Lecture Notes in Electrical Engineering

Volume 74

For further volume
http://www.springer.com/series/7818

Peter Benner · Michael Hinze ·
E. Jan W. ter Maten
Editors

Model Reduction for Circuit Simulation

Springer

Editors
Prof. Dr. Peter Benner
Max Planck Institute for Dynamics of Complex Technical Systems
Sandtorstr. 1
39106 Magdeburg
Germany
e-mail: benner@mpi-magdeburg.mpg.de

Prof. Michael Hinze
Dept. Mathematik
Universität Hamburg
Bundesstr. 55
20146 Hamburg
Germany
e-mail: michael.hinze@uni-hamburg.de

Dr. E. Jan W. ter Maten
Department of Mathematics & Computer Science
Eindhoven University of Technology
Den Dolech 2, HG 8.02, P.O. Box 513
5600 MB Eindhoven
The Netherlands
e-mail: jan.ter.maten@nxp.com

ISSN 1876-1100 e-ISSN 1876-1119

ISBN 978-94-007-0088-8 e-ISBN 978-94-007-0089-5

DOI 10.1007/978-94-007-0089-5

Springer Dordrecht Heidelberg London New york

Cover design: eStudio Calamar, Berlin/Figueres

Printed on acid-free paper

Springer is part of Springer Science+Business Media (www.springer.com)

Preface

Simulation plays a major role in computer aided design of integrated circuits (ICs). Mathematical models describe the dynamical processes and interactions of electrical devices. Verification of a circuit's behavior by means of solving these model equations in time and frequency domain is a mandatory task in the design process. The structures' sizes are decreasing, the packing density increases and so do the driving frequencies. This requires to use refined models and to take into account secondary, parasitic effects. The very high dimensional problems that emerge in this way may be solvable with the help of computer algebra in an unreasonable amount of time only. Clearly, this conflicts with the short time-to-market demands in industry.

Model order reduction (MOR) presents a way out of this dilemma. Redundancies are resolved, less relevant quantities are replaced by the most significant ones. In this way, the problem's complexity is reduced, keeping the main characteristics. Solving lower dimensional problems one can get statements on the circuit's performance more quickly.

This book surveys the state of the art in the challenging research field of MOR for integrated circuits (ICs), and also addresses future research directions in this area. Special emphasis is put on aspects stemming from miniaturization to nano-scale. Contributions cover complexity reduction using e.g., balanced truncation, Padé-approximation/Krylov-techniques, or POD approaches. For semiconductor applications, a focus is on generalizing current techniques to differential algebraic equations, on including design parameters, on preserving stability and other physical properties as well as model structures, and on including nonlinearities by means of piecewise linearizations along solution trajectories (TPWL) and interpolation techniques for nonlinear parts. Furthermore, the influence of inter-connect and powergrids on the physical properties of the device is considered, and also top-down system design approaches in which detailed block descriptions are combined with behavioral models. MOR faces the same requests as those appearing in Response Surface Modeling (RSM) techniques for robust design, with emphasis on parameterization, parameter screening, and nonlinearity. Further topics consider MOR and the combination of approaches from optimization and statistics, and the inclusion of Partial Differential Equation (PDE) models with emphasis on MOR for the resulting partial differential algebraic systems.

It is a proven fact that semiconductor industries see MOR as a key tool for simulations of nano-technology designs, with high demand of well educated experts in the field. This implies that timely education is needed. The current number of books in this area is very limited, so that this volume helps to fill a gap in providing the state of the art material, and to stimulate further research in the area of MOR for ICs. Furthermore, the methods which currently are being developed have also relevance in other application areas such as mechanical multibody systems, mechatronics, micro- and nano-technology, chemical and biological processes, etc.

Furthermore, this book reflects and documents the vivid interaction between three active research projects in this area, namely the EU-Marie Curie Action ToK project O-MOORE-NICE[1] (members in Belgium, The Netherlands and Germany), the EU-Marie Curie Action RTN-project COMSON[2] (members in The Netherlands, Italy, Germany, and Romania), and the German federal project *System Reduction for Nanoscale IC Design (SyreNe).*[3]

The material collected in this book reflects the contributions to the workshop "Model Reduction in Circuit Simulation", held at University of Hamburg, October 30–31, 2008. The workshop was jointly organized by the abovementioned projects SyreNe and O-Moore-Nice!, supported by the Federal Ministry of Education and Research (BMBF) of Germany and the EU FP6 programme "Marie Curie Industry Host Fellowships", respectively.

Part I of the book consists of overview papers of tutorial value corresponding to the invited presentations. Chapter 1 by Wilhelmus H. A. Schilders describes the need for novel MOR techniques in the electronics industry and thus provides a good motivation for the the technical papers to follow. In Chap. 2, Roland Freund derives the mathematical equations describing linear RCL circuits (which are the main tool for modeling interconnect and other parasitic effects in IC design) and presents Páde and Páde-type approximation methods respecting the structure of typical mathematical RCL models. A different approach to MOR is taken by Tatjana Stykel in Chap. 3 where the application of balancing-related methods from systems theory to circuit equations is surveyed. Special emphasis is put there on the exploitation of the topological structure implied by the electrical network. In Chap. 4, Sanda Lefteriu and Athanasios C. Antoulas pave the way to macro-modeling using measured data rather than starting from model equations. The main principle employed here is (tangential) interpolation using the Loewner matrix framework.

In Part II, contributed papers are collected. Chapters 5 and 7 provide different MOR methods and techniques to serve parameterized problems. In Chap. 5,

[1] www.tu-chemnitz.de/mathematik/industrie_technik/projekte/omoorenice/.

[2] www.comson.org for details.

[3] See www.syrene.org for details.

surrogate models (or RSMs) based on table models and neural network models are built that are valid for a range of parameters. These are applied to solve multi-input, multi-output (MIMO) problems. In Chap. 7, parameterized response functions are using expansions into bivariate rational basis functions. The coefficients are determined by optimizing a special cost function.

The Chaps. 6, 13 and 16 consider techniques to overcome problems associated with having massive numbers of inputs and outputs (or terminals, ports). In Chap. 6, recycling of Krylov sub spaces is applied to sequentially solve problems with changing right-hand sides occurring in an approach based on superposition of single input systems. Several recycling variants are studied for the Generalized Conjugate Residual method as basic linear solver. In Chap. 13, the sources are approximated by a finite expansion in basis functions. This allows for replacing current sources by a reduced number of controlled current sources. The network with the replaced sources remains passive and reciprocal. In Chap. 16, the situation where the large number of ports originates from validation problems that include many parasitics is considered. Extended SVD-based MOR (ESVDMOR) uses SVD on the input and output super matrices of the moments of the transfer function. A variant uses truncated SVD. Both approaches lead a system with virtual input and output operators and reduced number of terminals. For both approaches, stability, passivity, and reciprocity can be preserved under reasonable assumptions.

The Chaps. 8 and 9 address two different problems with circuits involving inductors. In Chap. 8, the transfer function at a node that is a candidate to be eliminated within the TICER method is first checked for the contribution coming from an attached inductor element. When passivity is preserved, RC couplings replace the first order inductor effect. After this the TICER step can be completed. In Chap. 9, the phase shift effect in a periodic steady state solution of an oscillator is determined by defining appropriate perturbation vectors covering the effects from other parts of the circuit. This allows to efficiently solve a balun circuit. The simulation time is further reduced by applying MOR (balanced truncation) to the RC models for the transmission lines coupling the oscillators.

In Chap. 10, POD-based model order reduction for semiconductors in electrical networks is investigated, where the semiconductor is modeled by the drift diffusion equations. The greedy approach is used to construct POD models which are valid over certain parameter ranges (frequencies). Numerical investigations for the reduction of a 4-diode rectifier network indicate that POD surrogate models depend on the position of the semiconductor in the network.

Chapter 11 deals with periodic linear systems as they arise in harmonic balance. It extends the notion of causal and non causal reachability and observability Gramians to discrete sequences of periodic descriptor systems (for instance arising after linearising around a periodic steady state solution). Periodic projected Lyapunov equations can be defined as well as (non-)causal Hankel singular values. This framework allows to extend MOR by balanced truncation to this class of problems.

In Chap. 12, synthesis of reduced models is considered: how to formulate those as netlist using RLC components and to reduce the amount of controlled sources. Foster synthesis only applies to partial fraction expansion of the transfer function. It is shown how MOR techniques that preserve the main structure of the equations and the input-output connectivity, formulated for problems in impedance form (current in, voltage out), can also be applied to problems in admittance form (voltage in, current out). This allows studying a second method, unstamping via RLCSYN. After synthesizing the reduced impedance model, this can be converted to an admittance model.

Structure preserving port-Hamiltonian reduction is introduced in Chap. 14 for electrical circuits. Formulation of a port-Hamiltonian state space system can easily guarantee stability. Kalman decomposition results in a controllable/observable port-Hamiltonian system. Balancing preserves the port-Hamiltonian structure of the equations.

In Chap. 15, symbolic reduction and simplification of expressions in nonlinear circuit equations is combined with reduction of blocks by numerical MOR techniques. The method relies on a hierarchical structure of the circuit.

In Chap. 17, MOR techniques like POD and Trajectory Piece-Wise Linear (TPWL) for nonlinear problems are considered, together with variants (like adapted POD, DEIM). The effect of the chosen variant on the quality of the approximations is considered as well as the efficiency and the robustness with respect to variations of the input sources.

In Chap. 18 an approach to nonlinear balancing and MOR is considered. For linear systems the notion of sliding interval balancing (SIB) is introduced, and it is shown that subsystems conserve stability. The truncated SIB realization bounds the optimal approximation in an input-output sense. Nonlinear balancing is approached by applying SIB on the linear variational system.

Hamburg, August 2010

Peter Benner
Michael Hinze
E. Jan W. ter Maten

Acknowledgments

This book contains the proceedings of the workshop *Model Reduction for Circuit Simulation* at the University of Hamburg, October 30–31, 2008, organized jointly by the German Federal Ministry of Education and Research (BMBF) research network *System reduction for Nanoscale IC Design (Systemreduktion für IC Design in der Nanoelektronik, SyreNe)*[4] together with the EU Marie Curie ToK project *O-Moore-Nice!*[5] (MTKICT-2006-042477). Financial support from the Deutsche Forschungsgemeinschaft (DFG) Priority Programme 1253 *Optimization with Partial Differential Equations* and from the DFG Collaborative Research Center *Elektromagnetische Strömungsbeeinflussung in Metallurgie, Kristallzüchtung und Elektrochemie (SFB 609)* is gratefully acknowledged.

We would like to thank the referees whose peer reviews were very valuable for preparing this book.We further would like to thank the Fachbereich Mathematik of the University of Hamburg for the kind hospitality and the support during the workshop. Finally, we want to express our special thanks to Martin Kunkel for his expertise and his great efforts in supporting the review process and in producing this volume.

[4] www.syrene.org.

[5] www.tu-chemnitz.de/mathematik/industrie_technik/projekte/omoorenice/.

Contents

Part I
Invited Papers

Chapter 1
The Need for Novel Model Order Reduction Techniques in the Electronics Industry

Wil H. A. Schilders

Abstract In this paper, we discuss the present and future needs of the electronics industry with regard to model order reduction. The industry has always been one of the main motivating fields for the development of MOR techniques, and continues to play this role. We discuss the search for provably passive methods, as well as passivity enforcement methods that are currently being developed. Structure preservation is another important research topic, for which new concepts are being developed. This also holds for the calculation of delays in long interconnect lines, a topic that leads to an entirely new type of methods. Topics that are still in their infancy are model order reduction for parameterized and nonlinear problems. We will discuss what the needs of the industry are in all of these fields, show specific applications and what has been achieved so far. The paper is meant as a guideline for future research, not as a detailed survey of existing methods.

1.1 Introduction

Both in the area of dynamical systems and control, and in numerical analysis, a wealth of model order reduction (MOR) techniques have been developed over the years. Balanced truncation, Krylov subspace methods, proper orthogonal decomposition and other SVD-based methods are just a few classes of methods that have been described. The electronics industry has been one of the main providers of

W. H. A. Schilders (✉)
NXP Semiconductors, High Tech Campus 46, 5656 AE Eindhoven, The Netherlands
e-mail: wil.schilders@nxp.com; w.h.a.schilders@tue.nl

Department of Mathematics and Computer Science, TU Eindhoven, PO Box 513
5600 MD Eindhoven, The Netherlands

P. Benner et al. (eds.), *Model Reduction for Circuit Simulation*,
Lecture Notes in Electrical Engineering, 74, DOI: 10.1007/978-94-007-0089-5_1,

motivating examples, and indeed model order reduction has found widespread use in this industry.

The starting point for MOR in the electronics industry is usually attributed to a method termed asymptotic waveform evaluation (AWE) [33]. The underlying idea of this method is simple, approximating the moments of the transfer function of the system. The idea being that moments will decay, so that calculating a sufficient but finite number of moments will eventually lead to an accurate approximation of the transfer function. Soon after its publication, it was realized that the method suffers from numerical problems that are similar to those encountered when applying the power method to calculate eigenvalues and eigenvectors of a system. The columns of the coefficient matrix generated to calculate the solution will tend more and more towards the eigenvector corresponding to the largest eigenvalue, which means that we will soon find this matrix to be very ill-conditioned. This is exactly what is observed when applying AWE: after 8–10 iterations, the process stagnates, and no additional useful information about moments is being generated. This stagnation is graphically illustrated in Fig. 1.1.

In order to overcome these numerical difficulties associated with AWE, Feldmann and Freund [13] proposed the use of Krylov subspace type methods. Within the area of numerical linear algebra, such methods had been suggested in the 1950s, and shown to be capable of avoiding the ill-conditioning of the coefficient matrices encountered in the power method. In fact, Lanczos already used a similar technique to determine eigenvalues and eigenvectors of symmetric matrices. Again, the idea is fairly simple: rather than generating a sequence of vectors that quickly become (almost) linearly dependent, after each iteration an orthogonalisation step is performed. The effect is that the same subspace is produced, but now with an orthonormal basis. Stated differently, information already contained in previous iterates is projected out of the newly generated vectors, and only the new information is taken into account. Not surprisingly, this also provided the solution to the problems arising in AWE. The method suggested by Feldmann and Freund was termed

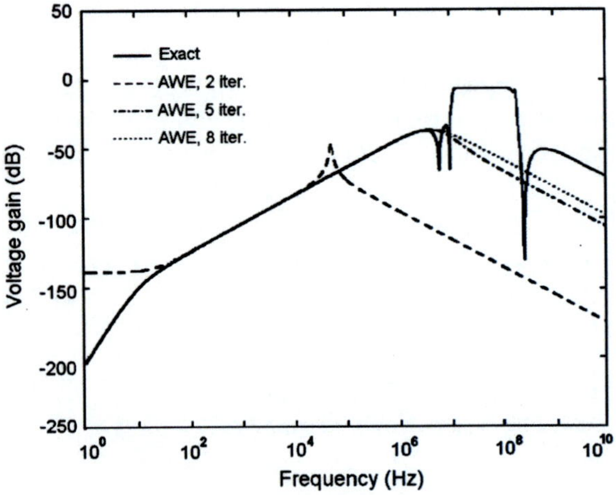

Fig. 1.1 Stagnation of AWE method [13]

Padé-via-Lanczos (PVL) [13], and shown to match the moments of the transfer function. It demonstrated that Krylov subspace methods, developed mainly in the area of numerical linear algebra, can also be used to perform model order reduction.

With the advent of PVL, the way was paved for new developments in MOR. In the past decade, the electronics industry has continued to formulate challenging research topics in the field of model order reduction. Although PVL eliminated the problem of ill-conditioning of the coefficient matrix associated with the calculation of the moments of the transfer function, it turned out to lead to non-passive approximations in some cases. Indeed, it is fair to expect that reduced order models for passive systems should also be passive. Thus, the invention of PVL was soon followed by the search for provably passive Krylov subspace methods. In 1998, the first of such methods was published (PRIMA [32], based on the Arnoldi process), soon followed by the Laguerre-type methods [27, 30]. PVL itself was also remedied, although the SyPVL and SymPVL methods [15] do not seem to have found widespread use; this also holds for Laguerre-type methods. Other work that should be mentioned here is that by Antoulas and Sorensen on passivity preservation based on methods using spectral zeros, in which forward and adjoint systems are treated together [3].

Since the beginning of the new century, developments in MOR for problems in the electronics industry have been stepped up. Engineers formulated more and more constraints and requirements, thereby formulating many interesting topics for researchers in the field of model order reduction. An important question was, for example, how to assess the accuracy of methods. Or, equivalently, how can we guarantee that the reduced order model generated is a sufficiently accurate approximation of the original transfer function. For methods based on Krylov subspaces, this question has not been answered so far, so that the iterative process is usually stopped when the difference between subsequent iterations is below a certain threshold. Methods used in the area of dynamical systems and control, most prominently those based on Hankel singular values, are equipped with good and reliable error estimates. The counterside is that these methods involve the solution of Lyapunov equations, which is limited to fairly small systems. Hence, research is now concentrating on extending the applicability of these methods to larger systems, often making use of the sparsity [7].

Another constraint that was formulated fairly recently is the preservation of the structure of systems. Again, this is a natural requirement. This has led to the development of classes of structure-preserving methods, of which SPRIM [17] and structure-preserving PMTBR [14] are examples. Although these methods have provided new ideas to perform MOR for electronic systems, new developments are still needed in this area. One of the problems is that the methods suggested so far destroy the structure of the incidence matrix that relates the different types of unknowns (such as currents and voltages), leading to a fully dense system of couplings. This also holds for the method described by Vandendorpe and Van Dooren [45] for interconnected systems, although, at first glance, this method appears to preserve the sparse interconnections. Also interesting in this context is the paper [16].

Despite all efforts, many unsolved problems remain, and the electronics industry is anxiously waiting for solutions. In this paper, the demands of the electronics industry with respect to model order reduction will be detailed. In Sect. 1.2, we briefly sketch where these needs arise, and place the problem in its context. We start by discussing, in Sect. 1.3, the problems of passivity and realization, the latter meaning that the mathematically reduced order model should be cast into a physically realizable model mimicking properties of the original model. In Sect. 1.4 it will be shown that structure preservation is still not in a state that is acceptable to the business, and that different concepts and methods are needed.

Another important problem is that of the reduction of networks with many inputs and outputs, the topic of Sect. 1.5 Krylov methods cannot handle such problems, and methods that make use of singular value decompositions are not applicable in most cases. We will also discuss methods for delay equations, a fairly recent development in MOR for long interconnect lines. Section 1.6 gives some details on the requirements associated with this problem.

Last, but certainly not least, the state of nonlinear and parameterized model order reduction techniques is far from mature enough to help the industry cope with their problems in behavioral and response surface modeling. Section 1.7 discusses some of the issues, and the state-oft-the-art in this important research area. Future developments will probably be concentrated in this area. Section 1.8 is the conclusion of this paper, with a summary of the current and future needs of the electronics industry.

It should be noted that the discussion in this paper will not be exhaustive; several other problems in the electronics industry lead to requirements for model order reduction techniques, and often these are the subject of research. Furthermore, the main emphasis will be on Krylov subspace methods, whence some readers may feel that solutions can also be provided by methods originating from the area of dynamical systems and control theory. Besides not being exhaustive, the discussion will also not be elementary, in the sense that some background knowledge by the reader is required, both in the electronics industry and in the area of model order reduction. For the latter, this background information can conveniently be obtained from the surveys found in the books [4, 8, 37, 43]. For the former, a wealth of books is available, as well as many websites.

1.2 Mathematical Problems in the Electronics Industry

The electronics industry is dealing with the design of computer chips. As is well known, the development over the years can be fairly well described by Moore's Law, as displayed in Fig. 1.2. It has various possible interpretations, but the bottom line is that every 18 months a new generation of chips is borne that is twice as fast as the previous generation. The dimensions are dramatically reduced from generation to generation, and the operating frequencies rise exponentially.

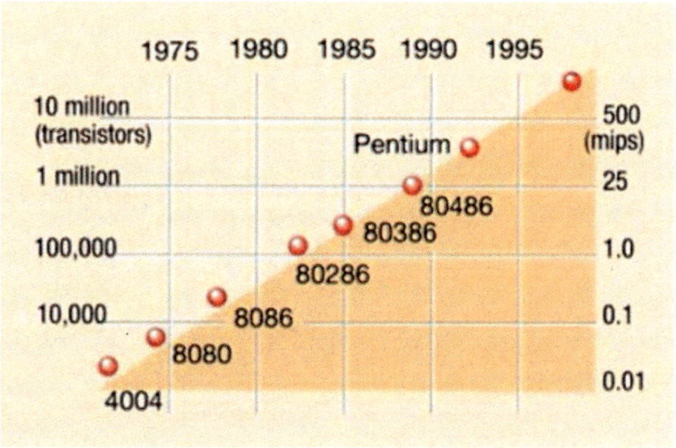

Fig. 1.2 Moore's law

Fig. 1.3 Illustration of the scale problem in the electronics industry

Dealing with computer chips means having to deal with complexity. Simulation of the behavior of semiconductor devices, electronic circuits and interconnect structures is a huge mathematical problem, in which we need to deal with scales that vary up to 10^{11}. This scale problem is illustrated in Fig 1.3. In the left picture, we see a close-up of the wiring in the interconnect structure, with details of 100 nanometers. The middle picture shows a 50 GHz integrated inductor, for which the scale is in the order of 500 μm. The picture on the right shows a close-up of the Intel Itanium chip (source: McKinley, 2002) for which approximately 10 km of interconnect are needed to connect all components.

Fortunately, we see a development in numerical methods that is similar to Moore's law. Since the advent of iterative methods for solving linear systems, occurring at the core of all simulation programmes, the size of systems that can be solved has increased exponentially. Most notable developments are preconditioned conjugate gradient methods and the multigrid method. Figure 1.4 shows the development of these methods in a period of 30 years (see also [49]).

When considering the mathematical problems in more detail, we should distinguish various levels, as indicated by the scale issue discussed before. At the top

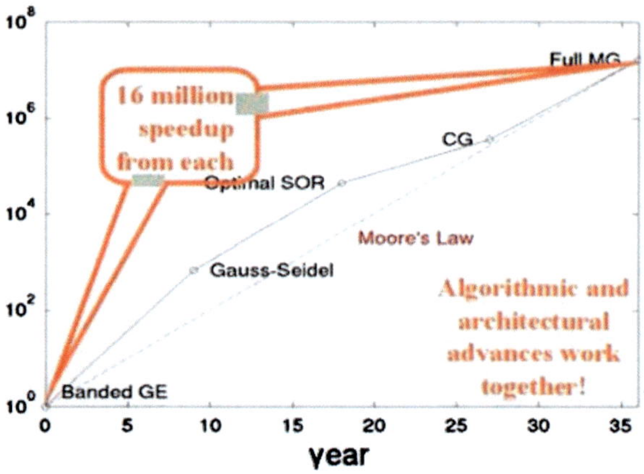

Fig. 1.4 "Moore's law" for numerical methods in solving linear systems

level, we need to deal with electronic circuit simulation which encompasses large (millions) discrete networks containing resistors, capacitors, inductors, diodes, transistors, and other components. The discretisation of this system is fairly simple, as it obeys Kirchhoff's laws. The models used for active devices, such as diodes and transistors, are extremely nonlinear, meaning that non-standard methods are needed to solve the associated nonlinear systems. The linear systems generated in this process often have a hierarchical structure and are sparse, but may also be indefinite. Transient simulations are also non-trivial, due to the sudden changes occurring due to externally applied signals. AC analysis for high frequencies requires the solution of large complex linear systems. Pole-zero analysis and harmonic balance are also worth mentioning, but there are some comments to be made here. Normal AC analysis has a matrix $\omega C + G$ in which $\omega = i \cdot 2\pi f$. Clearly, here the size of the system equals the size of C and G (i.e. defined by the design). In Harmonic Balance more harmonics are involved in one coupled system. So here the number of harmonics influences the size of the system. In nonlinear problems high frequency effects may result in the need to increase the number of harmonics. In Periodic AC (PAC), i.e. studying small perturbations after linearizing around a PSS solution, two frequencies have to be taken into account: that of the PSS solution and the one of the perturbation of the right hand side. Usually, a high frequency PSS solution needs many time points for a proper time-domain representation and to be able to study the effects of the perturbations to sum and difference frequencies $f_{\text{osc}} \pm f_{\text{pert}}$. The reader is referred to [9, 26], where MOR is studied for systems arising in PAC analysis.

Not surprisingly, the development of robust and efficient methods for electronic circuit simulation has taken a considerable number of years (since the 1970s). Even nowadays it is certainly not a trivial task, with new developments still being published, and new simulation software containing the latest advances in numerical methods being brought to the market. The additional benefits of using MOR is also a key topic in recent years.

At a lower level, we encounter the simulation of individual semiconductor devices. This is again a complicated mathematical problem. Often, one uses the model consisting of three coupled partial differential equations, together constituting the drift-diffusion model. The system is singularly perturbed, meaning that special discretization techniques are needed to obtain accurate solutions on relatively coarse meshes. The resulting discrete system of equations is again extremely nonlinear, requiring special numerical methods for their solution. Damped Newton methods were developed, as well as nonlinear block Gauss-Seidel methods, the latter being known in the field as Gummel's method. Nonlinear variable transformations have also been shown to provide successful solution techniques. In the 1980s and 1990s, many researchers have been working on this problem, both from academia and within large electronics companies like IBM, AT&T, and Philips. Although the set of methods for the drift-diffusion equations is by now fairly well accepted and stationary [39], this is certainly not the case for the extended models such as the energy transport model or the hydrodynamic set of equations [2].

Although electronic circuit simulation and semiconductor device simulation are now at a fairly mature level, new problems have occurred in the mean time. At high frequencies, electromagnetic effects in the interconnect structure (see Fig. 1.5) may start to influence the behavior of circuits and devices. This leads to delay of signals, malfunctioning devices, and undesired effects in the substrate. The Maxwell equations need to be solved to address these issues, either by using a full wave solution using methods like FDTD, FIT, BEM, and FVM, or approximate methods like PEEC [35]. The resulting extremely large systems must be reduced to acceptable size to be coupled with the underlying circuit. These are complex linear systems, which are difficult to solve at high frequencies. The eigenvalues depend on the frequency, and iterative methods have difficulties to solve such systems.

An important problem for the discussion in subsequent sections is that, even if we would be able to solve such large systems by iterative or direct methods, circuit

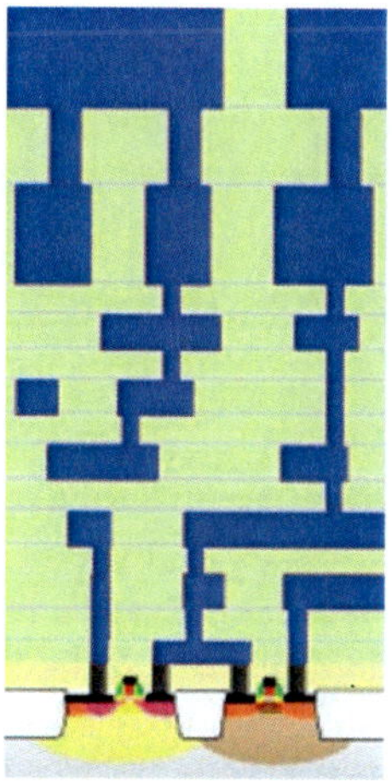

Fig. 1.5 Interconnect structure above the silicon containing the circuit and devices

simulation programmes will not be able to cope with these large systems. The problem is that a coupled simulation needs to be carried out, and the commonly used way to obtain this coupling is to translate the discretized electromagnetic system into a linear circuit, and solve this in conjunction with the original non-linear electronic circuit using a circuit simulator. As this is not a feasible approach, one needs to first reduce the electromagnetic system to an acceptable size. This is one of the reasons why model order reduction has received much attention within the electronics industry. Reducing the electromagnetic system is indeed possible, as we are only interested in the dominant electromagnetic effects that influence the underlying circuit. Therefore, not all detail contained in the full discretized system is needed, and the use of model order reduction is fully justified.

Model order reduction is not only used for the reduction of the electromagnetic system generated by the interconnect structure, but also by the substrate region below the individual semiconductor devices. This area of a chip also needs to be co-simulated with the electronic circuit, in order to account for signal propagations taking place within the substrate region. Often, this leads to fairly simple resistor or RC networks, the problem being that these systems are again very large. Hence, model order reduction is needed to reduce these substrate systems to acceptable size, so that they can also be co-simulated with the electronic circuit.

Having demonstrated the need for model order reduction, in the remainder of this paper we will address several issues that are encountered when performing this task. Once again, we stress that the discussion is not exhaustive. There are many papers describing the research on suitable MOR methods used in the electronics industry, and the reader is referred to these for more information.

1.3 Passivity and Realizability

As described in the introduction, the Padé-via-Lanczos method [13] was the starting point for many new developments in numerical methods for model order reduction. As we also discussed, it was soon realized that the method can generate non-passive reduced order models despite the fact that the original system was passive. The search for provably passive methods was started, and methods like PRIMA [32] and Laguerre-type methods [27, 30] are now routinely used. Thus, the problem appears to have been solved satisfactorily, and no further research is needed.

However, the latter is not entirely the case. The first problem encountered in practice is the possibility that the discretisation of the original problem does not guarantee a passive system. This may be caused in electromagnetic simulations, for example, by a mesh that is too coarse. Another example is when using the boundary element method for the simulation of printed circuit boards, when the fourfold integrals have not been calculated to sufficient accuracy (this is especially crucial for the diagonal elements, which contain singular integrals). In these cases, there is little hope that the reduced system will be passive.

A related problem is that the reduced order model is solely based upon experimentally obtained results, for example measurements of S-parameters. Again, due to inaccuracy in the data, the state space system constructed from these may have poles in the wrong halve of the complex plane, and be non-passive. Again, this leaves not much hope for passivity of the reduced order system that should capture the dominant behavior of the experimentally generated data (see also [28]).

It should be noted that a tendency is observed to increasingly perform MOR on data obtained fully or partially from experiments. Antoulas and his group [31] have done work on Loewner matrices and shifted Loewner matrices, that can conveniently be used to generate reduced order models based on available data. These methods split the data into two disjoint sets, and then perform a computational procedure using these so-called Loewner matrices. The data may come from measurements, and can also be combined with simulation results. In fact, one could also perform a huge number of simulations, and use these for the set of data. These Loewner approaches lead to linear state space models that interpolate exactly [5, 28].

From the foregoing it is clear that passivity is certainly not a closed chapter in the book of MOR! However, the emphasis has shifted from provably passive methods to techniques for passivity enforcement. This topic is receiving quite some attention in recent years. It is also an important topic, as it is essential that rational models be passive in order to avoid unstable time domain simulations such as shown in Fig. 1.6. Thus, several researchers are concentrating on passivity enforcement methods, some from the numerical analysis point of view, others in the area of dynamical systems and control. For example, in [19], building on earlier work (2004), a general class of a posteriori passivity enforcement algorithms is considered based on iteratively perturbing the Hamiltonian matrix. The authors present a frequency-weighting scheme leading to the definition of a modified Gramian that, when employed during passivity enforcement, effectively leads to control of the relative error.

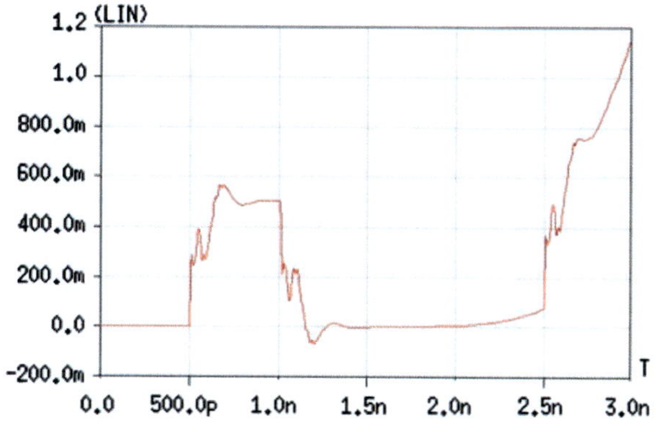

Fig. 1.6 Unstable time domain simulation for non-passive reduced order model of a low pass filter (see also [44])

A different technique is described in [23], where a fast approach for passivity enforcement of pole-residue models is obtained by perturbing the eigenvalues of the residue matrices, as opposed to the existing approach of perturbing matrix elements. This leads to large savings in computation time with only a small increase of the modeling error. This fast residue perturbation (FRP) approach is merged with the modal perturbation technique, leading to fast modal perturbation (FMP). Usage of FMP over FRP achieves to retain the relative accuracy of the admittance matrix eigenvalues. A complete approach is obtained by combining the passivity enforcement step with passivity assessment via the Hamiltonian matrix eigenvalues and a robust iteration scheme, giving a guaranteed passive model.

Also worth mentioning is the work in [36]. A nice and recent survey of passivity enforcement methods is provided in [20], and also at the recent SPI conference [25] interesting new developments were presented. There are, however, some remaining issues that are important from the electronics industry point of view. Most methods developed so far trade accuracy for passivity. In other words, the reduced order model can be made passive only at the expense of lower accuracy. This is not surprising, as the model was constructed in such a way that its transfer function gives a sufficiently accurate approximation of the transfer function of the original system. By changing the location of only a few poles, namely those causing the non-passivity, the accuracy may be seriously affected. A 5% change in the S-parameters (to obtain values not exceeding 1) will often lead to at least 5% loss of accuracy. This is not acceptable from an engineering point of view. The only solution for this problem appears to be an optimization approach that specifies the constraints, and changes the location and residues of all poles, until both passivity and sufficient accuracy are obtained. Techniques suggested so far avoid this optimization problem by performing iterations on a limited number of poles, but we feel it is necessary to invest in the more expensive approach. If this is achieved, then the chapter of passivity can be closed, and the engineering community within the electronics industry will also be much happier.

A topic that is closely related to passivity is that of realizability. A well known theorem tells us that passive systems are automatically realizable. In his thesis, Brune (1931) shows that any passive transfer function can be realized by some passive network. He also provides rules for the construction of such networks, a drawback being that he is forced to employ transformers with perfect coupling. From an engineering point of view, this is undesirable, and it is more attractive to use so-called Foster synthesis, dating back to Foster (1924).

Although this may, therefore, appear to be a problem solved long ago (see also [21] for a survey dating back to 1957), again this is not the case at closer inspection. Foster realization (or similar techniques) often leads to an extremely large number of resistive, capacitive and/or inductive elements. Thus, the size of the reduced order model is dramatically increased after realization. This is mainly due to the fact that the matrices in the reduced order model often have lost sparsity, so that all unknowns in the reduced order model are mutually coupled. Indeed, this dense structure is one of the stumbling blocks for efficient circuit synthesis of reduced order models. Other obstacles are that there may be negative values of resistors, capacitors and inductors, and, last but absolutely not least, that the methods often work only for single input single output (SISO) systems. For multiple input multiple output (MIMO) systems, the approaches often fail, or have not even been described for this case.

Recently, a method that makes use of "un-stamping" was published by Yang [50]. The method is termed RLCSYN, and heavily relies on block structure preserving MOR methods (see next section). This may indeed be the only way to avoid the mutual coupling between all elements in the reduced order system. So far, however, experiences with the method are lacking, and time will tell whether it is the solution to the aforementioned stumbling blocks.

A point of discussion in the electronics industry is, of course, whether or not one needs to realize/synthesize the reduced order models. The reasoning is simple: suppose we have an electronic circuit, and are able to reduce it using mathematical techniques. If we perform the intermediate step of realizing the reduced order model, we obtain a smaller circuit that can be treated by a circuit simulator. But the circuit simulator will immediately start "discretizing" the circuit, translating the equations back to the mathematical reduced order model. Hence, it appears unnecessary to make the effort of realizing and/or synthesizing a reduced order model. However, there is a need within the electronics industry. A synthesized reduced order model can provide more insight to engineers and designers than the reduce1d order model in mathematical form. In fact, as we shall see in Sect. 1.7, the compact models constructed by device physicists for transistors and diodes are translations of exactly these insights. Thus, there is a need in the industry for such realization and synthesis techniques. It would be even better if this could be done for parameterized reduced order models.....

1.4 Structure Preservation

When the discussions about passivity were drastically reduced after the publication of PRIMA and the Laguerre-type methods, a new topic emerged in the MOR community. In fact, this was initiated by the needs of the electronics industry, where it is a given that electronic circuits are formulated in terms of voltages and currents. Hence, the coefficient matrices in the state space systems can be regarded as consisting of blocks that correspond to these two types of unknowns. Unfortunately, conventional MOR methods do not make use of this knowledge. In fact, Krylov subspace methods like PVL immediately destroy this underlying structure, by mixing information that is of "current type" with that of "voltage type". This implies that the unknowns in the reduced order model do not have a physical interpretation.

Clearly, the foregoing is rather undesirable, and there is a clear need in the electronics industry to maintain the underlying structure of the problem within the context of MOR. This has led to new methods that are now termed structure-preserving MOR methods. The first to be published was that of Roland Freund, termed SPRIM [17]. The idea is simple, yet very clever: first, a Krylov subspace is generated using conventional MOR methods. However, rather than using the vectors generated as a basis for the Krylov subspace, Freund suggests to split each vector into two vectors, by splitting it into a vector containing only voltage unknowns, and a vector containing only current unknowns. In the vector containing the current unknowns, the voltage entries are all set to zero, and similarly for the vector containing the voltage unknowns. In this way, a basis is formed that consists of twice the number of basis vectors as the basis for the space constructed originally. The big advantage is, however, that it can now be shown that there will be no mixing of voltages and currents, meaning that it should be possible to assign a physical meaning to each of the vectors.

Unfortunately, despite the nice properties of SPRIM (such as matching double the number of moments, see [17]), the method was not felt by the industry to be the ultimate answer to the question of structure preservation. In fact, it was soon noticed that the incidence matrix, i.e. the matrix that relates voltages and currents, is destroyed completely. Originally, it is a very sparse matrix with at most two elements per row (1 and -1, relating branches and nodes), but using SPRIM it immediately turns into a dense matrix. This means that all currents and voltages in the reduced system are coupled, which is to be avoided in view of subsequent realization and synthesis (see Sect. 1.3).

Only recently, a remedy for the aforementioned problem has been presented. The method in [6] uses a clever transformation of variables before the reduction process starts, and then uses SPRIM. The transformation used is that which translates branch currents, arising in the frequently used MNA formulation of electronic circuits, into loop currents. Such transformations are also very helpful in methods for the solution of indefinite linear systems, as has been shown in [38]. In fact, long time ago, circuit simulation was usually done using the so-called "mesh method", which is precisely using loop currents rather than branch currents. With the advent of fast computers, the preference became MNA and using branch currents, but from a mathematical point of view, this is certainly not the preferred choice. In any case, Bai et al. [6] make use of the loop currents, the big advantage being that the incidence matrix becomes an identity matrix. That is, the (1,2)-block in the matrix will be identity, the corresponding (2,1) block is not affected. Hence, we immediately see a drawback of the method, namely it destroys the original symmetry.

The method proposed by Bai et al. [6] is certainly promising, and in a further publication it has been shown to be capable of addressing the realization problem discussed in the previous section. However, it is too early to state that all problems have been solved. There is still the clear need in the electronics industry to address the structure preservation issue, and it is the feeling of the author that other methods need to be explored. For example, when considering an electronic circuit, it usually contains more branches than nodes. This means that we have more unknowns of the type "current" than of the type "voltage". Structure preserving methods suggested so far generate the same number of basis vectors of both types. This seems unnatural. Adding a new basis vector for the voltages, should probably

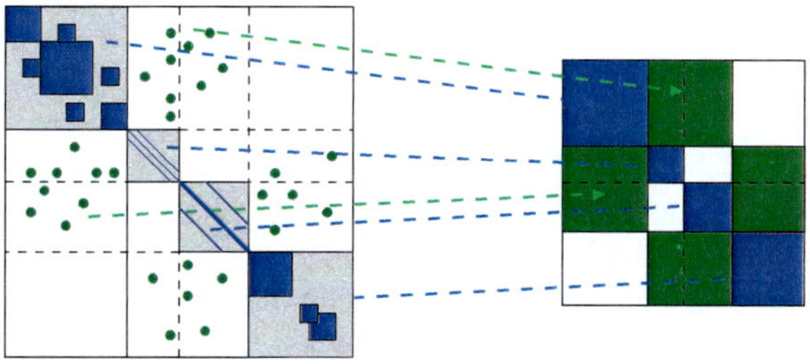

Fig. 1.7 Structure preserving MOR for coupled systems

lead to a number of basis vectors added for the currents. So that, in the end, having a full basis for the voltages at the same time leads to a full basis for the currents. In the presently used methods, this is not the case. Most probably, the answer needs to be found in graph theory, but currently it is not known how to do this.

In the foregoing, we have concentrated on electronic systems containing currents and voltages, reflecting a structure consisting of two types of unknowns. Of course, the structure may be much more refined. From a theoretical point of view, we could have interconnected systems as considered by Vandendorpe and Van Dooren (see [37], Chap. 14). More general (but very much related to that of Vandendorpe and Van Dooren) is the approach considered in [14]. The graphical interpretation shown in Fig. 1.7 clearly shows the idea: keep the structure of the original systems, and do not destroy it during the MOR phase. The picture also shows the drawback of the method developed so far, namely that the coupling blocks become dense after reduction. Here, a combination with the ideas of Bai et al. [6] could provide a solution. However, as before, we feel that other methods, possibly from graph theory, need to be investigated in order to arrive at a fully acceptable solution.

Summarizing, the problem of structure preservation remains unsolved, despite various useful attempts. The feeling within the electronics industry is that novel concepts and ideas are needed.

1.5 Reduction of MIMO Networks

Although not mentioned explicitly, so far the discussion has been limited to systems that are of single input single output type. Of course, several of the methods presented can also be applied to the MIMO case, but often this will reveal some hidden problems. One very clear disadvantage of Krylov subspace methods applied to MIMO systems is that the number of basis vectors added to the subspace is equal to the number of inputs and outputs of the system. In the SISO case, one vector is added at the time, but in the MIMO case, many vectors are added in each iteration. Thus, the dimension of Krylov subspaces quickly becomes large, in many cases even prohibitively large. In [27], this problem has been analyzed, and a block-Arnoldi algorithm was proposed that reduces the blocks of vectors that is added per iteration. Experiments conducted with this method show the advantage, but it is too marginal to be very effective in practical situations. Hence, new concepts need to be developed.

One new concept that has recently been investigated, and published in another chapter in this book, is the use of a mixture of graph theoretical and numerical methods to perform some kind of "preconditioning" of the original system. In fact, it is similar to the successive node elimination (SNE) technique [41]. Instead of applying Krylov subspace methods to the system, related to iterative solution methods for linear systems, the method employs a more direct approach of eliminating unknowns, corresponding to direct solution techniques for linear systems. We do not wish to repeat the information that is in the chapter elsewhere in this volume, and refer the reader to it for more details. What can be said is, however, that there is a very clear and relatively urgent need by the electronics industry to be able to drastically reduce such networks with many so-called "ports". As has been demonstrated in Sect. 1.2, both for the interconnect system and for the substrate, huge resistive (and sometimes RC networks) are generated that need to be co-simulated with the underlying electronic circuit. This is only feasible in present-day circuit simulators if these huge networks are reduced very substantially.

1.6 MOR for Delay Equations

A fairly recent development is concerned with delay extraction-based passive macromodeling, associated with multiconductor transmission line type interconnects. High-speed effects influencing a signal propagating on an interconnect could be multifold, such as delay, rise time degradation, attenuation, crosstalk, skin effect, overshoots, undershoots, ringing, and reflection. Concerning propagation delay, a signal traversing from one end of a transmission line to the other end takes a finite amount of time, resulting in the signal experiencing a certain amount of delay. In addition, the signal may encounter rise time degradation, where the rise time at the receiver end is larger than the rise time at the source end. Rise-time degradation further adds to the overall delay experienced by the signal.

Concerning signal reflection and the associated ringing, this can also severely distort signal propagation at higher frequencies. The prime cause of this is the discontinuity in characteristic impedance of the transmitting line, due to the impedance variation on a line taking place over a certain length. This can occur due to the change in the medium along the length of the signal trace, which may have to traverse several layers.

It turns out that conventional model order reduction techniques do not work well for this case. characterized by multiport (Y, Z, S, or the transfer function) tabulated parameters. In [10] an algorithm is described that approximates the logarithm of the transfer function using a low-order rational function. Subsequently, the DEPACT (delay extraction-based passive compact transmission line macromodel) algorithm is applied to obtain a passive and causal macromodel for SPICE simulation. This method leads to compact, low-order macromodels that are more reliable for time-domain simulations.

More recent developments can be found in [25], where several papers discuss the issue of MOR for delay equations. The most commonly used approach nowadays appears to be an adapted version of vector fitting. Rather than using only rational functions, it is suggested to have an expansion which is a mixture of exponential and rational terms. The exponential terms then account for the delay, and can account also for reflections. The "coefficients" of these exponential terms are then rational functions as used in the vector fitting approach.

Although simulations carried out so far suggest that the foregoing approach is the way to go, there is also the clear disadvantage that the methods are only suited for delay equations. Stated differently: if one includes the exponential terms, one can account for effects in long interconnect lines, but this expansion is not very suitable for short lines. If one does not include the exponential terms, the normal vector fitting approach is used, and one will find that a prohibitively large number of terms is needed in the expansion when used for long interconnect lines. Hence, the essential question is, whether a combined approach can be found that addresses both long and short interconnect lines. This is a clear need in present-day simulations of interconnect structures, also for extraction programmes [24].

1.7 Parameterized and Nonlinear MOR

As will have become clear from the foregoing sections, the state-of-the-art concerning MOR for linear problems is relatively well developed, although there are still many open questions and remaining problems. For parameterized and nonlinear problems, model order reduction techniques are still in their infancy.

When looking at parameterized MOR, the electronics industry has clear needs. Modern analog designs are often built using the concept of programmable cells (PCells, PCircuits) and the ROD/Skill programming language. All of these introduce parameters, either in the definition of the circuit sub blocks (Pcells), or the relation between sub blocks (ROD, relative object design). Often, these sub blocks are optimized with respect to a certain objective function on the so-called schematic level, but optimization needs to be performed again on the higher layout level, in order to account for parasitic effects that might take place due to interactions between the different components. Clearly, these problems cannot be addressed by an MOR technique that only uses numbers. We need to be able to perform MOR also when unknown parameters are in the system.

So far, several researchers have addressed this problem, but the techniques suggested are often not very general, or not capable of solving the aforementioned complex problems. Obvious extensions using assumptions on the linearity are often suggested, whereas what is needed are new concepts. Rather than using Taylor expansions in the parameters, leading to a multitude of cross terms, it is the

Fig. 1.8 Dependence of poles on parameters

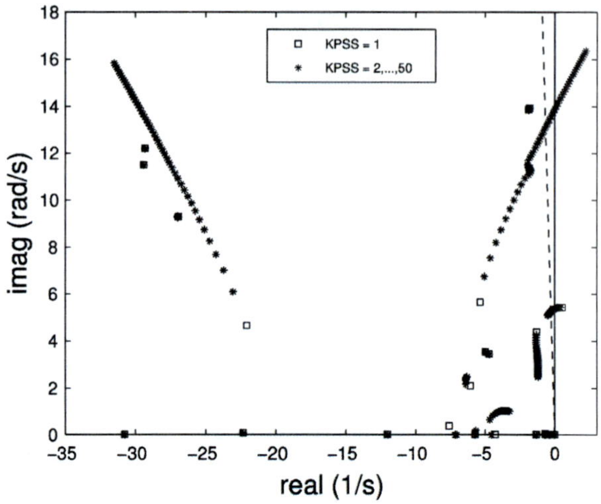

feeling of the author that we need a careful analysis of the poles of the transfer function in terms of the parameters. A development in this direction that is felt to be very promising is the sensitive pole algorithm (SPA), that shows the dependence of the location of poles on parameters. In Fig. 1.8 an example is shown (see [34]).

It is exactly this dependence that is important. For the Maxwell equations, for example, part of the eigenvalue spectrum appears to be independent of the parameter "frequency", whereas another part is heavily dependent on the frequency. Hence, it should be possible to develop MOR methods that make use of this knowledge, and only consider the poles that are most sensitive to changes of parameters. This would be the author's advice to researchers working in the area of parameterized MOR. The electronics industry would be extremely delighted when better methods, not as involved as the methods suggested so far, would "see the light". A potentially very interesting development is that using symbolic approximations. The reader is referred to [40, 48] for more information.

For nonlinear problems, the situation is similar to that of parameterized MOR. Some methods have been suggested, but often these are based on linearizations of the underlying problem. This can never be effective for general nonlinear problems, and is far from the needs of the electronics industry. In the area of Krylov subspace type methods, the trajectory piecewise linear method (TPWL) (Fig. 1.9) appears to be most popular; see also the chapter by Striebel in this volume. That chapter contains a wealth of information on the state-of-the-art of model order reduction for nonlinear problems [42]. In the area of dynamical systems and control, methods developed by Fujimoto and Scherpen [18] and Verriest [46] are promising, although so far the methods are limited to small nonlinear systems.

The question is, however, whether one can expect to be able to develop MOR methods for general nonlinear systems. In the author's opinion, larger nonlinear systems can only be tackled by using specific information about the underlying

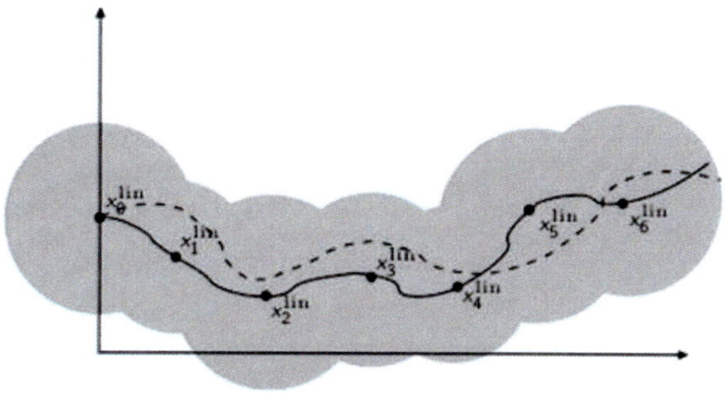

Fig. 1.9 Schematic illustration of the TPWL method

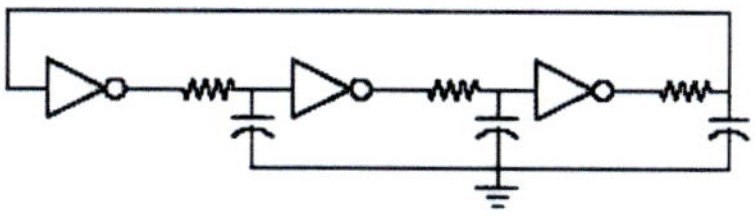

Fig. 1.10 Simple oscillator

nonlinear system. A nice example is provided by the simulation of voltage controlled oscillators. These are devices that are used frequently in electronic designs. Initiated by Roychowdhury [12], tremendous achievements have been obtain in the reduction of such systems. In Fig. 1.10 a simple oscillator schematic is shown, whereas in Fig. 1.11 the behaviour is shown in case it is in the so-called unlocked state.

The theory developed for oscillators is based on insight in the behaviour of such devices. The circuit describing these oscillators often consist of about 100 components, but their behaviour can be reduced to a single equation describing the phase noise. In fact, reduction to a single equation is possible as one knows that only the eigenvalue 1 plays a role in the final behaviour, all other modes will damp out. This example illustrates that it can be very effective to use detailed knowledge in order to obtain an adequate and effective reduction of the nonlinear system.

The electronics industry needs more of these specific ideas for the reduction of nonlinear components. In fact, physicists and device engineers have performed model order reduction already since the 1980s, but their activities are termed "compact modeling" rather than reduced order modeling. Initially, compact models were made in the form of so-called Gummel–Poon models (see Fig. 1.12).

More generally, the idea of compact modeling is to capture the full nonlinear behaviour of a semiconductor device into a model consisting of resistors, capacitors, inductors and simple nonlinear components like diodes, containing many parameters such as channel width, channel length, and doping profile. Compact modeling has become a discipline in itself, a good and solid reference being [29]. In recent years, efforts have mainly concentrated on compact models for the

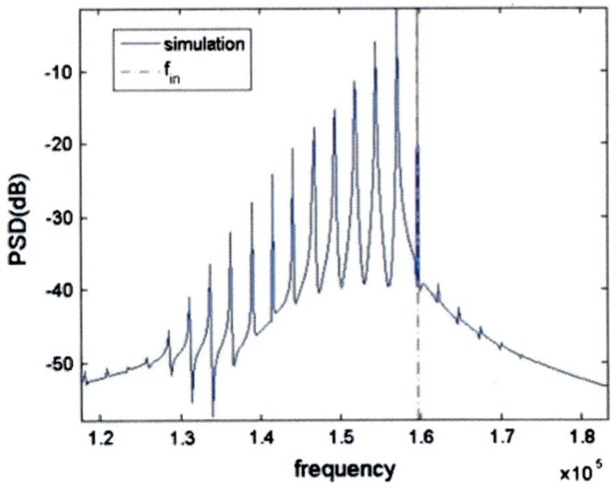

Fig. 1.11 Frequency domain behaviour of unlocked oscillator

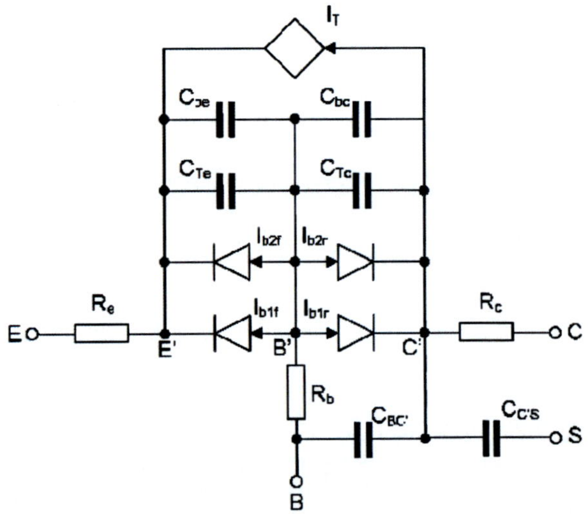

Fig. 1.12 Gummel–Poon model used in compact modeling of semiconductor devices

various generations of MOS transistors, building blocks of most electronic circuits nowadays. The Penn State Philips (PSP) model is the most recent model, now being accepted as the world standard [1].

It would be great if the MOR community could provide ideas for the automatic generation of such compact models. Thus far, such models have been constructed manually, using much insight and a huge number of experimental results being generated over years. In view of the latter, it is to be expected that successfully generated reduced order models must also make use of huge numbers (millions) of measured or simulated (via semiconductor device simulation) results.

A final remark is that the work presented in [11, 46, 47] is also of very much interest in the context of nonlinear MOR, and might provide the breakthroughs the industry is waiting for.

1.8 Summary: Present and Future Needs of the Electronics Industry

In this chapter, we have touched upon many subjects within the field of model order reduction as viewed from the electronics industry. The conclusion is that the area of linear problems is fairly well developed, but that many problems still remain, with new research topics being generated regularly. In the areas of parameterized and nonlinear model order reduction, many problems remain, and these areas still need many new ideas in order to get the status of being mature.

Specific topics that we identified in this paper, and are worth stressing again:

- Good and reliable error estimates for Krylov subspace based MOR techniques
- Passivity enforcement in such a way that accuracy is not traded for passivity
- More natural structure preserving methods that use different spaces for currents and voltages
- Efficient reduction methods for systems with multiple inputs and outputs
- An MOR approach that can address both long and short interconnect lines, i.e. is suitable for situations with and without delay
- Parameterized model order reduction based on the sensitive pole algorithm

The ultimate challenge, however, is to be able to automatically generate the nonlinear device models that can currently only be constructed by device physicists!

References

1. Aarts, A.C.T., Smit, G.D.J., Scholten, A.J., Klaassen, D.B.M.: A PSP-based small-signal MOSFET model for both quasi-static and nonquasi-static operations. IEEE Trans. El. Dev. **55**, 1424–1432 (2008)
2. Anile, A.M., Nikiforakis, N., Romano, V., Russo, G.: Discretization of semiconductor device problems (II). Chapter 5 of Handbook of Numerical Analysis Vol. XIII: special volume Numerical Methods in Electromagnetics. In: Schilders, W.H.A. and Ter Maten, E.J.W. (eds.), Elsevier, North-Holland (2005)
3. Antoulas, A.C., Sorensen, D.C.: Approximation of large-scale dynamical systems; an overview. Technical Report, Rice University (2001)
4. Antoulas, A.C.: Approximation of large-scale dynamical systems. SIAM series on Advances in design and control (2005)
5. http://www.ece.rice.edu/~aca/aca_monopoli_IV.pdf
6. Bai, Z., Li, R., Su, Y.: A unified Krylov projection framework for structure-preserving model reduction. In: Schilders, W.H.A., Van der Vorst, H.A., Rommes, J. (eds.) Model order

reduction: theory, research aspects and applications, Springer series Mathematics in Industry **13**, 75–93 (2008)

7. Benner, P., Li, J.-L., Penzl, T.: Numerical solution of large Lyapunov equations, Riccati equations, and linear-quadratic control problems. Numer. Lin. Algebra Appl. **15**, 755–777 (2008)
8. Benner, P., Mehrmann, V., Sorensen, D.C. (eds.): Dimension reduction of large-scale systems. Lecture Notes in Computational Science and Engineering **45**, Springer-Verlag, Berlin/Heidelberg (2005)
9. Benner, P., Hossain, M.-S., Stykel,T.: Model reduction of periodic descriptor systems using balanced truncation. These proceedings
10. Charest, A., Achar, R., Nakhla, M., Erdin, I.: Delay extraction-based passive macromodeling techniques for transmission line type interconnects characterized by tabulated multiport data. Analog Int. Circ. Sign. Proc. **60**, 13–25 (2009)
11. Chaturantabut, S., Sorensen, D.C.: Discrete empirical interpolation for nonlinear model reduction. Report TR09-05, Rice University (2009). http://www.caam.rice.edu/tech_reports/2009/TR09-05.pdf
12. Demir, A., Mehrotra, A., Roychowdhury, J.: Phase noise in oscillators: a unifying theory and numerical methods for characterization. IEEE Trans. Circ. Syst. **47**, 655–674 (2000)
13. Feldmann, P., Freund, R.: Efficient linear circuit analysis by Padé approximation via the Lanczos process. IEEE Trans. Computer-Aided Des. **14**, 639–649 (1995)
14. Fernandez Villena, J., Schilders, W.H.A., Silveira, L.M.: Block oriented model order reduction of interconnected systems. Int. J. Numer. Mod.: Electr. Netw. Dev. and Fields (to appear, 2009)
15. Freund, R.W., Feldmann, P.: The SyMPVL algorithm and its application to interconnect simulation. Proceedings of the International Conference on Simulation of Semiconductor Processes and Devices, 113–116 (1997)
16. Freund, R.W.: On Padé-type model order reduction of J-Herimitian linear dynamical systems. Linear Algebra Appl. **429**, 2451–2464 (2008)
17. Freund, R.W.: Structure-preserving model order reduction of RCL circuit equations. In: Schilders, W.H.A., Van der Vorst, H.A., Rommes, J. (eds.) Model order reduction: theory, research aspects and applications, Springer series Mathematics in Industry **13**, 49–73 (2008)
18. Fujimoto, K., Scherpen, J.M.A.: Singular value analysis and balanced realizations for nonlinear systems. In: Schilders, W.H.A., Van der Vorst, H.A., Rommes, J. (eds.) Model order reduction: theory, research aspects and applications, Springer series Mathematics in Industry **13**, 251–272 (2008)
19. Grivet-Talocia, S., Ubolli, A.: Passivity enforcement with relative error control. IEEE Trans. Microwave Theory Tech. **55**, 2374–2383 (2007)
20. Grivet-Talocia, S., Ubolli, A.: A comparative study of passivity enforcement schemes for linear lumped macromodels. IEEE Trans. Adv. Pack. **31**, 673–683 (2008)
21. Guillemin, E.A.: Synthesis of Passive Networks. Wiley, New York (1957)
22. Gummel, H.K., Poon, R.C.: An integral charge control model of bipolar transistors. Bell Syst. Tech. J. **49**, 827–852 (1970)
23. Gustavsen, B.: Fast passivity enforcement for pole-residue models by perturbation of residue matrix eigenvalues. IEEE Trans. Power Deliv. **23**, 2278–2285 (2008)
24. http://www.siliconfrontline.com
25. http://www.univ-brest.fr/SPI/pages/page.php?num=1-1
26. Harutyunyan, D., Rommes, J., Ter Maten, J., Schilders, W.: Simulation of mutually coupled oscillators using nonlinear phase macromodels. IEEE Trans. Comp. Aided Des. Integr. Circ. Syst. **28**, 1456–1466 (2009)
27. Heres, P.J.: Robust and efficient Krylov subspace methods for model order reduction. PhD Thesis, TU Eindhoven, The Netherlands (2005)
28. Ionutiu, R., Lefteriu, S., Antoulas, A.C.: Comparison of model reduction methods with applications to circuit simulation. In: Proceedings SCEE-2006 Conference, Ciuprina, G., Ioan, D. (eds.) Springer Verlag, Berlin, 3–24 (2008)

29. Klaassen, F.M., De Graaff, H.C.: Compact transistor modeling for circuit design. Springer-Verlag, Wien, New York (1990)
30. Knockaert, L., De Zutter, D.: Laguerre-SVD reduced-order modeling. IEEE Trans. Microw. Theory Tech. **48**, 1469–1475 (2000)
31. Mayo, A.J., Antoulas, A.C.: A framework for the solution of the generalized realization problem. Lin. Alg. Appl. **425**, 634–662 (2007)
32. Odabasioglu, A., Celik, M.: PRIMA: passive reduced-order interconnect macromodeling algorithm. IEEE Trans. Comput-Aided Des. **17**, 645–654 (1998)
33. Pillage, L.T., Rohrer, R.A.: Asymptotic waveform evaluation for timing analysis. IEEE Trans. Comput-Aided Des. **9**, 352–366 (1990)
34. Rommes, J., Martins, N.: Computing large-scale system eigenvalues most sensitive to parameter changes, with applications to power system small-signal stability. Power Syst. **23**, 434–442 (2008)
35. Ruehli, A.E.: Equivalent circuit models for three dimensional multiconductor systems. IEEE Trans. Microw. Theory (1974)
36. Saraswat, D., Achar, R., Nakhla, M.S.: Global passivity enforcement algorithm for macromodels of interconnect subnetworks characterized by tabulated data. IEEE Trans. VLSI Syst. **13**, 819–832 (2005)
37. Schilders, W.H.A., Van der Vorst, H.A., Rommes, J. (eds.): Model order reduction: theory, research aspects and applications, Springer series Mathematics in Industry **13** (2008)
38. Schilders, W.H.A.: Solution of indefinite linear systems using an LQ decomposition for the linear constraints. Lin. Alg. Appl. **431**, 381–395 (2009)
39. Schilders, W.H.A.: Numerical methods for semiconductor device simulation. Springer Verlag, Heidelberg (to appear, 2009)
40. Schmidt, O., Halfmann, Th., Lang, P.: Coupling of numerical and symbolic techniques for model order reduction in circuit design. These proceedings (2009)
41. Schrik, E., Van der Meijs, N.P.: Comparing two YΔ based methodologies for realizable model reduction. In: ProRISC IEEE 14th Annual Workshop on Circuits, Systems and Signal Processing, 148–152 (2003)
42. Striebel, M., Rommes, J.: Model order reduction of nonlinear systems in circuit simulation: status and applications. This volume (2009)
43. Tan, S., He, L.: Advanced model order reduction techniques in VLSI design. Cambridge University Press (2008)
44. Ugryumova, M., Schilders, W.H.A.: Stability and passivity of the super node algorithm for EM modeling of IC's. In: Proceedings of the SCEE-2008 Conference, Springer-Verlag (to appear)
45. Vandendorpe, A., Van Dooren, P.: Model reduction of interconnected systems. In: Schilders, W.H.A., Van der Vorst, H.A. and Rommes, J. (eds.) Model order reduction: theory, research aspects and applications, Springer series Mathematics in Industry **13**, 305–321 (2008)
46. Verriest, E.I.: Time variant balancing and nonlinear balanced realizations. In: Schilders, W.H.A., Van der Vorst, H.A., Rommes, J. (eds.) Model order reduction: theory, research aspects and applications, Springer series Mathematics in Industry **13**, 213–250 (2008)
47. Verriest, E.I.: An approach to nonlinear balancing and MOR. These proceedings (2009)
48. Vladislavleva, E., Smits, G.F., Den Hertog, D.: Order of nonlinearity as a complexity measure for models generated by symbolic regression via Pareto genetic programming. IEEE Trans. Evol. Comp. **13**, 333–349 (2009)
49. Voller, V.R., Porté-Agel, F.: Moore's law and numerical modeling. J. Comp. Phys. **179**, 698–703 (2002)
50. Yang, F., Zeng, X., Su, Y., Zhou, D.: RLC equivalent circuit synthesis method for structure-preserved reduced-order model of interconnect in VLSI. Commun. Comput. Phys. **3**, 376–396 (2008)

Chapter 2
The SPRIM Algorithm for Structure-Preserving Order Reduction of General RCL Circuits

Roland W. Freund

Abstract In recent years, order-reduction techniques based on Krylov subspaces have become the methods of choice for generating macromodels of large-scale multi-port RCL networks that arise in VLSI circuit simulation. A popular method of this type is PRIMA. Its main features are provably passive reduced-order models and a Padé-type approximation property. On the other hand, PRIMA does not preserve other structures inherent to RCL circuits, which makes it harder to synthesize the PRIMA models as actual circuits. For the special case of RCL circuits without voltage sources, SPRIM was introduced as a structure-preserving variant of PRIMA that overcomes many of the shortcomings of PRIMA and at the same time, is more accurate than PRIMA. The purpose of this paper is twofold. First, we review the formulation of the equations characterizing general RCL circuits as descriptor systems. Second, we describe an extension of SPRIM to the case of general RCL circuits with voltage and current sources. We present some properties of the general SPRIM algorithm and report results of numerical experiments.

Keywords Krylov subspace · Structure preservation · Reduced-order model · Passivity · Electronic circuit · Interconnect

2.1 Introduction

The problem of order reduction of linear dynamical systems has a long history and many powerful methods for generating provably good reduced-order models have been developed. Most of this work was motivated by problems in control theory,

Roland W.Freund (✉)
Department of Mathematics, University of California at Davis,
One Shields Avenue, Davis, CA 95616, USA
e-mail: freund@math.ucdavis.edu

P. Benner et al. (eds.), *Model Reduction for Circuit Simulation*,
Lecture Notes in Electrical Engineering, 74, DOI: 10.1007/978-94-007-0089-5_2,

where, typically, the sizes of the dynamical systems to be reduced are relatively small to begin with. In the early 1990s, the continued miniaturization ('Moore's Law') of VLSI circuits lead to the need for efficient order reduction of large-scale linear dynamical systems with ever-increasing state-space dimension. The linear dynamical systems to be reduced in VLSI circuit simulation are RCL models of the VLSI circuit's interconnect system. These RCL models are relatively 'simple', but large, electronic networks with resistors, capacitors, inductors, voltage sources, and current sources as their only elements. Unlike the typical order-reduction problems in control theory, the large state-space dimension of these RCL circuits made the use of most of the more powerful reduction methods prohibitive until recently. Furthermore, RCL circuits are described by systems of differential-algebraic equations, resulting in so-called descriptor systems. Many of the existing order-reduction methods could not be used directly for descriptor systems at that time. Motivated by this 'new' application in VLSI circuit simulation, starting in the early 1990s, there has been tremendous interest in developing techniques for order reduction of large-scale descriptor systems.[1]

Reduced-order modeling techniques based on Padé or Padé-type approximation were quickly recognized to be powerful tools for the problems arising in VLSI circuit simulation. The first such technique was asymptotic waveform evaluation (AWE) [27], which uses explicit moment matching. More recently, the attention has moved to reduced-order models generated by means of Krylov-subspace algorithms, which avoid the typical numerical instabilities of explicit moment matching; see, e.g., the survey papers [12–14].

PVL [8, 9] and its multi-port version MPVL [10] use variants of the Lanczos process [23] to stably compute reduced-order models that represent Padé or matrix-Padé approximations [4] of the circuit transfer function. SyPVL [18] and its multi-port version SyMPVL [11, 19, 20] are versions of PVL and MPVL, respectively, that are tailored to RCL circuits. By exploiting the symmetry of RCL transfer functions, the computational costs of SyPVL and SyMPVL are only half of those of general PVL and MPVL.

Reduced-order modeling techniques based on the Arnoldi process [2], which is another popular Krylov-subspace algorithm, were first proposed in [7, 24–26, 31]. Arnoldi-based reduced-order models are defined by a certain Padé-type approximation property, rather than Padé approximation, and as a result, in general, they are not as accurate as a Padé-based model of the same size. In fact, Arnoldi-based models are known to match only half as many moments as Lanczos-based models; see [13, 24, 25, 31].

In many applications, in particular in VLSI interconnect analysis, the reduced-order model is used as a substitute for the full-blown original model in higher-level simulations. In such applications, it is very important for the reduced-order model to maintain the passivity properties of the original circuit. In [3, 19, 20], it is shown

[1] See Chap. 3 for recent progress in the adaptation of control-theoretic methods to large-scale descriptor systems.

that SyMPVL is passive for RC, RL, and LC circuits. However, the Padé-based reduced-order model that characterizes SyMPVL cannot be guaranteed to be passive for general RCL circuits. On the other hand, in [24–26], it was proved that the Arnoldi-based reduction technique PRIMA produces passive reduced-order models for general RCL circuits. PRIMA employs a block version of the Arnoldi process and then obtains reduced-order models by projecting the matrices defining the RCL transfer function onto the Arnoldi basis vectors. While PRIMA generates provably passive reduced-order models, it does not preserve other structures, such as reciprocity or the block structure of the circuit matrices, inherent to RCL circuits. This has motivated the development of the reduction technique SPRIM [15], which overcomes these disadvantages of PRIMA. In particular, SPRIM generates provably passive and reciprocal macromodels of multi-port RCL circuits. Furthermore, SPRIM models match twice as many moments as the corresponding PRIMA models obtained with identical computational work.

SPRIM was originally proposed in [15] for the special case of RCL circuits without voltage sources. The purpose of this paper is twofold. First, we review the formulation of the equations characterizing general RCL circuits as descriptor systems. Second, we describe an extension of SPRIM to the case of general RCL circuits with voltage and current sources. We present some properties of the general SPRIM algorithm and report results of numerical experiments.

The remainder of this paper is organized as follows. In Sect. 2.2, we review the equations describing general RCL circuits and the formulation of these equations as descriptor systems. In Sect. 2.3, we present some basic facts about generating reduced-order models of descriptor systems by means of Krylov subspace-based projection. In Sect. 2.4, we describe the SPRIM algorithm for the case of general RCL circuits. In Sect. 2.5, we make some comments about how to reduce the number of voltage sources before applying SPRIM to the remaining RCL network. In Sect. 2.6, we report the results of some numerical experiments. Finally, in Sect. 2.7, we mention some open problems and make some concluding remarks.

Throughout this paper the following notation is used. The set of real and complex numbers is denoted by $\mathbb{R}$ and $\mathbb{C}$, respectively. For (real or complex) matrices $M = \left[m_{jk}\right]$, we denote by $M^T = \left[m_{kj}\right]$ the *transpose* of M, and by $M^H := \left[\overline{m_{kj}}\right]$ the *Hermitian* (or *complex conjugate*) of M. The identity matrix is denoted by I and the zero matrix by 0; the actual dimensions of I and 0 will always be apparent from the context. The notation $M \succeq 0$ ($M \succ 0$) is used to indicate that a real or complex square matrix M is *Hermitian positive semidefinite* (*positive definite*). If all entries of the matrix $M \succeq 0$ ($M \succ 0$) are real, then M is said to be *symmetric positive semidefinite* (*positive definite*).

2.2 RCL Circuit Equations

We consider general RCL circuits driven by voltage and current sources. In this section, we review the formulation of such RCL circuits as linear time-invariant differential-algebraic equations (DAEs).

2.2.1 The Lumped-Element Approach

The lumped-element approach uses directed graphs to model electronic circuits; see, e.g., [6, 28, 30, 32]. The edges $\mathcal{E}$ of the graph correspond to the electronic elements of the circuit and the nodes $\mathcal{N}$ of the graph correspond to the interconnections of the electronic elements. In this subsection, we review this approach for the case of general RCL circuits driven by voltage and current sources.

The elements of such a general RCL circuit are its resistors, capacitors, inductors, voltage sources, and current sources. With each such element we associate an edge $e \in \mathcal{E}$, written as an ordered pair of nodes:

$$e = (n_1, n_2).$$

Here, $n_1, n_2 \in \mathcal{N}$ are the nodes of the graph representing the interconnections of the element to other circuit elements. We call n_1 the *tail node* of e and n_2 the *head node* of e. Note that the *direction* of e is from n_1 to n_2. For circuit elements for which the direction of the electric current through the element is known beforehand, the direction of e is chosen accordingly. For all other elements, an arbitrary direction of e is assigned. If the computed electric current through such an element is nonnegative, then the current flow is in the direction of e; otherwise, the actual current flow is against the direction of e.

The resulting directed graph $\mathcal{G} = (\mathcal{N}, \mathcal{E})$ can be described by its *incidence matrix* the rows and columns of which correspond to the nodes $n_j \in \mathcal{N}$ and edges $e_k \in \mathcal{E}$, respectively. To this end, we set

$$a_{jk} = \begin{cases} 1 & \text{if edge } e_k \text{ leaves node } n_j, \\ -1 & \text{if edge } e_k \text{ enters node } n_j, \\ 0 & \text{otherwise.} \end{cases}$$

The rows of this matrix add up to the zero row, and thus the matrix is rank deficient. In order to avoid this redundancy, we label one of the nodes as the *ground node* of the circuit and delete the corresponding row. We call the resulting matrix $\mathcal{A}$. It has $|\mathcal{N}| - 1$ rows and $|\mathcal{E}|$ columns, where $|\mathcal{N}|$ and $|\mathcal{E}|$ denote the number of nodes and edges of the graph $\mathcal{G}$, respectively. If $\mathcal{G}$ (viewed as an undirected graph) is connected, then the matrix $\mathcal{A}$ has full row rank:

$$\operatorname{rank} \mathcal{A} = |\mathcal{N}| - 1.$$

Note that $\mathcal{G}$ is indeed connected for any real electronic circuit, and thus, this condition of full row rank of $\mathcal{A}$ is always satisfied.

The matrix $\mathcal{A}$ allows an elegant formulation of the Kirchhoff's laws for the given RCL circuit. To this end, we denote by $i_{\mathcal{E}}$ the vector the entries of which are the currents along the edges $\mathcal{E}$, by $v_{\mathcal{E}}$ the vector the entries of which are the voltages across the edges $\mathcal{E}$, and by v the vector the entries of which are the voltages at the nodes $\mathcal{N}$, except for the ground node at which the voltage is zero. We remark that $i_{\mathcal{E}}$ and $v_{\mathcal{E}}$ are vectors of length $|\mathcal{E}|$ and v is a vector of length $|\mathcal{N}| - 1$. *Kirchhoff's current laws* (KCLs) can then be stated compactly as follows:

$$\mathcal{A} i_{\mathcal{E}} = 0. \tag{2.1}$$

Kirchhoff's voltage laws (KVLs) have the following compact formulation:

$$\mathcal{A}^T v = v_{\mathcal{E}}. \tag{2.2}$$

To obtain a complete characterization of the given RCL circuit, the so-called *branch constitutive relations* (BCRs) for the actual circuit elements need to be added to the Kirchhoff's laws (2.1) and (2.2). To formulate the BCRs, it is convenient to assume that the edges $e_k \in \mathcal{E}$ are numbered according to element type in the following order: resistors, capacitors, inductors, voltage sources, and current sources. The matrix $\mathcal{A}$ and the vectors $i_{\mathcal{E}}$ and $v_{\mathcal{E}}$ can thus be partitioned as follows:

$$\mathcal{A} = [\mathcal{A}_r \quad \mathcal{A}_c \quad \mathcal{A}_l \quad \mathcal{A}_v \quad \mathcal{A}_i], \quad i_{\mathcal{E}} = \begin{bmatrix} i_r \\ i_c \\ i_l \\ i_v \\ i_i \end{bmatrix}, \quad v_{\mathcal{E}} = \begin{bmatrix} v_r \\ v_c \\ v_l \\ v_v \\ v_i \end{bmatrix}. \tag{2.3}$$

Here, the subscripts r, c, l, v, and i refer to resistors, capacitors, inductors, voltage sources, and current sources, respectively. Using the partitions (2.3), the Kirchhoff's laws (2.1) and (2.2) can be written as follows:

$$\begin{aligned} &\mathcal{A}_r\, i_r + \mathcal{A}_c\, i_c + \mathcal{A}_l\, i_l + \mathcal{A}_v\, i_v + \mathcal{A}_i\, i_i = 0, \\ &\mathcal{A}_r^T v = v_r, \quad \mathcal{A}_c^T v = v_c, \quad \mathcal{A}_l^T v = v_l, \quad \mathcal{A}_v^T v = v_v, \quad \mathcal{A}_i^T v = v_i. \end{aligned} \tag{2.4}$$

Furthermore, the BCRs for the resistors, capacitors, and inductors can be stated in the following compact form:

$$v_r(t) = R i_r(t), \quad i_c(t) = C \frac{d}{dt} v_c(t), \quad v_l(t) = L \frac{d}{dt} i_l(t). \tag{2.5}$$

Here, R and C are diagonal matrices, the diagonal entries of which are the resistances of the resistors and the capacitances of the capacitors, respectively, and in particular, $R \succ 0$ and $C \succ 0$. The matrix L contains the inductances between the inductors as its entries. If mutual inductances are included, then L is a full matrix; otherwise, L is also a diagonal matrix. In both cases, $L \succ 0$.

Therefore, the matrices in (2.5) always satisfy

$$R \succ 0, \quad C \succ 0, \quad \text{and} \quad L \succ 0. \tag{2.6}$$

2.2.2 RCL Circuit Equations as Integro-DAEs

Equations 2.4 and 2.5, together with suitable initial conditions for some initial time t_0, completely characterize the time behavior of the RCL circuit. Here,

$$v_v(t) \quad \text{and} \quad i_i(t), \quad t \geq t_0, \tag{2.7}$$

are given functions, which describe the voltages and currents provided by the voltage and current sources, respectively. The functions

$$v(t), \quad v_r(t), \quad v_c(t), \quad v_l(t), \quad v_i(t), \quad i_r(t), \quad i_c(t), \quad i_l(t), \quad i_v(t) \tag{2.8}$$

are the unknown solutions of the Eqs. 2.4 and 2.5. These equations can be greatly simplified by using the BCRs (2.5) and the formulae for v_r, v_c, and v_l in (2.4) to eliminate all but the unknown functions $v(t)$ and $i_v(t)$.

To this end, we now assume that $t_0 = 0$ and that

$$i_l(0) = 0. \tag{2.9}$$

The third relation in (2.5) can then be rewritten in the form

$$i_l(t) = L^{-1} \int_0^t v_l(\tau) d\tau. \tag{2.10}$$

Using (2.10), the first two relations of (2.5), and the formulae for v_r, v_c, and v_l in (2.4), it follows that

$$i_r(t) = R^{-1} \mathcal{A}_r^T v(t), \quad i_c(t) = C \mathcal{A}_c^T \frac{d}{dt} v(t), \quad i_l(t) = L^{-1} \mathcal{A}_l^T \int_0^t v(\tau) d\tau. \tag{2.11}$$

Inserting these relations into (2.4), we obtain the coupled system of equations

$$\begin{aligned} E_{11} \frac{d}{dt} v(t) - A_{11} v(t) + \mathcal{A}_l L^{-1} \mathcal{A}_l^T \int_0^t v(\tau) d\tau + \mathcal{A}_v i_v(t) &= -\mathcal{A}_i i_i(t), \\ -\mathcal{A}_v^T v(t) &= -v_v(t) \end{aligned} \tag{2.12}$$

for the unknown functions $v(t)$ and $i_v(t)$. Here, we have set

$$E_{11} := \mathcal{A}_c C \mathcal{A}_c^T \quad \text{and} \quad A_{11} := -\mathcal{A}_r R^{-1} \mathcal{A}_r^T. \tag{2.13}$$

The system (2.12) is completed by adding initial values

$$v(0) \quad \text{and} \quad i_v(0) \tag{2.14}$$

for the solutions $v(t)$ and $i_v(t)$ at initial time $t_0 = 0$.

We remark that the system (2.12) with initial conditions (2.14) represents an *integro-differential-algebraic equation* (integro-DAE). This is the most compact formulation of the equations describing RCL circuits. By solving the integro-DAE (2.12), we obtain the vector $v(t)$ of nodal voltages and the vector $i_v(t)$ of currents through the voltage sources. The remaining circuit quantities, namely

$$v_r(t), \quad v_c(t), \quad v_l(t), \quad v_i(t), \quad i_r(t), \quad i_c(t), \quad \text{and} \quad i_l(t),$$

are then readily computed using the KVLs in (2.4), the first two BCRs in (2.5), and (2.10).

2.2.3 RCL Circuit Equations as Descriptor Systems

While integro-DAEs of the type (2.12) are the most compact formulation of RCL circuit equations, they are not directly amenable to Krylov subspace-based reduction techniques. Instead, we rewrite (2.13) as a descriptor system. This can be done by treating the vector $i_l(t)$ of inductor currents also as an unknown function, along with the functions $v(t)$ and $i_v(t)$. We denote by

$$x(t) := \begin{bmatrix} v(t) \\ i_l(t) \\ i_v(t) \end{bmatrix} \tag{2.15}$$

the resulting new *state-space vector* of unknowns. Furthermore, we define vectors of input and output functions

$$u(t) := \begin{bmatrix} -i_i(t) \\ v_v(t) \end{bmatrix} \quad \text{and} \quad y(t) := \begin{bmatrix} v_i(t) \\ -i_v(t) \end{bmatrix}. \tag{2.16}$$

Recall from (2.12) that the entries of the *input vector* $u(t)$ are all given functions. The *output vector* $y(t)$ is readily obtained from the state-space vector (2.15). Indeed, using the relation $v_i(t) = \mathcal{A}_i^T v(t)$ from (2.4), it follows that

$$y(t) = B^T x(t), \quad \text{where} \quad B := \begin{bmatrix} \mathcal{A}_i & 0 \\ 0 & 0 \\ 0 & -I \end{bmatrix}. \tag{2.17}$$

We now re-introduce $i_l(t)$ into (2.12). To this end, we employ the formula for $i_l(t)$ stated in (2.11) and the differentiated form thereof:

$$L \frac{d}{dt} i_l(t) = \mathcal{A}_l^T v(t).$$

The resulting equivalent version of the system (2.12) can be stated as follows:

$$\begin{aligned} E_{11}\frac{d}{dt}v(t) - A_{11}v(t) + \mathcal{A}_l i_l(t) + \mathcal{A}_v i_v(t) &= -\mathcal{A}_i i_i(t), \\ L\frac{d}{dt}i_l(t) - \mathcal{A}_l^T v(t) &= 0, \\ -\mathcal{A}_v^T v(t) &= -v_v(t). \end{aligned} \tag{2.18}$$

Finally, setting

$$E := \begin{bmatrix} E_{11} & 0 & 0 \\ 0 & L & 0 \\ 0 & 0 & 0 \end{bmatrix} \quad \text{and} \quad A := \begin{bmatrix} A_{11} & -\mathcal{A}_l & -\mathcal{A}_v \\ \mathcal{A}_l^T & 0 & 0 \\ \mathcal{A}_v^T & 0 & 0 \end{bmatrix}, \tag{2.19}$$

and adding the relation (2.17) and (2.18), we obtain the following formulation of the RCL circuit equations as a *descriptor system*:

$$\begin{aligned} E\frac{d}{dt}x(t) &= Ax(t) + Bu(t), \\ y(t) &= B^T x(t). \end{aligned} \tag{2.20}$$

The system (2.20) is completed by the initial conditions from (2.9) and (2.14):

$$x(0) = x_0 := \begin{bmatrix} v(0) \\ 0 \\ i_v(0) \end{bmatrix}.$$

In the following, we denote by N and m the length of the state-space vector $x(t)$ and the length of the input vector $u(t)$ of (2.20), respectively. The number N is called the *state-space dimension* of the descriptor system (2.20). We remark that m is also the length of the vector-valued output vector $y(t)$ of (2.20). In particular, A and E are $N \times N$ matrices, and B is an $N \times m$ matrix. Note that, in view of (2.16), m is the total number of voltage and current sources in the RCL circuit.

The formulation (2.20) is essential for using Krylov subspace techniques for model order reduction of RCL circuits. Furthermore, the matrices A and E in (2.20) need to be such that the *matrix pencil*

$$sE - A, \quad s \in \mathbb{C}, \tag{2.21}$$

is *regular*, i.e., the matrix $sE - A$ is singular only for finitely many values of $s \in \mathbb{C}$.

Regularity of the matrix pencil (2.21) is equivalent to certain rank conditions involving the subblocks of the matrix $\mathcal{A}$ in (2.3). Indeed, the matrix pencil (2.21) is regular if, and only if, the matrix

$$\mathcal{A}_v \quad \text{has full column rank} \tag{2.22}$$

and the matrix

$$\begin{bmatrix} \mathcal{A}_r & \mathcal{A}_c & \mathcal{A}_l & \mathcal{A}_v \end{bmatrix} \quad \text{has full row rank.} \tag{2.23}$$

For an elementary proof of this characterization of regularity, we refer the reader to [17, Theorem 1].

The rank conditions (2.22) and (2.23) have simple interpretations in terms of the RCL circuit described by the descriptor system (2.20). Condition (2.22) means that the subcircuit consisting of only the voltage sources of the given RCL circuit has no closed (undirected) loops. Condition (2.23) means that the (undirected) graph corresponding to the subcircuit obtained from the given RCL circuit by deleting all current sources is still connected. Both these conditions are satisfied for any practically relevant RCL circuit, and thus from now on, we always assume that the matrix pencil $sE - A$ associated with the descriptor system (2.20) is regular. Furthermore, we stress that the occurrence of a singular matrix pencil is a strong indication that some error was made in the design or the modeling of the RCL circuit described by (2.20).

2.2.4 Passivity

Any RCL circuit is *passive*, i.e., it consumes energy (provided by the voltage and current sources), but does not generate energy.

One of the possible mathematical characterizations of passivity uses the concept of the transfer function of a descriptor system. Recall that the matrix pencil (2.21) is assumed to be regular. Therefore, we can define an $m \times m$-matrix-valued function by setting

$$H(s) := B^T(sE - A)^{-1}B, \quad s \in \mathbb{C}. \tag{2.24}$$

Note that H is a rational function, with possible poles at those finitely many values of $s \in \mathbb{C}$ for which the matrix $sE - A$ is singular. The function (2.24) is called the *transfer function* of the descriptor system (2.20).

The function (2.24), H, is said to be *positive real* if it has no poles in the right-half

$$\mathbb{C}_+ := \{s \in \mathbb{C} \mid \operatorname{Re} s > 0\}$$

of the complex plane and

$$H(s) + (H(s))^H \succeq 0 \quad \text{for all} \quad s \in \mathbb{C}_+.$$

It is well known that a system described by (2.20) is passive if, and only if, the transfer function of (2.20) is positive real, see, e.g., [1].

Since RCL circuits are passive, it thus follows that the transfer functions of the circuit equations stated as descriptor systems (2.20) are guaranteed to be positive real. An alternative proof of this fact uses the following simple theorem, which can be found as Theorem 13 in [13].

Theorem 1 *Let $A, E \in \mathbb{R}^{N \times m}$ and $B \in \mathbb{R}^{N \times m}$. Assume that*

$$E = E^T \succeq 0, \quad A + A^T \preceq 0, \tag{2.25}$$

and that the matrix pencil $sE - A$ is regular. Then, the function (2.24), H, is positive real.

In view of this theorem, we only need to verify that the matrices A and E defined in (2.19) satisfy the conditions (2.25). Note that by (2.6) and (2.13), we have

$$E_{11} = E_{11}^T \succeq 0, \quad L = L^T \succ 0, \quad \text{and} \quad A_{11} = A_{11}^T \preceq 0.$$

Together with (2.19) it follows that

$$E = E^T \succeq 0 \quad \text{and} \quad A + A^T = \begin{bmatrix} 2A_{11} & 0 & 0 \\ 0 & 0 & 0 \\ 0 & 0 & 0 \end{bmatrix} \preceq 0.$$

Hence the transfer functions of RCL circuit equations stated as descriptor systems (2.20) are indeed positive real.

2.3 Projection-Based Order Reduction

In this section, we review some basic facts about generating reduced-order models of descriptor systems (2.20) by means of projection. From now on, we always assume that the system (2.20) describes a given RCL circuit. In particular, the matrices A and E are of the form (2.19), and the matrix B is of the form (2.17).

2.3.1 Reduced-Order Models

A general *reduced-order model* of (2.20) is a descriptor system of the same form as (2.20), but with a reduced state-space dimension $n < N$, instead of N. Thus a reduced-order model is of the form

$$\begin{aligned} E_n \frac{d}{dt} \tilde{x}(t) &= A_n \tilde{x}(t) + B_n u(t), \\ \tilde{y}(t) &= B_n^T \tilde{x}(t), \end{aligned} \tag{2.26}$$

where $A_n, E_n \in \mathbb{R}^{n \times n}$ and $B_n \in \mathbb{R}^{n \times m}$. Note that the input vector $u(t)$ is the same as in (2.20). In particular, the number m is unchanged from (2.20). The output vector $\tilde{y}(t)$ of (2.26) is only an approximation of the original output vector $y(t)$ of (2.20). In fact, the problem of order reduction is to find a sufficiently large reduced state-space dimension n and matrices A_n, E_n, and B_n such that the output vector of the reduced-order model (2.26) is a 'sufficiently good' approximation of the output vector of the original system (2.20).

Provided that the matrix pencil

$$sE_n - A_n, \quad s \in \mathbb{C}, \tag{2.27}$$

associated with (2.26) is regular, we can define a transfer function as before:

$$H_n(s) := B_n^T (sE_n - A_n)^{-1} B_n, \quad s \in \mathbb{C}. \tag{2.28}$$

In terms of transfer functions, the problem of order reduction is to find a sufficiently large reduced state-space dimension n and matrices A_n, E_n, and B_n such that the transfer function (2.28), H_n, of the reduced-order model is a 'sufficiently good' approximation of the transfer function (2.24), H of the original system:

$$H_n(s) \approx H(s) \quad \text{in 'some sense'}. \tag{2.29}$$

2.3.2 *Order Reduction Via Projection*

A very basic approach to constructing reduced-order models is to employ *projection*. Let

$$V_n \in \mathbb{R}^{N \times n} \quad \text{with} \quad \operatorname{rank} V_n = n \tag{2.30}$$

be given. Then, by simply setting

$$A_n := V_n^T A V_n, \quad E_n := V_n^T E V_n, \quad \text{and} \quad B_n := V_n^T B, \tag{2.31}$$

one obtains a reduced-order model (2.26) that can be viewed as a projection of the N-dimensional state space of the original system onto the n-dimensional subspace spanned by the columns of the matrix V_n. In particular, projection employs an ansatz of the form

$$\tilde{x}(t) = V_n x(t)$$

for the state-space vector $\tilde{x}(t)$ of the reduced-order model (2.26). Recall that $x(t)$ denotes the state-space vector of the original system (2.20).

There are two appealing aspects of the projection approach:

1. Reduced-order models obtained by means of projection trivially preserve passivity of the original system; see, e.g., [24–26]. Indeed, the only additional condition on the matrix (2.30), V_n, is that the resulting matrix pencil (2.27) is regular. Recall that A and E satisfy the semidefiniteness properties (2.25). The definitions of A_n and E_n in (2.31) readily imply that these matrices also satisfy (2.25). Therefore, by Theorem 1, the transfer function (2.28), H_n is positive real, and thus the corresponding reduced-order model (2.26) is passive.

2. By choosing the matrix V_n such that the subspace spanned by its columns contains a certain Krylov subspace, one obtains reduced-order models for which (2.28), H_n, is a Padé-type approximation of the original transfer function (2.24), H; see Sect. 2.3.4 below.

2.3.3 Block Krylov Subspaces

Suppose we want to evaluate the transfer function (2.24), H, of the descriptor system (2.20) at some point $s_0 \in \mathbb{C}$ at which the matrix $s_0E - A$ is nonsingular. The most efficient way of obtaining $H(s_0)$ is to first solve the system of linear equations

$$(s_0E - A)X(s_0) = B \tag{2.32}$$

for $X(s_0)$ and then compute

$$H(s_0) = B^T X(s_0).$$

The coefficient matrix $s_0E - A$ of (2.32) is large, but sparse, and as it is the case for all practical circuit matrices, a sparse LU factorization of this matrix can be computed with typically little fill-in; see, e.g., [30, 32].

The basic idea behind Krylov subspace-based order reduction for electronic circuits is to reuse the LU factorization, which was computed to obtain $H(s_0)$, to generate the information contained in the leading Taylor coefficients of H expanded about s_0. To this end, we rewrite the transfer function (2.24) as follows:

$$H(s) = B^T\big(s_0E - A + (s - s_0)E\big)^{-1}B = B^T\big(I + (s - s_0)M\big)^{-1}R, \tag{2.33}$$

where

$$M := (s_0E - A)^{-1}E \quad \text{and} \quad R := (s_0E - A)^{-1}B. \tag{2.34}$$

In view of (2.33), the Taylor expansion of H about s_0 is given by

$$H(s) = \sum_{j=0}^{\infty}(-1)^j B^T M^j R \quad (s - s_0)^j. \tag{2.35}$$

The leading Taylor coefficients of H can thus be obtained by computing inner products of the columns of the matrix B and the leading columns of the *block Krylov matrix*

$$\begin{bmatrix} R & MR & M^2R & \cdots & M^{j-1}R & \cdots \end{bmatrix}. \tag{2.36}$$

Let $N_{\max}(\leq N)$ denote the rank of this matrix. Then for $\hat{n} = 1, 2, \ldots, N_{\max}$, the $\hat{n}$th *block Krylov subspace* (induced by M and R) is defined as the $\hat{n}$-dimensional subspace of $\mathbb{C}^N$ spanned by the first $\hat{n}$ linearly independent columns of the block Krylov matrix (2.36). In the following, $\mathcal{K}_{\hat{n}}(M,R)$ denotes this $\hat{n}$th block Krylov subspace. Note that, by construction, $\mathcal{K}_{\hat{n}}(M,R)$ contains the necessary information to generate at least the first

$$\left\lfloor \frac{\hat{n}}{m} \right\rfloor \tag{2.37}$$

Taylor coefficients of the expansion (2.35) of H about s_0. In the generic case, the integer (2.37) is the exact number of Taylor coefficients that can be obtained from $\mathcal{K}_{\hat{n}}(M,R)$. However, in certain degenerate cases, one can obtain even more coefficients; we refer the reader to [16, 17] for a complete characterization of the exact number of coefficients.

We remark that for actual numerical computations, this definition of $\mathcal{K}_{\hat{n}}(M,R)$ is useless. Instead, one employs a suitable Krylov subspace algorithm to generate a numerically well-behaved basis of $\mathcal{K}_{\hat{n}}(M,R)$. One such algorithm is the band Arnoldi process described in [14]. It produces orthonormal basis vectors for the subspaces $\mathcal{K}_{\hat{n}}(M,R)$. We remark that Krylov subspace algorithms involve the matrix M only in the form of matrix-vector products Mv. For the computation of these, the matrix M never needs to be formed explicitly. Instead, in view of the definition of M in (2.34), each matrix-vector product Mv can be obtained via one sparse multiplication with E and two sparse triangular solves with the LU factors of the matrix $s_0E - A$.

Finally, we stress that for general $s_0 \in \mathbb{C}$, $\mathcal{K}_{\hat{n}}(M,R)$ is a subspace of $\mathbb{C}^N$. If we restrict s_0 to be real, then the matrices M and R are real and $\mathcal{K}_{\hat{n}}(M,R)$ is a subspace of $\mathbb{R}^N$.

2.3.4 *Krylov Subspace-Based Projection*

We now employ the block Krylov subspaces defined in Sect. 2.3.3 to choose suitable projection matrices V_n.

Recall from (2.30) that the matrices V_n are assumed to be real. For the sake of generating reduced-order models, this assumption is not essential, and in fact, complex $N \times n$ matrices can be used as well. However, in order to obtain passive models, the matrices V_n need to be real. To guarantee this condition, from now on we assume that

$$s_0 \in \mathbb{R}, \tag{2.38}$$

so that $\mathcal{K}_{\hat{n}}(M,R)$ is a subspace of $\mathbb{R}^N$. At the end of this subsection, we include some remarks about how to proceed in the case of truly complex $s_0 \in \mathbb{C} \setminus \mathbb{R}$.

For Krylov subspace-based projection, we choose the matrix (2.30), V_n, such that

$$\mathcal{K}_{\hat{n}}(M,R) \subseteq \text{colspan } V_n. \tag{2.39}$$

Since, by construction, $\mathcal{K}_{\hat{n}}(M,R)$ has dimension $\hat{n}$ and, by (2.30), V_n is assumed to have rank n, the condition (2.39) implies that

$$\hat{n} \leq n. \tag{2.40}$$

An obvious choice for V_n is a matrix whose columns form a basis of $\mathcal{K}_{\hat{n}}(M,R)$, such as the vectors generated by the band Arnoldi process. In this case, we have $\hat{n} = n$ in (2.40).

Let A_n, E_n, and B_n be the matrices (2.31), and let H_n be the transfer function (2.28) of the corresponding reduced-order model (2.26). The main result of Krylov subspace-based projection then states that H_n is a *Padé-type approximation* of the original transfer function (2.24), H, in the following sense:

$$H_n(s) = H(s) + \mathcal{O}\Big((s - s_0)^{j(\hat{n})}\Big), \quad \text{where} \quad j(\hat{n}) \geq \left\lfloor \frac{\hat{n}}{m} \right\rfloor. \tag{2.41}$$

Moreover, in the generic case, $j(\hat{n}) = \lfloor \hat{n}/m \rfloor$ in (2.41). The property (2.41) is well known. For example, it was established for various special cases in [5, 21, 24]. Proofs for the general case be found in [13, 16]. We remark that the matrix-valued coefficients of the Taylor expansion of the transfer function (2.24), H, about the expansion point s_0 are often called *moments* and that the Padé-type approximation (2.41) is also referred to as *moment matching*.

If instead of (2.38), the expansion point s_0 is chosen to be truly complex, then $\mathcal{K}_{\hat{n}}(M,R)$ is a complex subspace and the projection matrix V_n in (2.39) will be complex as well in general. One possibility of keeping the projection matrix real is to replace the complex matrix V_n satisfying (2.39) by the real matrix

$$[\,\text{Re } V_n \quad \text{Im } V_n\,]. \tag{2.42}$$

The obvious disadvantage of this approach is that the dimension of the reduced-order model is doubled. Furthermore, in general, the matrix (2.42) is not guaranteed to have full column rank, and so before using (2.42) as a projection matrix, one would need to check for and possibly delete any linearly dependent columns of (2.42). On the other hand, the transfer function of the resulting reduced-order model will satisfy a Padé-type property of the form (2.41) for both s_0 and the complex conjugate expansion point $\overline{s_0}$; see, e.g., [22].

2.3.5 Structure Preservation

Recall from (2.19) and (2.17) that the matrices A, E, and B exhibit certain block structures reflecting the fact that the descriptor system (2.20) describes an RCL circuit. As long as the expansion point s_0 is chosen to be real, the reduced-order model generated via projection preserves the passivity of the original RCL circuit, but not these block structures of the data matrices. In fact, in general V_n will be a dense matrix, and then the data matrices (2.31) of the reduced-order model will be dense matrices as well.

For example, consider the 'minimal' choice of the matrix V_n in (2.39), i.e., $\hat{n} = n$ and V_n is chosen as any matrix whose columns span $\mathcal{K}_{\hat{n}}(M, R)$. The resulting order reduction method is mathematically equivalent to PRIMA [24–26]. However, in general the data matrices of the PRIMA reduced-order model are dense and do not preserve the block structures of the original matrices A, E, and B. In the next section, we describe how these structures can indeed be preserved by taking advantage of the fact that in (2.39) we can choose a matrix V_n with $n > \hat{n}$.

The reader may ask why preservation of the block structures of the matrices A, E, and B is important. There are two reasons:

1. In practice, one would like to synthesize the system described by the reduced-order model (2.26) as an actual physical electronic circuit. It is well known that passivity of the reduced-order model is sufficient to guarantee the existence of such a synthesized circuit; see, e.g., [1]. However, passivity alone is not enough to guarantee synthesis as an RCL circuit, and in general, some other circuit elements are needed. Additional properties, such as *reciprocity*, of the reduced-order model are necessary to ensure synthesis as an RCL circuit. While preserving the block structures of the data matrices of the original descriptor system alone is not enough to guarantee synthesis as an RCL circuit, it ensures reciprocity and significantly increases the probability that the reduced-order model can be synthesized as an actual RCL circuit.
2. Preserving the block structure also doubles the number of Taylor coefficients that are matched in the Padé-type approximation property of the reduced-order model. More precisely, the structure-preserving SPRIM algorithm matches twice as many coefficients as the non-structure-preserving PRIMA algorithm; see Eq. 2.47 in Sect. 2.4.3 below.

2.4 The SPRIM Algorithm

The SPRIM (*Structure-Preserving Reduced-order Interconnect Macromodeling*) algorithm was originally introduced in [15] for the special case of RCL circuits with only current sources, but no voltage sources. In this case, the third block row and column of A and E in (2.19) and the third block row of E in (2.17) are non-existent. This significantly simplifies both the implementation of SPRIM and the derivation of certain theoretical properties of SPRIM. In this section, we describe the SPRIM algorithm for the case of general RCL circuits with both voltage and current sources.

2.4.1 Preserving the Block Structures

The main computational step of SPRIM is the generation of a suitable basis for the $\hat{n}$th block Krylov subspace $\mathcal{K}_{\hat{n}}(M,R)$. This step is identical to what is done in PRIMA. Let $\hat{V}_{\hat{n}}$ be the resulting matrix, the columns of which form a basis of $\mathcal{K}_{\hat{n}}(M,R)$. PRIMA employs this matrix as the projection matrix to obtain the reduced-order data matrices (2.31). As pointed out before, in general, these PRIMA data matrices are dense and thus do not preserve the block structures of the original matrices A, E, and B.

Instead of using the matrix $\hat{V}_{\hat{n}}$ directly for the projection, SPRIM employs a modified version of this matrix that trivially leads to structure preservation. To this end, $\hat{V}_{\hat{n}}$ is first partitioned as follows:

$$\hat{V}_{\hat{n}} = \begin{bmatrix} V^{(1)} \\ V^{(2)} \\ V^{(3)} \end{bmatrix}. \tag{2.43}$$

Here, the block sizes correspond to the block sizes of A and E in (2.19). While $\hat{V}_{\hat{n}}$ has full column rank $\hat{n}$, the same is not necessarily true for the three subblocks $V^{(l)}$, $l = 1, 2, 3$, in (2.43). In particular, the third block, $V^{(3)}$ is of size $n_v \times \hat{n}$, where n_v denotes the number of voltage sources of the given RCL circuit. The number n_v is very small and usually $n_v < \hat{n}$. Therefore, $V^{(3)}$ typically does not have full column rank. In the actual implementation of SPRIM, we run a Gram-Schmidt algorithm on the rows of $V^{(3)}$ to determine a matrix $\tilde{V}^{(3)}$ the columns of which span (instead of spans) the same space as the columns of $V^{(3)}$, but has full column rank. The other two blocks usually have many more rows than columns. and these blocks are unlikely not to have full column rank. In the actual implementation of SPRIM, there is the option to check the column ranks of the first two blocks and replace them by matrices $\tilde{V}^{(1)}$ and $\tilde{V}^{(2)}$ of full columns rank. Next, we set up the actual projection matrix V_n as follows:

$$V_n := \begin{bmatrix} \tilde{V}^{(1)} & 0 & 0 \\ 0 & \tilde{V}^{(2)} & 0 \\ 0 & 0 & \tilde{V}^{(3)} \end{bmatrix}. \tag{2.44}$$

By construction, we have

$$\mathcal{K}_{\hat{n}}(M,R) = \text{colspan}\hat{V}_{\hat{n}} \subseteq \text{colspan } V_n. \tag{2.45}$$

Thus the matrix (2.44), V_n, satisfies the condition (2.39), which in turn guarantees the Padé-type approximation property (2.41). Furthermore, in view of the block structure of V_n, the data matrices (2.31) of the resulting reduced-order model obtained via projection with V_n have the same block structure as the original data matrices A, E, and B.

2.4.2 The Algorithm

The order reduction procedure outlined in the previous subsection is the SPRIM algorithm for general RCL circuits. SPRIM can be formulated as an actual algorithm in the following form.

Algorithm 2 (SPRIM for general RCL circuits)

- Input: matrices of the form

$$A = \begin{bmatrix} A_{11} & -\mathcal{A}_l & -\mathcal{A}_v \\ \mathcal{A}_l^T & 0 & 0 \\ \mathcal{A}_v^T & 0 & 0 \end{bmatrix}, \quad E = \begin{bmatrix} E_{11} & 0 & 0 \\ 0 & L & 0 \\ 0 & 0 & 0 \end{bmatrix}, \quad B = \begin{bmatrix} \mathcal{A}_i & 0 \\ 0 & 0 \\ 0 & -I \end{bmatrix},$$

  where $A_{11} \preceq 0, E_{11} \succeq 0$, and $L \succ 0$;
  an expansion point $s_0 \in \mathbb{R}$.

- Formally set

$$M = (s_0 E - A)^{-1} E, \quad R = (s_0 E - A)^{-1} B.$$

- Until $\hat{n}$ is large enough, run your favorite block Krylov subspace method (applied to M and R) to construct the columns of the basis matrix

$$\hat{V}_{\hat{n}} = \begin{bmatrix} v_1 & v_2 & \cdots & v_{\hat{n}} \end{bmatrix}$$

  of the $\hat{n}$th block Krylov subspace $\mathcal{K}_{\hat{n}}(M, R)$, i.e.,

$$\text{colspan } \hat{V}_{\hat{n}} = \mathcal{K}_{\hat{n}}(M, R).$$

- Let

$$\hat{V}_{\hat{n}} = \begin{bmatrix} V^{(1)} \\ V^{(2)} \\ V^{(3)} \end{bmatrix}$$

  be the partitioning of $V_{\hat{n}}$ corresponding to the block sizes of A and E.

- For l = 1, 2, 3 do:
  If $r_l := \text{rank } V^{(l)} < \hat{n}$, determine an $N \times r_l$ matrix $\tilde{V}^{(l)}$ with

$$\text{colspan } \tilde{V}^{(l)} = \text{colspan } V^{(l)} \quad \text{and} \quad \text{rank } \tilde{V}^{(l)} = r_l.$$

- Set

$$\tilde{A}_{11} = \left(\tilde{V}^{(1)}\right)^T A_{11} \tilde{V}^{(1)}, \quad \tilde{\mathcal{A}}_l = \left(\tilde{V}^{(1)}\right)^T \mathcal{A}_l \tilde{V}^{(2)}, \quad \tilde{\mathcal{A}}_v = \left(\tilde{V}^{(1)}\right)^T \mathcal{A}_v \tilde{V}^{(3)},$$

$$\tilde{E}_{11} = \left(\tilde{V}^{(1)}\right)^T E_{11} \tilde{V}^{(1)}, \quad \tilde{L} = \left(\tilde{V}^{(2)}\right)^T L \tilde{V}^{(2)}, \quad \tilde{\mathcal{A}}_i = \left(\tilde{V}^{(1)}\right)^T \mathcal{A}_i.$$

- Output: the data matrices

$$A_n = \begin{bmatrix} \tilde{A}_{11} & -\tilde{A}_l & -\tilde{A}_v \\ \tilde{A}_l^T & 0 & 0 \\ \tilde{A}_v^T & 0 & 0 \end{bmatrix}, \quad E_n = \begin{bmatrix} \tilde{E}_{11} & 0 & 0 \\ 0 & \tilde{L} & 0 \\ 0 & 0 & 0 \end{bmatrix}, \quad \text{and}$$

$$B_n = \begin{bmatrix} \tilde{A}_i & 0 \\ 0 & 0 \\ 0 & -\left(\tilde{V}^{(3)}\right)^T \end{bmatrix},$$

of the SPRIM reduced-order model

$$\begin{aligned} E_n \frac{d}{dt}\tilde{x}(t) &= A_n \tilde{x}(t) + B_n u(t), \\ \tilde{y}(t) &= B_n^T \tilde{x}(t). \end{aligned} \tag{2.46}$$

2.4.3 The Padé-Type Approximation Property of SPRIM

Recall that SPRIM is a Krylov subspace-based projection method. In particular, the matrix V_n used in the projection satisfies the condition (2.39), which implies the Padé-type approximation property (2.41). It turns out that SPRIM actually has a stronger Padé-type approximation property. More precisely, the transfer function H_n of the SPRIM reduced-order model (2.46) satisfies

$$H_n(s) = H(s) + \mathcal{O}\left(\left(s - s_0\right)^{2j(\hat{n})}\right), \quad \text{where} \quad j(\hat{n}) \geq \left\lfloor \frac{\hat{n}}{m} \right\rfloor, \tag{2.47}$$

instead of (2.41). The integer $j(\hat{n})$ is the same as in (2.41), and in the generic case, $j(\hat{n}) = \lfloor \hat{n}/m \rfloor$ in (2.47). The property means that at the expansion point s_0, the transfer function of the SPRIM reduced-order model matches twice as many leading Taylor coefficients as the theory of general Krylov subspace-based projection methods predicts.

This higher accuracy of SPRIM is a consequence of structure preservation. Projection methods that do not preserve the structures of the original data matrices, such as PRIMA, do not satisfy the more accurate Padé-type property (2.47).

The proof of (2.47) is relatively straightforward for the special case of RCL circuits without voltage sources; see [15]. In this case, the third block row and column of the matrices A and E and the third block row of the matrix B are empty, and one can easily make both matrices A and E symmetric by changing the sign of the second block row, without changing B. This symmetry, along with the corresponding symmetry of the data matrices of the SPRIM reduced-order models, can be used to establish (2.47). In the case of general RCL circuits, it is no longer possible to make A and E symmetric without changing B, and a different approach to proving (2.47) is required. The key observation here is that the matrices A and E are *J-symmetric*, i.e., they satisfy

$$JA = A^T J \quad \text{and} \quad JE = E^T J,$$

with respect to the indefinite matrix

$$J = \begin{bmatrix} I & 0 & 0 \\ 0 & -I & 0 \\ 0 & 0 & -I \end{bmatrix},$$

where the I's denote identity matrices of the same size as the diagonal blocks of A and E. Since the matrices A_n and E_n of the SPRIM reduced-order model have the same block structure as A and E, respectively, the matrices A_n and E_n are J_n-symmetric with respect to a 'reduced' version J_n of J. Finally, the projection matrix (2.44), V_n, employed in SPRIM is *compatible* with J and J_n in the sense that

$$JV_n = V_n J_n. \tag{2.48}$$

In [16], we developed a general theory of Krylov subspace-based projection of J-symmetric descriptor systems. More precisely, we showed that the stronger Padé-type approximation property (2.47) holds true, provided that the data matrices of the reduced-order models are J_n-symmetric and the compatibility condition (2.48) is satisfied. By applying this more general result [16, Theorem 9] to SPRIM, we obtain its stronger Padé-type approximation property (2.47).

2.5 Treatment of Voltage Sources

Recall from (2.19) that the third block row and column of the matrices A and E arise due to the presence of voltage sources in the given RCL circuit. Note that the size of the third block rows is $n_v \times N$ and the size of the third block columns is $N \times n_v$. Here, n_v denotes the number of voltage sources, which is usually very small. In SPRIM, the corresponding third block $V^{(3)}$ of the matrix (2.43), $\hat{V}_{\hat{n}}$ is usually rank deficient and thus needs to be replaced by a block $\tilde{V}^{(3)}$ of full column rank, see Algorithm 2.

In many cases, it is actually possible to treat all or at least some of the voltage sources separately and to apply the model reduction algorithm itself to a slightly smaller RCL circuit. The reason is that voltage sources are usually connected to the ground node of the circuit and they are connected to the remaining RCL circuit by a resistor. Such a case is illustrated in Fig. 2.1, which shows an RCL circuit with 4 such voltage sources connected to the ground node and by a resistor to the remaining RCL network. Recall from (2.16) that the given voltages of the voltage sources are part of the input vector $u(t)$ of the descriptor system (2.20) and from (2.16) that the unknown currents through the voltage sources and the voltages at

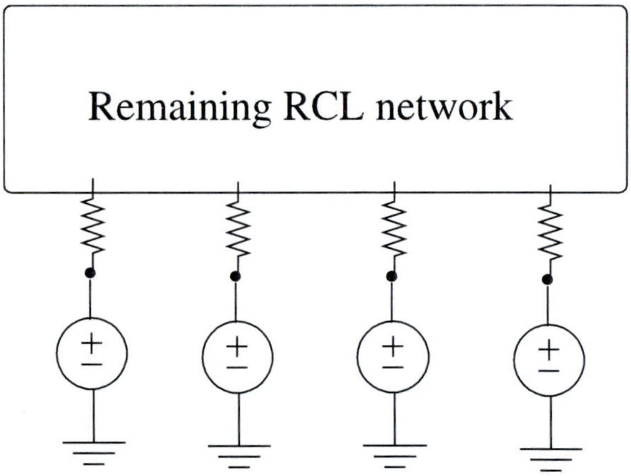

Fig. 2.1 An RCL circuit for which 4 voltage sources can be eliminated before model reduction is applied

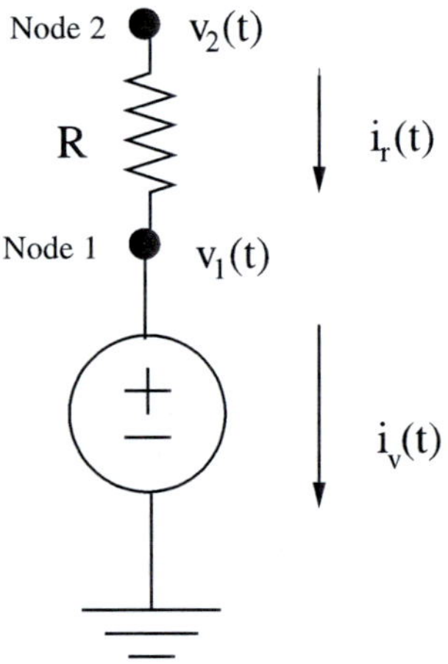

Fig. 2.2 A typical voltage source connected to the ground node and by a single resistor to the remaining RCL network

the circuit nodes are part of the unknown state-vector $x(t)$ of (2.20). For voltage sources connected to the remaining RCL network as in Fig. 2.1, there are some trivial relations between some of these circuit quantities, which can be used to reduce the state-space dimension of (2.20). More precisely, consider a typical voltage source connected to ground and to a resistor, as shown in Fig. 2.2. In this case, the voltage $v_1(t)$ at node 1 is equal to the voltage $v_v(t)$ provided by the voltage source:

$$v_1(t) = v_v(t). \tag{2.49}$$

By Kirchhoff's current law, the unknown current $i_v(t)$ through the voltage source is equal to the unknown current $i_r(t)$ through the resistor. Together with Ohm's law for resistors, it follows that

$$i_v(t) = i_r(t) = \frac{1}{R}(v_2(t) - v_1(t)), \tag{2.50}$$

where $v_2(t)$ denotes the unknown voltage at node 2 and R is the given resistance of the resistor. Using the two relations (2.49) and (2.50), we can eliminate the unknowns $v_1(t)$ and $i_v(t)$ from the circuit equation, thus reducing the state-space dimension by 2.

The above procedure can be carried out for all voltage sources that are connected to the ground node and by a single resistor to the remaining RCL network. If $n_v^{(e)}$ denotes the number of such voltage sources in the given RCL circuit, then the state-space dimension of the resulting descriptor system is $N^{(r)} := N - 2n_v^{(e)}$. Here, N is the state-space dimension of the original descriptor system (2.20).

It is relatively straightforward to carry out this elimination on the matrices of the original system (2.20). Here, we omit the full details and just state the final result. The given RCL circuit is again described by a descriptor system that now has the following form:

$$\begin{aligned} E^{(r)}\frac{d}{dt}x^{(r)}(t) &= A^{(r)}x^{(r)}(t) + B^{(r)}u(t), \\ y(t) &= Du(t) + \left(C^{(r)}\right)^T x^{(r)}(t). \end{aligned} \tag{2.51}$$

Here, $A^{(r)}$ and $E^{(r)}$ are $N^{(r)} \times N^{(r)}$ matrices corresponding to the remaining RCL network. These matrices have the same block structure as the matrices in (2.20), but with the sizes of each first and third block row and column reduced by $n_v^{(e)}$. In particular, in the case that all voltage sources have been eliminated, then $A^{(r)}$ and $E^{(r)}$ have no third block rows and columns at all. The state-vector $x^{(r)}(t)$ of (2.51) is obtained from the state-vector $x(t)$ of (2.20) by deleting the voltages at the nodes between the eliminated voltage sources and the directly connected resistors and the currents through the eliminated voltage sources. The input and output vectors of (2.51) are the same as in (2.16). We now partition these vectors as follows:

$$u(t) = \begin{bmatrix} -i_i(t) \\ v_v^{(e)}(t) \\ v_v^{(r)}(t) \end{bmatrix} \quad \text{and} \quad y(t) = \begin{bmatrix} v_i(t) \\ -i_v^{(e)}(t) \\ -i_v^{(r)}(t) \end{bmatrix}. \tag{2.52}$$

Here, the superscripts "(e)" and "(r)" refer to eliminated and remaining voltages sources, respectively. Finally, in (2.51), $D \in \mathbb{R}^{m \times m}$ and $B^{(r)}, C^{(r)} \in \mathbb{R}^{N^{(r)} \times m}$ are matrices of the following form:

$$\begin{aligned} B^{(r)} &= \begin{bmatrix} \mathcal{A}_i & B_{12} & 0 \\ 0 & 0 & 0 \\ 0 & 0 & -I \end{bmatrix}, \quad C^{(r)} = \begin{bmatrix} \mathcal{A}_i & -B_{12} & 0 \\ 0 & 0 & 0 \\ 0 & 0 & -I \end{bmatrix}, \quad \text{and} \\ D &= \begin{bmatrix} 0 & 0 & 0 \\ 0 & R_1^{-1} & 0 \\ 0 & 0 & 0 \end{bmatrix}. \end{aligned} \tag{2.53}$$

Here, R_1 denotes the diagonal matrix the entries of which are the resistances of the resistors connected directly to the eliminated voltage sources, and the partitions in (2.53) are conforming with the partitions of the input and output vectors (2.52).

To obtain a reduced-order model of the descriptor system (2.51), we can again employ SPRIM. The data matrices of the reduced system are obtained analogous to (2.31), together with the additional relations

$$C_n := V_n^T C^{(r)} \quad \text{and} \quad D_n := D.$$

It is easy to see that SPRIM applied to the descriptor system (2.51) preserves the structures of all the data matrices.

2.6 Numerical Examples

In this section, we present some results for two classes of circuit examples. In all cases, we ran the SPRIM algorithm for increasing values of the dimension $\hat{n}$ of the underlying block Krylov subspaces and stopped when the transfer function of the SPRIM reduced-order model had converged to the transfer function of the unreduced original descriptor system. Here, convergence is monitored over a relevant frequency range of interest of the form

$$s = \mathtt{i}\omega, \qquad \omega_{\min} \le \omega \le \omega_{\max},$$

where $\mathtt{i} = \sqrt{-1}$. For the value of $\hat{n}$ at convergence of SPRIM, we also generated the corresponding PRIMA reduced-order model produced from the same $\hat{n}$-dimensional block Krylov subspace. This is a fair comparison since generating basis vectors for this subspace is the dominating computational cost for both SPRIM and PRIMA. In all cases, we plot the absolute values of $H(s)$ and $H_n(s)$ (for both SPRIM and PRIMA) over the frequency range of interest.

The first example is a variant of the so-called PEEC circuit [29]. It only has state-space dimension $N = 308$, but due to its many poles and zeros close to the frequency range of interest, its transfer function has many features. This variant of the PEEC circuit has two current sources and one voltage source, and thus $m = 3$. The expansion point $s_0 = \pi \times 10^{10}$ was used. In this case, the Krylov dimension $\hat{n} = 90$ is needed to achieve convergence. Figure 2.3 depicts the absolute values of the (1,1)-component of the 3×3-matrix-valued transfer functions. Clearly, for $\hat{n} = 90$ PRIMA has not converged yet. Figure 2.4 shows a close-up of the subrange where the PRIMA and SPRIM reduced-order models differ the most. Figures 2.5 and 2.6 display the corresponding plots for the (1,3)-component of the 3×3-matrix-valued transfer functions.

The second example (referred to as "package example") is a larger RCL circuit with state-space dimension $N = 1841$. This circuit has 8 current sources and 8 voltage sources, and thus $m = 16$. Its transfer function is 16×16-matrix-valued

Fig. 2.3 PEEC example, (1,1)-component of transfer functions

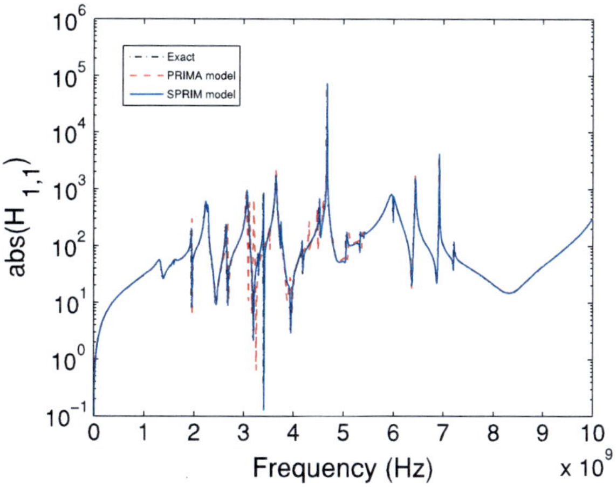

Fig. 2.4 PEEC example, close-up of (1,1)-component of transfer functions

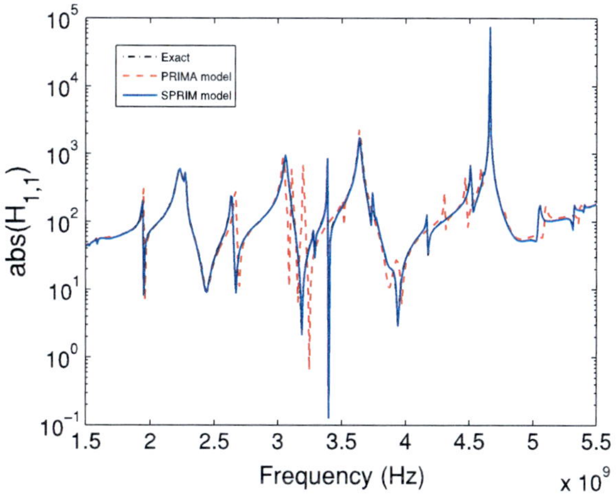

and has 256 components. The expansion point $s_0 = 2\pi \times 10^{10}$ was used. For this example, the Krylov dimension $\hat{n} = 128$ is needed to achieve convergence. Figures 2.7 and 2.8 depict the absolute values of the (8,1)-component and the (9,9)-component of the transfer functions. Note that for $\hat{n} = 128$ PRIMA has not converged yet.

The 8 voltage sources of the package example are all of the type shown in Fig. 2.2, and so all voltage sources can be eliminated using the approach outlined in Sect. 2.5. We have applied SPRIM and PRIMA to the resulting descriptor system (2.51) of state-space dimension $N^{(r)} = 1841-16 = 1825$. As before, the Krylov dimension $\hat{n} = 128$ is needed to achieve convergence. Figure 2.9 shows the absolute values of the (16,9)-component of the transfer functions.

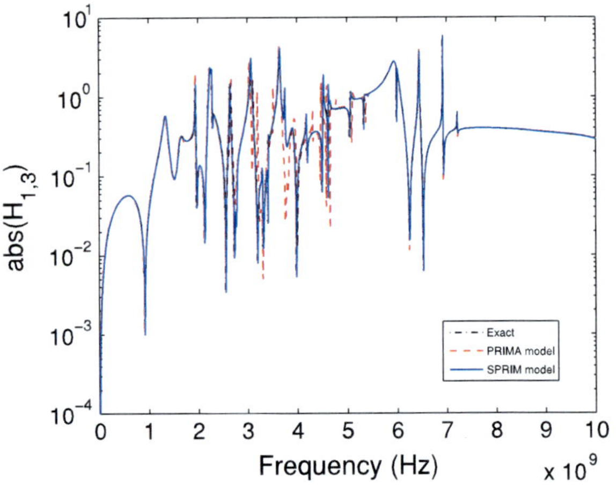

Fig. 2.5 PEEC example, (1,3)-component of transfer functions

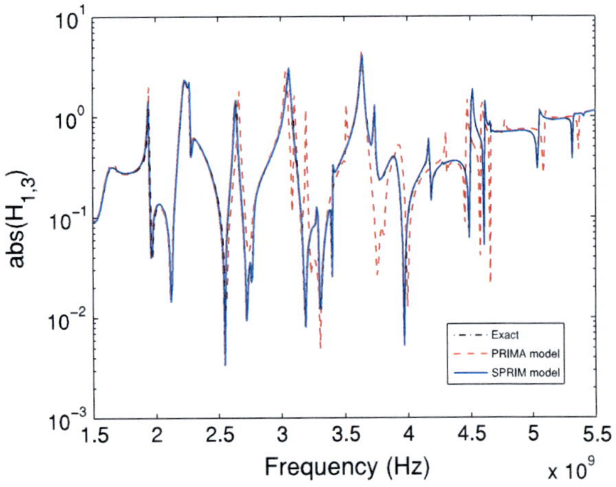

Fig. 2.6 PEEC example, close-up of the (1,3)-component of the transfer functions

2.7 Concluding Remarks

In this paper, we reviewed the formulation of general RCL circuits as descriptor systems and described the SPRIM reduction algorithm for general RCL circuits.

While there has been a lot of progress in Krylov subspace-based order reduction of large-scale RCL circuits in recent years, there are still many open problems. All state-of-the-art structure-preserving methods, such as SPRIM, first generate a basis matrix of the underlying Krylov subspace and then employ explicit projection using some suitable partitioning of the basis matrix to obtain a structure-preserving reduced-order model. In particular, there are two major problems with the use of

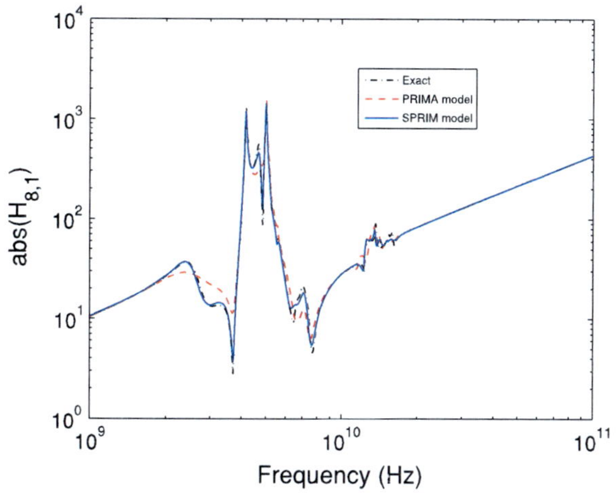

Fig. 2.7 Package example, (8,1)-component of transfer functions

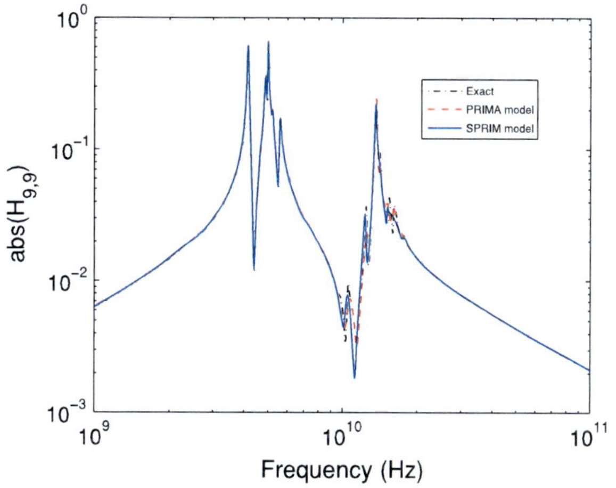

Fig. 2.8 Package example, (9,9)-component of transfer functions

such explicit projections. First, it requires the storage of the basis matrix, which becomes prohibitive in the case of truly large-scale linear dynamical systems. Second, the approximation properties of the resulting structure-preserving reduced-order models are not optimal, and they show that the available degrees of freedom are not fully used in general. It would be highly desirable to have structure-preserving reduction methods that do not involve explicit projection and would thus be applicable in the truly large-scale case. Other unresolved issues include the automatic and adaptive choice of suitable expansion points s_0 and robust and reliable stopping criteria and error bounds.

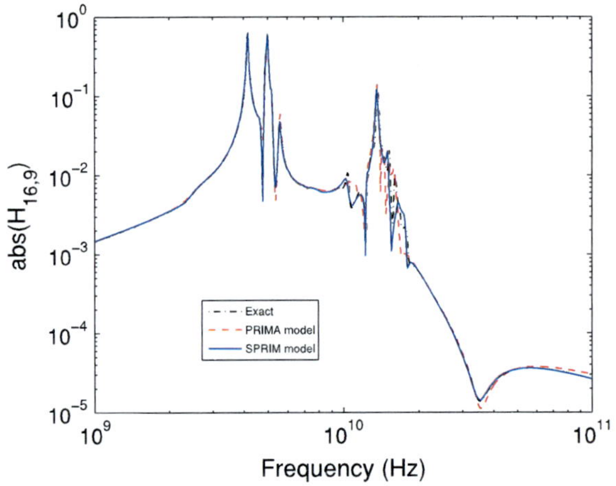

Fig. 2.9 Package example with voltage sources eliminated, (16,9)-component of transfer functions

Acknowledgements This work was supported in part by the National Science Foundation through Grant DMS-0613032. The author is grateful to the two anonymous referees for their careful reading of the original manuscript and their constructive comments that helped to improve the presentation of the paper.

References

1. Anderson, B.D.O., Vongpanitlerd, S.: Network Analysis and Synthesis. Prentice-Hall, Englewood Cliffs, New Jersey (1973)
2. Arnoldi, W.E.: The principle of minimized iterations in the solution of the matrix eigenvalue problem. Quart. Appl. Math. **9**, 17–29 (1951)
3. Bai, Z., Feldmann, P., Freund, R.W.: How to make theoretically passive reduced-order models passive in practice. In: Proceedings of IEEE 1998 Custom Integrated Circuits Conference, pp. 207–210. IEEE, Piscataway (1998)
4. Baker, Jr., G.A., Graves-Morris, P.: Padé Approximants, 2nd edn. Cambridge University Press, New York (1996)
5. de Villemagne, C., Skelton, R.E.: Model reductions using a projection formulation. Int. J. Control **46**(6), 2141–2169 (1987)
6. Deo, N.: Graph Theory with Applications to Engineering and Computer Science. Prentice-Hall, Englewood Cliffs (1974)
7. Elfadel, I.M., Ling, D.D.: Zeros and passivity of Arnoldi-reduced-order models for interconnect networks. In: Proceedings of 34nd ACM/IEEE Design Automation Conference, pp. 28–33. ACM, New York (1997)
8. Feldmann, P., Freund, R.W.: Efficient linear circuit analysis by Padé approximation via the Lanczos process. In: Proceedings of EURO-DAC '94 with EURO-VHDL '94, pp. 170–175. IEEE Computer Society Press, Los Alamitos (1994)
9. Feldmann, P., Freund, R.W.: Efficient linear circuit analysis by Padé approximation via the Lanczos process. IEEE Trans. Comput. Aided Des. **14**, 639–649 (1995)
10. Feldmann, P., Freund, R.W.: Reduced-order modeling of large linear subcircuits via a block Lanczos algorithm. In: Proceedings 32nd ACM/IEEE Design Automation Conference, pp. 474–479. ACM, New York (1995)

11. Feldmann, P., Freund, R.W.: Interconnect-delay computation and signal-integrity verification using the SyMPVL algorithm. In: Proceedings of 1997 European Conference on Circuit Theory and Design, pp. 132–138. IEEE Computer Society Press, Los Alamitos (1997)
12. Freund, R.W.: Reduced-order modeling techniques based on Krylov subspaces and their use in circuit simulation. In: Datta, B.N. (ed.) Applied and Computational Control, Signals, and Circuits, vol. 1, pp. 435–498. Birkhäuser, Boston (1999)
13. Freund, R.W.: Krylov-subspace methods for reduced-order modeling in circuit simulation. J. Comput. Appl. Math. **123**(1–2), 395–421 (2000)
14. Freund, R.W.: Model reduction methods based on Krylov subspaces. Acta Numerica **12**, 267–319 (2003)
15. Freund, R.W.: SPRIM: structure-preserving reduced-order interconnect macromodeling. In: Technical Digest 2004 IEEE/ACM International Conference on Computer-Aided Design, pp. 80–87. IEEE Computer Society Press, Los Alamitos (2004)
16. Freund, R.W.: On Padé-type model order reduction of J-Hermitian linear dynamical systems. Linear Algebra Appl. **429**, 2451–2464 (2008)
17. Freund, R.W.: Structure-preserving model order reduction of RCL circuit equations. In: Schilders, W., van der Vorst, H., Rommes, J. (eds.) Model Order Reduction: Theory, Research Aspects and Applications, Mathematics in Industry, vol. 13, pp. 49–73. Springer-Verlag, Berlin (2008)
18. Freund, R.W., Feldmann, P.: Reduced-order modeling of large passive linear circuits by means of the SyPVL algorithm. In: Technical Digest 1996 IEEE/ACM International Conference on Computer-Aided Design, pp. 280–287. IEEE Computer Society Press, Los Alamitos (1996)
19. Freund, R.W., Feldmann, P.: The SyMPVL algorithm and its applications to interconnect simulation. In: Proceedings of 1997 International Conference on Simulation of Semiconductor Processes and Devices, pp. 113–116. IEEE, Piscataway (1997)
20. Freund, R.W., Feldmann, P.: Reduced-order modeling of large linear passive multi-terminal circuits using matrix-Padé approximation. In: Proceedings of Design, Automation and Test in Europe Conference 1998, pp. 530–537. IEEE Computer Society Press, Los Alamitos (1998)
21. Grimme, E.J.: Krylov projection methods for model reduction. Ph.D. thesis, Department of Electrical Engineering, University of Illinois at Urbana-Champaign, Urbana-Champaign (1997)
22. Kamon, M., Marques, N.A., Silveira, L.M., White, J.: Automatic generation of accurate circuit models of 3-D interconnect. IEEE Trans. Compon. Packag. Manuf. Technol—Part B **21**(3), 225–240 (1998)
23. Lanczos, C.: An iteration method for the solution of the eigenvalue problem of linear differential and integral operators. J. Res. Nat. Bur. Standards **45**, 255–282 (1950)
24. Odabasioglu, A.: Provably passive RLC circuit reduction. M.S. thesis, Department of Electrical and Computer Engineering, Carnegie Mellon University (1996)
25. Odabasioglu, A., Celik, M., Pileggi, L.T.: PRIMA: passive reduced-order interconnect macromodeling algorithm. In: Technical Digest 1997 IEEE/ACM International Conference on Computer-Aided Design, pp. 58–65. IEEE Computer Society Press, Los Alamitos (1997)
26. Odabasioglu, A., Celik, M., Pileggi, L.T.: PRIMA: passive reduced-order interconnect macromodeling algorithm. IEEE Trans. Comput. Aided Des. **17**(8), 645–654 (1998)
27. Pillage, L.T., Rohrer, R.A.: Asymptotic waveform evaluation for timing analysis. IEEE Trans. Comput. Aided Des. **9**, 352–366 (1990)
28. Pillage, L.T., Rohrer, R.A., Visweswariah, C.: Electronic circuit and system simulation methods. McGraw-Hill, Inc., New York (1995)
29. Ruehli, A.E.: Equivalent circuit models for three-dimensional multiconductor systems. IEEE Trans. Microw. Theory Tech. **22**, 216–221 (1974)
30. Ruehli, A.E. (ed.): Circuit Analysis, Simulation, and Design Part 1: General Aspects of Circuit Analysis and Design. North-Holland, Amsterdam (1986)

31. Silveira, L.M., Kamon, M., Elfadel, I., White, J.: A coordinate-transformed Arnoldi algorithm for generating guaranteed stable reduced-order models of RLC circuits. In: Technical Digest 1996 IEEE/ACM International Conference on Computer-Aided Design, pp. 288–294. IEEE Computer Society Press, Los Alamitos (1996)
32. Vlach, J., Singhal, K.: Computer Methods for Circuit Analysis and Design, 2nd edn. Van Nostrand Reinhold, New York (1994)

Chapter 3
Balancing-Related Model Reduction of Circuit Equations Using Topological Structure

Tatjana Stykel

Abstract In recent years, model order reduction has been recognized to be a powerful tool in analysis and simulation of integrated circuits. We consider balancing-related model reduction methods for differential-algebraic equations arising in circuit simulation. We show how positive real and bounded real balanced truncation can be used for passivity-preserving model reduction of circuit equations. These methods are based on balancing the solutions of projected Lur'e or Riccati matrix equations and admit computable error bounds. We also discuss efficient algorithms for solving such matrix equations that exploit the topological structure of circuit equations. Numerical experiments demonstrate the performance of the presented model reduction methods.

Keywords Model reduction · Balanced truncation · Circiut equations · Passivity · Matrix equations

3.1 Introduction

Modern integrated circuits have hundreds of millions of semiconductor devices whose feature size is nowadays reaching the nanometer range. These devices are placed on several layers and interconnected to each other by wires. Due to increased packing density and interconnect length, modelling of thermal and electromagnetic effects is highly required in order to verify that the heat conduction and internal electromagnetic field do not disturb signal propagation. Design of VLSI circuits with distributed elements is no longer possible without

T. Stykel (✉)
Institut für Mathematik, Technische Universität Berlin, MA 4-5, Straße des 17.Juni 136, 10623 Berlin, Germany
e-mail: stykel@math.tu-berlin.de

P. Benner et al. (eds.), *Model Reduction for Circuit Simulation*,
Lecture Notes in Electrical Engineering, 74, DOI: 10.1007/978-94-007-0089-5_3,

computer simulations that involve numerical solution of coupled systems of partial differential equations and differential-algebraic equations (DAEs). After spatial discretization, such systems have very large state space dimension that makes the analysis and simulations unacceptably time consuming and expensive. In this context, model order reduction is of great importance.

A general idea of model order reduction is to approximate the large-scale system by a much smaller model that captures the input–output behavior of the original system to a required accuracy and also preserves essential physical properties such as stability and passivity. Many different model reduction approaches have been developed in computational fluid dynamics, control design and electrical and mechanical engineering, see [2, 15, 70] for recent books on this topic. One of the most used model reduction techniques in circuit simulation is *moment matching approximation* based on Krylov subspace methods, e.g., [4, 30, 38]. Although these methods are efficient for very large sparse problems, the resulting reduced-order systems have only locally good approximation properties. Another drawback of the moment matching methods is that stability and passivity are not necessarily preserved in the reduced-order models, so that usually post-processing is needed to realize these properties. Recently, passivity-preserving model reduction methods based on Krylov subspaces have been developed for structured systems arising in circuit simulation [31, 33, 45, 56] and also for general systems [3, 28, 41, 72]. However, none of these methods provides computable global error bounds.

Balanced truncation is another model reduction approach commonly used in control design. In order to capture specific system properties, different balancing techniques have been developed in the last thirty years for standard state space systems [26, 39, 53, 55, 60, 77] and also for DAEs [7, 14, 59, 63, 74]. In particular, passivity-preserving balanced truncation has been considered in [9, 13, 60, 63–65, 83]. An important property of balancing-related model reduction methods is the existence of computable error bounds. Unfortunately, these methods have a reputation for being very expensive since they involve solving (projected) Lyapunov and/or Riccati matrix equations. However, recent developments on iterative methods for such equations [16, 49, 57, 71, 75] show that balanced truncation methods can also be applied to large-scale problems.

In this paper, we give a brief survey on model reduction of circuit equations using balanced truncation and its relatives. In Sect. 3.2, we present some basic foundations from graph theory and network analysis required in the following. In Sect. 3.3, the balanced truncation model reduction approach for DAEs is described. Passivity-preserving model reduction methods for circuit equations based on positive real and bounded real balanced truncation are also considered. In Sect. 3.4, we discuss numerical solution of projected Lyapunov and Riccati equations with large-scale matrix coefficients. Section 3.5 contains some results of numerical experiments demonstrating the efficiency of the balancing-related model reduction techniques.

Throughout the paper, $\mathbb{R}^{n,m}$ and $\mathbb{C}^{n,m}$ denote the spaces of $n \times m$ real and complex matrices, respectively. The open left and right half-planes are denoted by

$\mathbb{C}_-$ and $\mathbb{C}_+$, respectively, and $i = \sqrt{-1}$. The matrices A^T and A^* denote, respectively, the transpose and the conjugate transpose of $A \in \mathbb{C}^{n,m}$, and $A^{-T} = (A^{-1})^T$. An identity matrix of order n is denoted by I_n or simply by I. We use rank(A) and *ker*(A) for the rank and the kernel of A, respectively. A matrix $A \in \mathbb{C}^{n,n}$ is *positive definite* (*semidefinite*), if $v^*Av > 0$ ($v^*Av \geq 0$) for all non-zero $v \in \mathbb{C}^n$. Note that positive (semi)definiteness of A does not require A to be Hermitian. For $A, B \in \mathbb{C}^{n,n}$, we write $A > B$ ($A \geq B$) if $A - B$ is positive definite (semidefinite).

3.2 Circuit Equations

In this section, we briefly describe the formulation of linear RLC circuits via DAEs and discuss their properties. For more details on graph theory and network analysis, we refer to [1, 22, 44, 79].

A general electrical circuit can be modelled as a *directed graph* $\mathfrak{G} = (\mathfrak{N}, \mathfrak{B})$ whose vertices $\mathfrak{n}_k \in \mathfrak{N}$ correspond to the nodes of the circuit and whose edges (branches) $\mathfrak{b}_{k_1,k_2} = \langle \mathfrak{n}_{k_1}, \mathfrak{n}_{k_2} \rangle \in \mathfrak{B}$ correspond to the circuit elements like capacitors, inductors and resistors. For the ordered pair $\mathfrak{b}_{k_1,k_2} = \langle \mathfrak{n}_{k_1}, \mathfrak{n}_{k_2} \rangle$, we say that $\mathfrak{b}_{k_1,k_2}$ leaves $\mathfrak{n}_{k_1}$ and enters $\mathfrak{n}_{k_2}$. An alternating sequence $(\mathfrak{n}_{k_1}, \mathfrak{b}_{k_1}, \mathfrak{n}_{k_2}, \ldots, \mathfrak{n}_{k_{s-1}}, \mathfrak{b}_{k_{s-1}}, \mathfrak{n}_{k_s})$ of vertices and edges in $\mathfrak{G}$ is called a *path* connecting $\mathfrak{n}_{k_1}$ and $\mathfrak{n}_{k_s}$ if $\mathfrak{b}_{k_j} = \langle \mathfrak{n}_{k_j}, \mathfrak{n}_{k_{j+1}} \rangle$ and $\mathfrak{n}_{k_i} \neq \mathfrak{n}_{k_j}$ for $2 \leq i < j \leq s$. A path is *closed* if $\mathfrak{n}_{k_1}$ and $\mathfrak{n}_{k_s}$ are the same, and *open* if they are different. A closed path is called a *loop*. A graph $\mathfrak{G}$ is called *connected* if for every two vertices there exists an open path connecting them. A *cutset* is a set of edges of a connected graph whose removal disconnects the graph, and this set is minimal with this property. A subgraph of the graph $\mathfrak{G}$ is called a *tree* if it has all nodes of $\mathfrak{G}$, is connected and does not contain loops.

Any directed graph $\mathfrak{G} = (\mathfrak{N}, \mathfrak{B})$ with $\mathfrak{N} = \{\mathfrak{n}_1, \ldots, \mathfrak{n}_{n_\eta+1}\}$ and $\mathfrak{B} = \{\mathfrak{b}_1, \ldots, \mathfrak{b}_{n_b}\}$ can be described by an *incidence matrix* $\mathbf{A}_0 = [a_{kl}] \in \mathbb{R}^{n_\eta+1, n_b}$ defined as

$$a_{kl} = \begin{cases} 1 & \text{if edge } \mathfrak{b}_l \text{ leaves vertex } \mathfrak{n}_k, \\ -1 & \text{if edge } \mathfrak{b}_l \text{ enters vertex } \mathfrak{n}_k, \\ 0 & \text{otherwise.} \end{cases}$$

In a connected graph, any n_η rows of $\mathbf{A}_0$ are linearly independent. Thus, deleting any row from $\mathbf{A}_0$ yields a full rank matrix $\mathbf{A} \in \mathbb{R}^{n_\eta, n_b}$ known as *reduced incidence matrix*. For circuits, the deleted row corresponds to a reference (or grounding) node.

We now consider a general linear RLC circuit that contains linear resistors, inductors, capacitors, independent voltage sources and independent current sources only. Such circuits are often used to model the interconnects, transmission lines and pin packages in VLSI networks. They arise also in the linearization of nonlinear circuit equations around DC operating points. RLC circuits are completely described by the graph-theoretic relations like Kirchhoff's current and voltage laws together with the branch constitutive relations that characterize the circuit elements. *Kirchhoff's current law* states that the sum of the currents along all edges leaving and entering any circuit node is zero. *Kirchhoff's voltage law* states that the sum of the voltages along the branches of any loop is zero. Let $j = [j_R^T, j_C^T, j_L^T, j_V^T, j_I^T]^T \in \mathbb{R}^{n_b}$ and $v = [v_R^T, v_C^T, v_L^T, v_V^T, v_I^T]^T \in \mathbb{R}^{n_b}$ denote the vectors of branch currents and branch voltages, respectively, and let the reduced incidence matrix $\mathbf{A} = [A_R, A_C, A_L, A_V, A_I]$ be partitioned accordingly, where the subscripts R, C, L, V and I stand for resistors, capacitors, inductors, voltage sources and current sources, respectively. Then Kirchhoff's current and voltage laws can be expressed in the compact form as

$$\mathbf{A}j = 0, \quad \mathbf{A}^T\eta = v, \tag{3.1}$$

respectively, where $\eta \in \mathbb{R}^{n_\eta}$ denotes the vector of potentials of all nodes excepting the reference node.

The *branch constitutive relations* for the linear capacitors, inductors and resistors are given by

$$\mathcal{C}\frac{d}{dt}v_C(t) = j_C(t), \quad v_L(t) = \mathcal{L}\frac{d}{dt}j_L(t), \quad v_R(t) = \mathcal{R}j_R(t), \tag{3.2}$$

where $\mathcal{C} \in \mathbb{R}^{n_C, n_C}$, $\mathcal{L} \in \mathbb{R}^{n_L, n_L}$ and $\mathcal{R} \in \mathbb{R}^{n_R, n_R}$ are the *capacitance*, *inductance* and *resistance matrices*, respectively. These matrices are often diagonal and their diagonal entries are the capacitances, inductances and resistances of the capacitors, resistors and inductors, respectively. However, the diagonal structure gets lost in case of mutually coupled elements. If $\mathcal{R}$ and $\mathcal{L}$ are nonsingular, then $\mathcal{G} = \mathcal{R}^{-1}$ and $\mathcal{S} = \mathcal{L}^{-1}$ are the *conductance* and *susceptance matrices*, respectively.

Using relations (3.1) and (3.2), the behaviour of a linear RLC circuit can be described via modified nodal analysis (MNA) [79] by the following system of DAEs

$$\begin{aligned} E\dot{x}(t) &= Ax(t) + Bu(t), \\ y(t) &= Cx(t) + Du(t), \end{aligned} \tag{3.3}$$

where

$$\begin{aligned} &E = \begin{bmatrix} A_C\mathcal{C}A_C^T & 0 & 0 \\ 0 & \mathcal{L} & 0 \\ 0 & 0 & 0 \end{bmatrix}, \quad A = \begin{bmatrix} -A_R\mathcal{G}A_R^T & -A_L & -A_V \\ A_L^T & 0 & 0 \\ A_V^T & 0 & 0 \end{bmatrix}, \\ &C = \begin{bmatrix} -A_I^T & 0 & 0 \\ 0 & 0 & -I \end{bmatrix} = B^T, \quad D = 0, \end{aligned} \tag{3.4}$$

$$x(t) = \begin{bmatrix} \eta(t) \\ j_L(t) \\ j_V(t) \end{bmatrix}, \quad u(t) = \begin{bmatrix} j_I(t) \\ v_V(t) \end{bmatrix}, \quad y(t) = -\begin{bmatrix} v_I(t) \\ j_V(t) \end{bmatrix}.$$

The number of state variables $n = n_\eta + n_L + n_V$ is called the *order* of system (3.3), and $m = n_I + n_V$ is the number of inputs and outputs. In the following we will assume that the circuit is *well-posed* in the sense that it has neither V-loops nor I-cutsets. These assumptions can be written in terms of the incidence matrices as follows:

(A1) The matrix A_V has full column rank, i.e., $\operatorname{rank}(A_V) = n_V$.

(A2) The matrix $A_{CLRV} = [A_C, A_L, A_R, A_V]$ has full row rank, i.e., $\operatorname{rank}(A_{CLRV}) = n_\eta$.

We will also assume that

(A3) $\mathcal{C}$, $\mathcal{G}$ and $\mathcal{L}$ are positive definite.

Assumptions (A1)–(A3) together guarantee that the matrix pencil $\lambda E - A$ is *regular*, i.e., $\det(\lambda E - A) \neq 0$ for some $\lambda \in \mathbb{C}$, see [32]. In this case, we can define a transfer matrix $\mathbf{G}(s) = C(sE - A)^{-1}B + D$ that describes the input–output relation of (3.3) in the frequency domain. The transfer function $\mathbf{G}$ is called *proper* if $\lim_{s\to\infty} \mathbf{G}(s) < \infty$, and *improper*, otherwise. If $\lim_{s\to\infty} \mathbf{G}(s) = 0$, then $\mathbf{G}$ is called *strictly proper*.

Any regular pencil $\lambda E - A$ can be reduced into the *Weierstrass canonical form*

$$E = T_l \begin{bmatrix} I_{n_f} & 0 \\ 0 & E_\infty \end{bmatrix} T_r, \quad A = T_l \begin{bmatrix} A_f & 0 \\ 0 & I_{n_\infty} \end{bmatrix} T_r, \tag{3.5}$$

where T_l and T_r are the left and right nonsingular transformation matrices, and E_∞ is nilpotent with index of nilpotency μ, see [34]. The eigenvalues of A_f are the finite eigenvalues of $\lambda E - A$, and E_∞ corresponds to an eigenvalue at infinity. The number μ is called the *index* of $\lambda E - A$ and also of the DAE system (3.3). Index concept plays an important role in the analysis and numerical solution of DAEs, e.g., [19, 20, 37, 46, 67]. The following proposition characterizes the index of the MNA equations (3.3), (3.4).

Proposition 1 [27] *Let E and A be as in* (3.4) *and let* (A1)–(A3) *be fulfilled.*

1. *The index of the pencil $\lambda E - A$ is at most two.*
2. *The index of $\lambda E - A$ is equal to zero if and only if*

$$n_V = 0, \quad \operatorname{rank}(A_C) = n_\eta. \tag{3.6}$$

3. *The index of $\lambda E - A$ is equal to one if and only if*

$$\operatorname{rank}(Q_C^T A_V) = n_V, \quad \operatorname{rank}[A_C, A_R, A_V] = n_\eta, \tag{3.7}$$

where Q_C is a projector onto $\ker(A_C^T)$.

Considering the topological structure of the circuit, the conditions (3.6) imply that the circuit does not contain voltage sources and the circuit graph contains a capacitive tree. Furthermore, the first condition in (3.7) implies that the circuit does not contain CV-loops except for C-loops, whereas the second condition in (3.7) means that the circuit does not contain LI-cutsets.

Using the Weierstrass canonical form (3.5), the MNA system (3.3), (3.4) can be decoupled into the *slow* subsystem

$$\dot{x}_1(t) = A_f x_1(t) + B_f u(t), \tag{3.8a}$$

$$y_1(t) = C_f x_1(t), \tag{3.8b}$$

and the *fast* subsystem

$$E_\infty \dot{x}_2(t) = x_2(t) + B_\infty u(t), \tag{3.9a}$$

$$y_2(t) = C_\infty x_2(t), \tag{3.9b}$$

where $T_r x(t) = [(x_1(t))^T, (x_2(t))^T]^T, y(t) = y_1(t) + y_2(t)$ and

$$T_l^{-1} B = \begin{bmatrix} B_f \\ B_\infty \end{bmatrix}, \quad C T_r^{-1} = [C_f, C_\infty]. \tag{3.10}$$

Equation 3.8a with the initial condition $x_1(0) = x_1^0$ has a unique solution

$$x_1(t) = e^{tA_f} x_1^0 + \int_0^t e^{(t-\tau)A_f} B_f u(\tau) d\tau$$

for any integrable input u and any initial vector $x_1^0 \in \mathbb{R}^{n_f}$. Since the index μ of system (3.3), (3.4) does not exceed two, a unique solution of Eq. (3.9a) is given by

$$x_2(t) = -B_\infty u(t) - E_\infty B_\infty \dot{u}(t).$$

This representation shows that for the existence of a continuously differentiable solution x of (3.3), (3.4), it is necessary that the input function u is μ times continuously differentiable. Moreover, the initial condition $x(0) = x_0$ has to be consistent, i.e., $x_0 = T_r^{-1}[(x_1^0)^T, (x_2^0)^T]^T$ must satisfy

$$x_2^0 = -B_\infty u(0) - E_\infty B_\infty \dot{u}(0).$$

If the initial vector x_0 is inconsistent or the input u is not sufficiently smooth, then the solution of the MNA system (3.3), (3.4) may have impulsive modes [20].

3.2.1 Stability

Stability is a qualitative property of dynamical systems which describes the behaviour of their solutions under small perturbations in the initial data. For the linear time-invariant DAE system (3.3), stability can be characterized in terms of the finite eigenvalues of the pencil $\lambda E - A$, e.g., [24]. System (3.3) is *stable* if all the finite eigenvalues of $\lambda E - A$ lie in the closed left half-plane and the eigenvalues on the imaginary axis are semi-simple, i.e., they have the same algebraic and geometric multiplicity. System (3.3) is *asymptotically stable* if the pencil $\lambda E - A$ is *c-stable*, i.e., all its finite eigenvalues lie in the open left half-plane. The following proposition gives the topological conditions for the asymptotic stability of the MNA equations (3.3), (3.4).

Proposition 2 [66] *Let the matrices E and A be as in* (3.4) *and let* (A1)–(A3) *be fulfilled. Assume that* $\mathcal{C}$ *and* $\mathcal{L}$ *are symmetric and one of the following two pairs of topological conditions holds:*

1. $\operatorname{rank}[A_L,\ A_V] = n_L + n_V, \quad \operatorname{rank}[A_R, A_V] = n_\eta,$ (3.11)

2. $\operatorname{rank}[A_C, A_L, A_V] = n_C + n_L + n_V, \quad \operatorname{rank}[A_L, A_R, A_V] = n_\eta.$ (3.12)

Then the MNA system (3.3), (3.4) *is asymptotically stable.*

Conditions (3.11) are equivalent to the absence of LV-loops and CLI-cutsets (except maybe for LI-cutsets), whereas (3.12) implies that the circuit does not contain CLV-loops (except maybe for CV-loops) and CI-cutsets.

If system (3.3) is asymptotically stable, then the $\mathbb{H}_\infty$*-norm* of its transfer function $\mathbf{G}$ is defined as $\|\mathbf{G}\|_{\mathbb{H}_\infty} = \sup_{\omega\in\mathbb{R}} \|\mathbf{G}(i\omega)\|$, where $\|\cdot\|$ denotes the spectral matrix norm.

3.2.2 Passivity and Positive Realness

Passivity is a most basic property of circuit equations. Generally speaking, passivity means that the system does not produce energy. More precisely, system (3.3) is *passive* if

$$\int_0^t u(\tau)^T y(\tau)d\tau \geq 0 \tag{3.13}$$

for all $t \geq 0$ and all admissible u such that $u^T y$ is locally integrable. For a circuit element with a voltage v and a current j, condition (3.13) implies that the storage energy of this element defined as

$$\int_0^t v(\tau)^T j(\tau)d\tau$$

is always nonnegative. Thus, capacitors, resistors and inductors with nonnegative element values are passive. Furthermore, interconnection of a finite number of passive circuit components yields a passive network [1].

It is well known in network theory [1] that the DAE system (3.3) is passive if and only if its transfer function $\mathbf{G}(s) = C(sE - A)^{-1}B + D$ is *positive real*, i.e., $\mathbf{G}$ is analytic in $\mathbb{C}_+$ and $\mathbf{G}(s) + \mathbf{G}(s)^* \geq 0$ for all $s \in \mathbb{C}_+$. Using the Weierstrass canonical form (3.5) and (3.10), the transfer function of (3.3) can be additively decomposed as

$$\mathbf{G}(s) = \mathbf{G}_{sp}(s) + M_0 + sM_1 + \cdots + s^{\mu-1}M_{\mu-1},$$

where $\mathbf{G}_{sp}(s) = C_f(sI - A_f)^{-1}B_f$ is the strictly proper part of $\mathbf{G}$, $M_0 = D - C_\infty B_\infty$ and $M_k = -C_\infty E_\infty^k B_\infty$ for $k \geq 1$. One can show that $\mathbf{G}$ is positive real if and only if its proper part $\mathbf{G}_p(s) = \mathbf{G}_{sp}(s) + M_0$ is positive real, $M_1 = M_1 \geq 0$ and $M_k = 0$ for $k > 1$, see [1].

The following proposition gives sufficient conditions for system (3.3), (3.4) to be stable and passive.

Proposition 3 *If Assumptions* (A1)–(A3) *are fulfilled and the matrices* $\mathcal{C}$ *and* $\mathcal{L}$ *are symmetric, then the MNA system* (3.3), (3.4) *is stable and passive.*

Proof The facts that the pencil $\lambda E - A$ in (3.4) has no finite eigenvalues in $\mathbb{C}_+$ and the transfer function $\mathbf{G}(s) = C(sE - A)^{-1}B$ of (3.3), (3.4) is positive real have been proved in [61, 66], respectively. Analogously, we can show that $(sE - A)^{-1}$ is also positive real. Hence, the purely imaginary eigenvalues of $\lambda E - A$ are semi-simple [1, Theorem 2.7.2]. Thus, the MNA system (3.3), (3.4) is stable and passive. □

3.2.3 Contractivity and Bounded Realness

An important class of dynamical systems are contractive systems. System (3.3) is called *contractive* if

$$\int_0^t (\|u(\tau)\|^2 - \|y(\tau)\|^2)d\tau \geq 0 \tag{3.14}$$

for all $t \geq 0$ and all admissible u such that u and y are both square integrable. The integral in (3.14) expresses the difference between the input and output energy of the system. One can show that (3.3) is contractive if and only if its transfer function $\mathbf{G}$ is *bounded real*, i.e., $\mathbf{G}$ is analytic in $\mathbb{C}_+$ and $I - \mathbf{G}(s)^*\mathbf{G}(s) \geq 0$ for all $s \in \mathbb{C}_+$, see [1]. For the asymptotically stable system (3.3), contractivity is equivalent to the condition $\|\mathbf{G}\|_{\mathbb{H}_\infty} \leq 1$ that justifies the name 'contractive'. Note that the bounded real transfer function is necessarily proper.

Positive real and bounded real square transfer functions are related to each other via a *Moebius transformation* defined as

$$\mathscr{M}(\mathbf{G})(s) = (I - \mathbf{G}(s))(I + \mathbf{G}(s))^{-1}.$$

The transfer function $\mathbf{G}$ is positive real if and only if the Moebius-transformed function $\hat{\mathbf{G}}(s) = \mathscr{M}(\mathbf{G})(s)$ is bounded real [1]. For system (3.3) with nonsingular $I + D$, the function $\hat{\mathbf{G}}(s)$ can be represented as $\hat{\mathbf{G}}(s) = \hat{C}(s\hat{E} - \hat{A})^{-1}\hat{B} + \hat{D}$, where

$$\hat{E} = E, \quad \hat{A} = A - B(I + D)^{-1}C, \quad \hat{B} = -\sqrt{2}B(I + D)^{-1},$$
$$\hat{C} = \sqrt{2}(I + D)^{-1}C, \quad \hat{D} = (I - D)(I + D)^{-1}.$$

For the MNA matrices as in (3.4), we have

$$\hat{E} = \begin{bmatrix} A_C \mathcal{C} A_C^T & 0 & 0 \\ 0 & \mathcal{L} & 0 \\ 0 & 0 & 0 \end{bmatrix}, \quad \hat{A} = \begin{bmatrix} -A_R \mathcal{G} A_R^T - A_I A_I^T & -A_L & -A_V \\ A_L^T & 0 & 0 \\ A_V^T & 0 & -I \end{bmatrix},$$
$$\hat{B} = \sqrt{2}\begin{bmatrix} A_I & 0 \\ 0 & 0 \\ 0 & I \end{bmatrix} = -\hat{C}^T, \quad \hat{D} = I. \tag{3.15}$$

It has been shown in [64] that under Assumptions (A1)–(A3) the pencil $\lambda\hat{E} - \hat{A}$ in (3.15) is of index at most two. It is equal to one if and only if $\operatorname{rank}[A_C, A_R, A_I, A_V] = n_\eta$. This condition means that the circuit does not contain L-cutsets.

3.2.4 Reciprocity

Another relevant property of circuit equations is reciprocity. We call a matrix $S \in \mathbb{R}^{m,m}$ a *signature* if S is diagonal and $S^2 = I_m$. System (3.3) is *reciprocal* with an external signature $S_{\text{ext}} \in \mathbb{R}^{m,m}$ if its transfer function satisfies $\mathbf{G}(s) = S_{\text{ext}}\mathbf{G}(s)^T S_{\text{ext}}$ for all $s \in \mathbb{C}$. The following proposition shows that the symmetry of $\mathcal{C}$, $\mathcal{L}$ and $\mathcal{G}$ guarantees the reciprocity of system (3.3), (3.4).

Proposition 4 [62] *Let Assumptions* (A1)–(A3) *be fulfilled and let the matrices* $\mathcal{C}$, $\mathcal{L}$ and $\mathcal{G}$ *be symmetric. Then the MNA system* (3.3), (3.4) *is reciprocal with the external signature* $S_{\text{ext}} = \operatorname{diag}(I_{n_I}, -I_{n_V})$.

3.3 Balancing-Related Model Reduction

The aim of model order reduction for circuit equations is to approximate the DAE system (3.3), (3.4) with a reduced-order model

$$\begin{aligned} \tilde{E}\dot{\tilde{x}}(t) &= \tilde{A}\tilde{x}(t) + \tilde{B}u(t), \\ \tilde{y}(t) &= \tilde{C}\tilde{x}(t) + \tilde{D}u(t), \end{aligned} \tag{3.16}$$

where $\tilde{E}, \tilde{A} \in \mathbb{R}^{\ell,\ell}$, $\tilde{B} \in \mathbb{R}^{\ell,m}$, $\tilde{C} \in \mathbb{R}^{m,\ell}$, $\tilde{D} \in \mathbb{R}^{m,m}$ and $\ell \ll n$. It is required for the approximate system (3.16) to preserve physical properties like stability, passivity and reciprocity. Such a system can then be synthesized as an electrical circuit in an standard netlist format, e.g., [62, 84] and Chap. 12 of this volume, that is often required in the industrial circuit simulators. It is also important to have a small approximation error $\tilde{y} - y$ or $\tilde{\mathbf{G}} - \mathbf{G}$, where $\tilde{\mathbf{G}}(s) = \tilde{C}(s\tilde{E} - \tilde{A})^{-1}\tilde{B} + \tilde{D}$. In the ideal case, we would like to have a computable error bound that allows us to approximate (3.3) to a given accuracy and makes model reduction fully automatic.

Most of the model reduction methods for linear dynamical systems are based on the projection of the system onto lower dimensional subspaces. In this case, the system matrices of the reduced-order model (3.16) have the form

$$\tilde{E} = W^T E T, \quad \tilde{A} = W^T A T, \quad \tilde{B} = W^T B, \quad \tilde{C} = CT, \tag{3.17}$$

where the projection matrices $W, T \in \mathbb{R}^{n,\ell}$ determine the subspaces of interest. For example, in modal model reduction the columns of W and T span, respectively, the left and right deflating subspaces of the pencil $\lambda E - A$ corresponding to the dominant eigenvalues, e.g., [25, 50]. In the moment matching approximation, one chooses the projection matrices W and T whose columns form the bases of certain Krylov subspaces associated with (3.3), e.g., [4, 30].

3.3.1 Balanced Truncation Model Reduction

Balanced truncation also belongs to the projection-based model reduction techniques. This method consists in transforming the dynamical system into a balanced form whose controllability and observability Gramians are both equal to a diagonal matrix. Then a reduced-order model (3.16), (3.17) is obtained by projecting (3.3) onto the subspaces corresponding to the dominant diagonal elements of the balanced Gramians. This idea goes back to [54] and has been extended over the years in different directions by many authors, e.g., [11, 14, 26, 35, 39, 52, 53, 55, 58, 60, 74, 77].

For standard state space systems with $E = I$, the balanced truncation model reduction method makes use of the dual Lyapunov equations

$$AG_c + G_c A^T = -BB^T, \quad A^T G_o + G_o A = -C^T C.$$

If all eigenvalues of the matrix A have negative real part, then these equations have unique symmetric, positive semidefinite solutions G_c and G_o known as the *controllability* and *observability Gramians*, respectively. One can show that all eigenvalues of the product $G_c G_o$ are real and nonnegative. The square roots of these eigenvalues, denoted by σ_j, are called the *Hankel singular values* of system (3.3) with $E = I$. Such a system is *balanced* if $G_c = G_o = \mathrm{diag}(\sigma_1, \ldots, \sigma_n)$. If the system is controllable and observable, then the Gramians G_c and G_o are both positive definite. In this case, there exists a balancing state space transformation such that the Gramians of the transformed system become equal and diagonal with the Hankel singular values on the diagonal. Then the reduced-order model is obtained by truncating the states corresponding to the small Hankel singular values. Such states are simultaneously difficult to reach and to observe, since they have a small impact on the energy transfer from input to output, see [35, 53] for details.

The balanced truncation model reduction approach can be extended to system (3.3) with $E \neq I$. If E is nonsingular, then the Gramians are defined as unique symmetric, positive semidefinite solutions of the generalized Lyapunov equations

$$AG_cE^T + EG_cA^T = -BB^T, \quad A^TG_oE + E^TG_oA = -C^TC, \tag{3.18}$$

provided the pencil $\lambda E - A$ is c-stable. However, for singular E, these equations cannot be used any more to determine the Gramians for the DAE system (3.3). As the following example shows, the generalized Lyapunov equations (3.18) with singular E may not have solutions even if $\lambda E - A$ is c-stable. Moreover, if the solutions of (3.18) exist, they are always nonunique, see [73] for detailed discussions.

Example 1 Consider the simple RL circuit shown in Fig. 3.1. This circuit is described by the DAE system (3.3) with

$$E = \begin{bmatrix} 0 & 0 & 0 & 0 \\ 0 & 0 & 0 & 0 \\ 0 & 0 & \mathcal{L} & 0 \\ 0 & 0 & 0 & 0 \end{bmatrix}, \quad A = \begin{bmatrix} 0 & 0 & -1 & 1 \\ 0 & -1/\mathcal{R} & 1 & 0 \\ 1 & -1 & 0 & 0 \\ -1 & 0 & 0 & 0 \end{bmatrix},$$

$$B^T = [0, \; 0, \; 0, \; -1] = C.$$

The pencil $\lambda E - A$ has only one finite eigenvalue $\lambda = -\mathcal{R}/\mathcal{L} < 0$. However, the generalized Lyapunov equations (3.18) are not solvable.

An extension of the balanced truncation method to DAEs based on projected Lyapunov equations has been presented in [52, 74]. Unlike the standard state space case, the DAE system (3.3) has two pairs of the Gramians, one pair for the slow subsystem (3.8a, b) and the other pair for the fast subsystem (3.9a, b). If (3.3) is

Fig. 3.1 A simple RL circuit

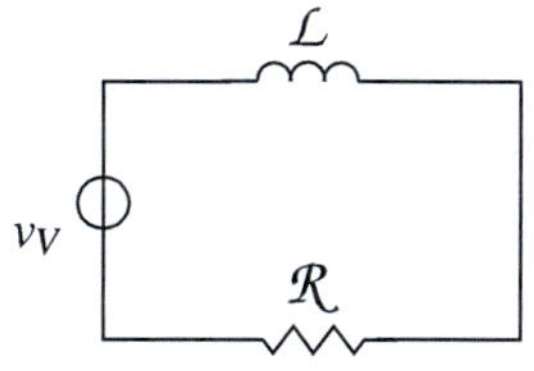

asymptotically stable, then the *proper controllability* and *observability Gramians* G_{pc} and G_{po} of (3.3) are defined as unique symmetric, positive semidefinite solutions of the *projected generalized continuous-time Lyapunov equations*

$$EG_{pc}A^T + AG_{pc}E^T = -P_l BB^T P_l^T, \quad G_{pc} = P_r G_{pc} P_r^T, \tag{3.19}$$

$$E^T G_{po} A + A^T G_{po} E = -P_r^T C^T C P_r, \quad G_{po} = P_l^T G_{po} P_l, \tag{3.20}$$

where P_l and P_r are the spectral projectors onto the left and right deflating subspaces of the pencil $\lambda E - A$ corresponding to the finite eigenvalues along the left and right deflating subspaces corresponding to the eigenvalue at infinity. Using the Weierstrass canonical form (3.5), these projectors can be represented as

$$P_r = T_r^{-1}\begin{bmatrix} I & 0 \\ 0 & 0 \end{bmatrix} T_r, \quad P_l = T_l \begin{bmatrix} I & 0 \\ 0 & 0 \end{bmatrix} T_l^{-1}.$$

Furthermore, the *improper controllability* and *observability Gramians* G_{ic} and G_{io} of system (3.3) are defined as unique symmetric, positive semidefinite solutions of the *projected generalized discrete-time Lyapunov equations*

$$AG_{ic}A^T - EG_{ic}E^T = Q_l BB^T Q_l^T, \quad G_{ic} = Q_r G_{ic} Q_r^T, \tag{3.21}$$

$$A^T G_{io} A - E^T G_{io} E = Q_r^T C^T C Q_r, \quad G_{io} = Q_l^T G_{io} Q_l, \tag{3.22}$$

where $Q_l = I - P_l$ and $Q_r = I - P_r$ are the complementary projectors. Note that unlike generalized Lyapunov equations considered in [42, 48, 76], the existence and uniqueness results for the projected Lyapunov equations (3.19)–(3.22) can be stated independently of the index of the pencil $\lambda E - A$, see [73].

Using the proper and improper Gramians, we can define the proper and improper Hankel singular values that characterize the importance of state variables in the slow and fast subsystems (3.8), and (3.9), respectively. Let n_f be the dimension of the deflating subspaces of $\lambda E - A$ corresponding to the finite eigenvalues. Then the *proper Hankel singular values* σ_j of system (3.3) are defined as the square roots of the largest n_f eigenvalues of the matrix $G_{pc}E^T G_{po}E$, and the *improper Hankel singular values* θ_j are defined as the square roots of the largest $n_\infty = n - n_f$ eigenvalues of the matrix $G_{ic}A^T G_{io}A$. We assume that the proper and improper Hankel singular values are ordered decreasingly. System (3.3) is *balanced* if the Gramians satisfy

$$\begin{aligned} G_{pc} = G_{po} &= \mathrm{diag}(\Sigma, 0) \quad \text{with} \quad \Sigma = \mathrm{diag}(\sigma_1, \ldots, \sigma_{n_f}), \\ G_{ic} = G_{io} &= \mathrm{diag}(0, \Theta) \quad \text{with} \quad \Theta = \mathrm{diag}(\theta_1, \ldots, \theta_{n_\infty}). \end{aligned}$$

States of the balanced system corresponding to the small proper Hankel singular values are less involved in the energy transfer from inputs to outputs, and, therefore, they can be truncated without changing the system properties significantly. Furthermore, we can remove the states of the balanced system corresponding to the zero improper Hankel singular values. Such states are uncontrollable and unobservable at infinity and do not influence the input–output relation. However, if we truncate the states that correspond to the non-zero improper Hankel singular values, even if they are small, then the approximation may be inaccurate. These states are subject to constraints, and their elimination may lead to undesirable disturbances in the approximate system and physically meaningless results.

Example 2 Consider the DAE system (3.3) with

$$E = \begin{bmatrix} 0 & 1 & 0 \\ 0 & 0 & 1 \\ 0 & 0 & 0 \end{bmatrix}, \quad A = I_3, \quad B = \begin{bmatrix} 10 \\ 0.1 \\ 0 \end{bmatrix}, \quad C^T = \begin{bmatrix} 0.04 \\ 30 \\ 1 \end{bmatrix}. \tag{3.23}$$

Since E is nilpotent, this system has only the improper Hankel singular values given by $\theta_1 = 3.4$, $\theta_2 = 4.7 \times 10^{-6}$, $\theta_3 = 0$. The truncation of the state corresponding to the Hankel singular value $\theta_3 = 0$ results in the reduced-order model

$$\begin{aligned} \begin{bmatrix} 1.18 & 1.18 \\ -1.18 & -1.18 \end{bmatrix} \dot{\tilde{x}}(t) &= \begin{bmatrix} 10^3 & 0 \\ 0 & 10^3 \end{bmatrix} \tilde{x}(t) + \begin{bmatrix} 1.84 \times 10^3 \\ 2.25 \times 10^{-3} \end{bmatrix} u(t), \\ \tilde{y}(t) &= [1.84 \times 10^3, -2.25 \times 10^{-3}] \tilde{x}(t). \end{aligned} \tag{3.24}$$

Figure 3.2a shows the output functions of the original and the reduced-order systems with the input $u(t) = \sin(t)$. They coincide since both systems have the same transfer function. However, if we truncate one more state corresponding to the second Hankel singular value, which is relatively small, we obtain the standard state space system

$$\dot{\tilde{x}}(t) = 850\tilde{x}(t) + 1567u(t), \quad \tilde{y}(t) = 1.84\tilde{x}(t). \tag{3.25}$$

This system is unstable, and, as Fig. 3.2b demonstrates, its output has nothing in common with the output of the original system.

We summarize the balanced truncation model reduction method for DAEs in Algorithm 1. For this method, we have the following a priori error bound.

$$\|\tilde{y} - y\|_{\mathbb{L}_2} \leq \|\tilde{\mathbf{G}} - \mathbf{G}\|_{\mathbb{H}_\infty} \|u\|_{\mathbb{L}_2} \leq 2(\sigma_{\ell_f+1} + \cdots + \sigma_{n_f}) \|u\|_{\mathbb{L}_2}$$

that allows an adaptive choice of the order of the approximate model. Furthermore, the resulting reduced-order system (3.16) is asymptotically stable and its index does not exceed the index of (3.3). If $\mathcal{C}$, $\mathcal{L}$ and $\mathcal{G}$ in (3.4) are symmetric, i.e., (3.3), (3.4) is reciprocal with the external signature S_{ext} as in Proposition 4, then the reduced-order model computed by Algorithm 1 is also reciprocal with the same signature. Unfortunately, the Lyapunov-based balanced truncation method does not, in general, ensure the preservation of passivity. However, for special reciprocal circuits such as RC and RL networks, Algorithm 1 can be modified for computing a passive reduced-order model, see [65] for details.

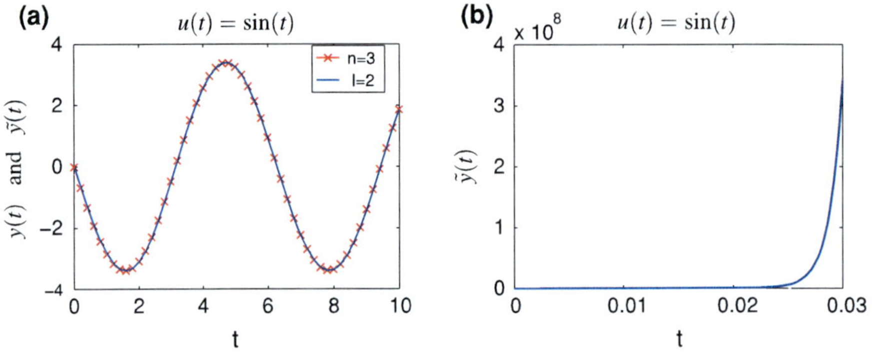

Fig. 3.2 **a** The output functions of the original system (3.3), (3.23) and the reduced-order system (3.24); **b** the output of the reduced-order system (3.25). In both cases, the input is $u(t) = \sin(t)$

Algorithm 1 Balanced truncation model reduction for DAEs.
Given $\mathbf{G} = (E, A, B, C, D)$, compute a reduced-order model $\tilde{\mathbf{G}} = (\tilde{E}, \tilde{A}, \tilde{B}, \tilde{C}, \tilde{D})$.

1. Compute the Cholesky factors R_p and L_p of the proper Gramians $G_{pc} = R_p R_p^T$ and $G_{po} = L_p L_p^T$ that satisfy the projected Lyapunov equations (3.19) and (3.20), respectively.
2. Compute the Cholesky factors R_i and L_i of the improper Gramians $G_{ic} = R_i R_i^T$ and $G_{io} = L_i L_i^T$ that satisfy the projected Lyapunov equations (3.21) and (3.22), respectively.
3. Compute the singular value decomposition $L_p^T E R_p = [U_1, U_2] \text{diag}(\Sigma_1, \Sigma_2) [V_1, V_2]^T$, where the matrices $[U_1, U_2]$ and $[V_1, V_2]$ have orthonormal columns, $\Sigma_1 = \text{diag}(\sigma_1, \ldots, \sigma_{\ell_f})$ and $\Sigma_2 = \text{diag}(\sigma_{\ell_f+1}, \ldots, \sigma_{n_f})$.
4. Compute the singular value decomposition $L_i^T A R_i = U_3 \Theta V_3^T$, where U_3 and V_3 have orthonormal columns and $\Theta = \text{diag}(\theta_1, \ldots, \theta_{\ell_\infty})$ is nonsingular.
5. Compute the reduced-order system $(\tilde{E}, \tilde{A}, \tilde{B}, \tilde{C}, \tilde{D}) = (W^T E T, W^T A T, W^T B, C T, D)$ with $W = [L_p U_1 \Sigma_1^{-1/2}, L_i U_3 \Theta^{-1/2}]$ and $T = [R_p V_1 \Sigma_1^{-1/2}, R_i V_3 \Theta^{-1/2}]$.

3.3.2 Positive Real Balanced Truncation

In this section, we describe passivity-preserving model reduction for general RCL circuits based on positive real balancing.

Passivity of the DAE system (3.3) can be characterized via the *projected positive real Lur'e equations*

$$\begin{aligned} AXE^T + EXA^T &= -K_c K_c^T, \quad X = P_r X P_r^T \geq 0, \\ EXC^T - P_l B &= -K_c J_c^T, \quad M_0 + M_0^T = J_c J_c^T \end{aligned} \tag{3.26}$$

and

$$\begin{aligned} A^T Y E + E^T Y A &= -K_o^T K_o, \quad Y = P_l^T Y P_l \geq 0, \\ E^T Y B - P_r^T C^T &= -K_o^T J_o, \quad M_0 + M_0^T = J_o^T J_o \end{aligned} \tag{3.27}$$

with unknowns $X \in \mathbb{R}^{n,n}$, $K_c \in \mathbb{R}^{n,m}$, $J_c \in \mathbb{R}^{m,m}$ and $Y \in \mathbb{R}^{n,n}$, $K_o \in \mathbb{R}^{m,n}$, $J_o \in \mathbb{R}^{m,m}$, respectively. Such equations are known in the literature also as Kalman–Yakubovich–Popov equations [40]. Similarly to [64, Theorem 4.1], one can show that if the MNA system (3.3), (3.4) is passive, then (3.26) and (3.27) are solvable. Conversely, solvability of the projected Lur'e equations (3.26) and (3.27) together with the conditions $M_1 = M_1^T \geq 0$ and $M_k = 0$ for $k > 1$ implies that (3.3) is passive.

Remark 1 Note that for a general DAE system, passivity alone does not guarantee the existence of the solution of the projected Lur'e equations. For such a system, in addition, R-minimality conditions

$$\operatorname{rank}[\lambda E - A,\ B] = n, \quad \operatorname{rank}[\lambda E^T - A^T,\ C^T] = n \quad \text{for all } \lambda \in \mathbb{C}$$

(or other weaker conditions) have to be assumed [23, 29, 47, 62].

The projected Lur'e equations (3.26) and (3.27) have, in general, many symmetric solutions X and Y that can be ordered with respect to the Loewner ordering in the set of symmetric matrices. The minimal solutions X_{pr} and Y_{pr} that satisfy $0 \leq X_{pr} \leq X$ and $0 \leq Y_{pr} \leq Y$ for all symmetric solutions X and Y of (3.26) and (3.27), respectively, are called the *positive real controllability* and *observability Gramians* of (3.3). System (3.3) is called *positive real balanced* if $X_{pr} = Y_{pr} = \operatorname{diag}(\Xi, 0)$ with $\Xi = \operatorname{diag}(\xi_1, \ldots, \xi_{n_f})$. The values ξ_j ordered decreasingly are called the *positive real characteristic values* of (3.3). Similarly to Lyapunov-based balanced truncation, the reduced-order system (3.16) can be computed by projecting onto the subspaces corresponding to the dominant positive real characteristic values and non-zero improper Hankel singular values. Note that if system (3.3) has a proper transfer function, then solving the projected discrete-time Lyapunov equations (3.21) and (3.22) can be avoided. The positive real balanced truncation method for such a system is summarized in Algorithm 2. It can be shown that the resulting reduced-order system is passive, and we have the error bound

$$\|\tilde{\mathbf{G}} - \mathbf{G}\|_{\mathbb{H}_\infty} \leq 2\|(M_0 + M_0^T)^{-1}\| \|\mathbf{G}_0\|_{\mathbb{H}_\infty} \|\tilde{\mathbf{G}}_0\|_{\mathbb{H}_\infty} (\xi_{\ell_f+1} + \cdots + \xi_{n_f}) \tag{3.28}$$

with $\mathbf{G}_0 = \mathbf{G} + M_0^T$ and $\tilde{\mathbf{G}}_0 = \tilde{\mathbf{G}} + M_0^T$, see [9, 63]. Note that this error bound requires the computation of the $\mathbb{H}_\infty$-norm of $\mathbf{G}_0$, which is expensive for large-scale systems. If ℓ_f is chosen in Algorithm 2 such that

$$4\|(M_0+M_0^T)^{-1}\|\|\tilde{\mathbf{G}}_0\|_{\mathbb{H}_\infty}(\xi_{\ell_f+1}+\cdots+\xi_{n_f})<1,$$

then bound (3.28) can be simplified to

$$\|\tilde{\mathbf{G}}-\mathbf{G}\|_{\mathbb{H}_\infty}\leq 4\|(M_0+M_0^T)^{-1}\|\|\tilde{\mathbf{G}}_0\|^2_{\mathbb{H}_\infty}(\xi_{\ell_f+1}+\cdots+\xi_{n_f}), \tag{3.29}$$

where only the evaluation of the $\mathbb{H}_\infty$-norm of the reduced-order system $\tilde{\mathbf{G}}_0$ is required.

Algorithm 2 Positive real balanced truncation for DAEs with a proper transfer function.
Given passive $\mathbf{G}=(E,A,B,C,D)$, compute a reduced-order system $\tilde{\mathbf{G}}=(\tilde{E},\tilde{A},\tilde{B},\tilde{C},\tilde{D})$.

1. Compute the matrix $M_0=C(s_0E-A)^{-1}Q_lB+D$ $s_0\in(0,\infty)$.
2. Compute the Cholesky factors R and L of the positive real Gramians $X_{pr}=RR^T$ and $Y_{pr}=LL^T$ that are the minimal solutions of the projected positive real Lur'e equations (3.26) and (3.27).
3. Compute the singular value decomposition $L^TER=[U_1,U_2]\text{diag}(\Xi_1,\Xi_2)[V_1,V_2]^T$, where the matrices $[U_1,U_2]$ and $[V_1,V_2]$ have orthonormal columns, $\Xi_1=\text{diag}(\xi_1,\ldots,\xi_{\ell_f})$ and $\Xi_2=\text{diag}(\xi_{\ell_f+1},\ldots,\xi_{n_f})$.
4. Compute the reduced-order system

$$\tilde{E}=\begin{bmatrix} I & 0\\ 0 & 0\end{bmatrix},\quad \tilde{A}=\begin{bmatrix} W_1^TAT_1 & 0\\ 0 & I,\end{bmatrix},\quad \tilde{B}=\begin{bmatrix} W_1^TB\\ B_2\end{bmatrix},\quad \tilde{C}=[CT_1,C_2],\quad \tilde{D}=D,$$

where $W_1=LU_1\Xi_1^{-1/2}$, $T_1=RV_1\Xi_1^{-1/2}$, and B_2 and C_2 are chosen such that $D-M_0=C_2B_2$.

If $D_0=M_0+M_0^T$ is nonsingular, then the projected Lur'e equations (3.26) and (3.27) can be written as the *projected positive real Riccati equations*

$$AXE^T+EXA^T+(EXC^T-P_lB)D_0^{-1}(EXC^T-P_lB)^T=0,\quad X=P_rXP_r^T, \tag{3.30}$$

$$A^TYE+E^TYA+(B^TYE-CP_r)^TD_0^{-1}(B^TYE-CP_r)=0,\quad Y=P_l^TYP_l. \tag{3.31}$$

The numerical solution of these equations will be discussed in Sect. 3.4.2. The major difficulty in solving these equations is that the spectral projectors P_r and P_l are required. They can be computed by the matrix chain approach from [51]. In the large-scale setting, however, it would be beneficial to have an explicit representation for P_r and P_l as it has been done in [75] for some other structured DAEs arising in computational fluid dynamics and multibody systems. Such a representation in terms of the incidence matrices is currently under investigation.

3.3.3 Passivity-Preserving Model Reduction Via Bounded Real Balanced Truncation

Another passivity-preserving model reduction approach presented first in [63] is based on the bounded real balanced truncation model reduction method applied to the Moebius-transformed system $\hat{\mathbf{G}} = \mathcal{M}(\mathbf{G}) = (\hat{E}, \hat{A}, \hat{B}, \hat{C}, \hat{D})$ as in (3.15). For the MNA equations (3.3), (3.4), where $\mathcal{G}$ is positive definite and both $\mathcal{L}$ and $\mathcal{C}$ are symmetric and positive definite, it has been shown in [64] that the *projected bounded real Lur'e equations*

$$\begin{aligned} &\hat{A}X\hat{E}^T + \hat{E}X\hat{A}^T + \hat{P}_l\hat{B}\hat{B}^T\hat{P}_l^T = -K_cK_c^T, \quad X = \hat{P}_rX\hat{P}_r^T \geq 0, \\ &\hat{E}X\hat{C}^T - \hat{P}_l\hat{B}\hat{M}_0^T = -K_cJ_c^T, \quad I - \hat{M}_0\hat{M}_0^T = J_cJ_c^T \end{aligned} \tag{3.32}$$

and

$$\begin{aligned} &\hat{A}^TY\hat{E} + \hat{E}^TY\hat{A} + \hat{P}_r^T\hat{C}^T\hat{C}\hat{P}_r = -K_o^TK_o, \quad Y = \hat{P}_l^TY\hat{P}_l \geq 0, \\ &-\hat{E}^TY\hat{B} + \hat{P}_r^T\hat{C}^T\hat{M}_0 = -K_o^TJ_o, \quad I - \hat{M}_0^T\hat{M}_0 = J_o^TJ_o \end{aligned} \tag{3.33}$$

are solvable for $X \in \mathbb{R}^{n,n}, K_c \in \mathbb{R}^{n,m}$, $J_c \in \mathbb{R}^{m,m}$ and $Y \in \mathbb{R}^{n,n}$, $K_o \in \mathbb{R}^{m,n}$, $J_o \in \mathbb{R}^{m,m}$, respectively. Here, $\hat{P}_r$ and $\hat{P}_l$ are the spectral projectors onto the right and left deflating subspaces of $\lambda\hat{E} - \hat{A}$ corresponding to the finite eigenvalues along the right and left deflating subspaces corresponding to the eigenvalue at infinity, and $\hat{M}_0 = \lim_{s\to\infty} \hat{C}(s\hat{E} - \hat{A})^{-1}\hat{B} + \hat{D}$. The minimal solutions X_{br} and Y_{br} satisfying $0 \leq X_{br} \leq X$ and $0 \leq Y_{br} \leq Y$ for all symmetric solutions X and Y of (3.32) and (3.33), respectively, are called the *bounded real controllability* and *observability Gramians* of system $\hat{\mathbf{G}}$. This system is *bounded real balanced* if $X_{br} = Y_{br} = \operatorname{diag}(\Gamma, 0)$ with $\Gamma = \operatorname{diag}(\gamma_1, \ldots, \gamma_{n_f})$. The values γ_j ordered decreasingly are called the *bounded real characteristic values* of $\hat{\mathbf{G}}$. Truncating the states of $\hat{\mathbf{G}}$ corresponding to small γ_j and applying the Moebius transformation to the obtained contractive reduced-order model, we get a passive reduced-order system $\tilde{\mathbf{G}}$. The resulting passivity-preserving model reduction method for circuit equations is presented in Algorithm 3. For this method, we have the following a priori error bound

$$\|\tilde{\mathbf{G}} - \mathbf{G}\|_{\mathbb{H}_\infty} \leq \frac{\|I + \mathbf{G}\|_{\mathbb{H}_\infty}^2 (\gamma_{\ell_f+1} + \cdots + \gamma_{n_f})}{I - \|I + \mathbf{G}\|_{\mathbb{H}_\infty} (\gamma_{\ell_f+1} + \cdots + \gamma_{n_f})},$$

provided $\|I + \mathbf{G}\|_{\mathbb{H}_\infty} (\gamma_{\ell_f+1} + \cdots + \gamma_{n_f}) < 1$, see [63]. Furthermore, if we choose ℓ_f in Algorithm 3 such that $2\|I + \tilde{\mathbf{G}}\|_{\mathbb{H}_\infty} (\gamma_{\ell_f+1} + \cdots + \gamma_{n_f}) < 1$, then we obtain the a posteriori error bound

$$\|\tilde{\mathbf{G}} - \mathbf{G}\|_{\mathbb{H}_\infty} \le 2\|I + \tilde{\mathbf{G}}\|_{\mathbb{H}_\infty}^2 (\gamma_{\ell_f+1} + \cdots + \gamma_{n_f}) \tag{3.34}$$

that is inexpensive to compute.

Algorithm 3 Passivity-preserving model reduction based on bounded real balanced truncation.
Given passive $\mathbf{G} = (E, A, B, C, 0)$, compute a reduced-order model $\tilde{\mathbf{G}} = (\tilde{E}, \tilde{A}, \tilde{B}, \tilde{C}, 0)$.

1. Compute the Moebius-transformed system $\hat{\mathbf{G}} = (\hat{E}, \hat{A}, \hat{B}, \hat{C}, \hat{D})$ as in (3.15).
2. Compute the matrix $\hat{M}_0 = \hat{D} + \hat{C}(s_0\hat{E} - \hat{A})^{-1}\hat{Q}_l\hat{B}$ for some $s_0 \in (0, \infty)$.
3. Compute the Cholesky factors $\hat{R}$ and $\hat{L}$ of the bounded real Gramians $X_{br} = \hat{R}\hat{R}^T$ and $Y_{br} = \hat{L}\hat{L}^T$ that are the minimal solutions of the projected bounded real Lur'e equations (3.32) and (3.33).
4. Compute the singular value decomposition $\hat{L}^T\hat{E}\hat{R} = [U_1, U_2]\mathrm{diag}(\Gamma_1, \Gamma_2)[V_1, V_2]^T$, where the matrices $[U_1, U_2]$ and $[V_1, V_2]$ have orthonormal columns, $\Gamma_1 = \mathrm{diag}(\gamma_1, \ldots, \gamma_{\ell_f})$ and $\Gamma_2 = \mathrm{diag}(\gamma_{\ell_f+1}, \ldots, \gamma_{n_f})$.
5. Compute the reduced-order system

$$\tilde{E} = \begin{bmatrix} I & 0 \\ 0 & 0 \end{bmatrix}, \quad \tilde{A} = \frac{1}{2}\begin{bmatrix} 2\hat{W}_1^T A\hat{T}_1 & ;\sqrt{2}\hat{W}_1^T B\hat{C}_2 \\ -\sqrt{2}\hat{B}_2 C\hat{T}_1 & ;2I - \hat{B}_2\hat{C}_2 \end{bmatrix},$$

$$\tilde{B} = \begin{bmatrix} \hat{W}_1^T B \\ -\hat{B}_2/\sqrt{2} \end{bmatrix}, \quad \tilde{C}^T = \begin{bmatrix} (C\hat{T}_1)^T \\ \hat{C}_2^T/\sqrt{2} \end{bmatrix},$$

where $\hat{W}_1 = \hat{L}U_1\Gamma_1^{-1/2}$, $\hat{T}_1 = \hat{R}V_1\Gamma_1^{-1/2}$, and $\hat{B}_2$ and $\hat{C}_2$ are chosen such that $I - \hat{M}_0 = \hat{C}_2\hat{B}_2$.

It should be noted that for DAE systems with a proper transfer function, Algorithms 2 and 3 are equivalent in the sense that they provide reduced-order models with the same transfer function.

Using the topological structure of circuit equations, the matrix $\hat{M}_0$ and the projector $\hat{P}_r$ can be computed in explicit form

$$\hat{M}_0 = \begin{bmatrix} I - 2A_I^T Q_C H_0^{-1} Q_C^T A_I & 2A_I^T Q_C H_0^{-1} Q_C^T A_V \\ -2A_V^T Q_C H_0^{-1} Q_C^T A_I & -I + 2A_V^T Q_C H_0^{-1} Q_C^T A_V \end{bmatrix}, \tag{3.35}$$

$$\hat{P}_r = \begin{bmatrix} H_5(H_4H_2 - I) & H_5H_4A_LH_6 & 0 \\ 0 & H_6 & 0 \\ -A_V^T(H_4H_2 - I) & -A_V^T H_4A_LH_6 & 0 \end{bmatrix}, \tag{3.36}$$

where

$$
\begin{aligned}
&H_0 = Q_C^T(A_R \mathcal{G} A_R^T + A_I A_I^T + A_V A_V^T)Q_C + Q_{RIV-C}^T Q_{RIV-C},\\
&H_1 = P_{CRIV}^T P_{CRIV} + Q_{CRIV}^T A_L \mathcal{S} A_L^T Q_{CRIV},\\
&H_2 = A_R \mathcal{G} A_R^T + A_I A_I^T + A_V A_V^T + A_L \mathcal{S} A_L^T Q_{CRIV} H_1^{-1} Q_{CRIV}^T A_L \mathcal{S} A_L^T,\\
&H_3 = A_C \mathcal{C} A_C^T + Q_C^T H_2 Q_C, \qquad H_4 = Q_C H_3^{-1} Q_C^T,\\
&H_5 = Q_{CRIV} H_1^{-1} Q_{CRIV}^T A_L \mathcal{S} A_L^T - I, \qquad H_6 = I - \mathcal{S} A_L^T Q_{CRIV} H_1^{-1} Q_{CRIV}^T A_L,\\
&Q_{CRIV} \text{ is a projector onto} \ker([A_C, A_R, A_I, A_V]^T), \qquad P_{CRIV} = I - Q_{CRIV},\\
&Q_{RIV-C} \text{ is a projector onto} \ker([A_R, A_I, A_V]^T Q_C),
\end{aligned}
$$

see [64] for details. Furthermore, the left projector is given by $\hat{P}_l = S_{\text{int}} \hat{P}_{r,t}^T S_{\text{int}}$, where $S_{\text{int}} = \text{diag}(I_{n_\eta}, -I_{n_L}, -I_{n_V})$ and $\hat{P}_{r,t}$ is as $\hat{P}_r$ with $\mathcal{G}$, $\mathcal{S}$ and $\mathcal{C}$ replaced by $\mathcal{G}^T$, $\mathcal{S}^T$ and $\mathcal{C}^T$, respectively. The projectors Q_C, Q_{CRIV} and $Q_{RIV\text{-}C}$ can easily be computed in sparse form using graph search algorithms like breadth-first-search [44]. Although $\hat{M}_0$ and $\hat{P}_r$ look very complex, their computation is inexpensive if the sparsity of the incidence matrices and Q-projectors is exploited. Due to the space limitation, we omit detailed discussions.

If the circuit contains neither CVI-loops except for C-loops nor LVI-cutsets except for L-cutsets, i.e., if $\text{rank}(Q_C^T[A_I, A_V]) = n_I + n_V$ and $Q_{RC}^T[A_I, A_V] = 0$, where Q_{RC} is a projector onto $\ker([A_R, A_C]^T)$, then both $\hat{D}_c = I - \hat{M}_0 \hat{M}_0^T$ and $\hat{D}_o = I - \hat{M}_0^T \hat{M}_0$ are nonsingular [64], and the projected Lur'e equations (3.32) and (3.33) can be written as the *projected bounded real Riccati equations*

$$
\begin{aligned}
&\hat{A} X E^T + E X \hat{A}^T + 2\hat{P}_l B B^T \hat{P}_l^T + 2(E X C^T - \hat{P}_l B \hat{M}_0^T)\hat{D}_c^{-1}(E X C^T - \hat{P}_l B \hat{M}_0^T)^T = 0,\\
&X = \hat{P}_r X \hat{P}_r^T,
\end{aligned}
\tag{3.37}
$$

and

$$
\begin{aligned}
&\hat{A}^T Y E + E^T Y \hat{A} + 2\hat{P}_r^T C^T C \hat{P}_r + 2(B^T Y E - \hat{M}_0^T C \hat{P}_r)^T \hat{D}_o^{-1}(B^T Y \hat{E} - \hat{M}_0^T C \hat{P}_r) = 0,\\
&Y = \hat{P}_l^T Y \hat{P}_l.
\end{aligned}
\tag{3.38}
$$

The numerical solution of these equations will be considered in Sect. 3.4.2.

3.3.4 Balanced Truncation for Reciprocal Circuit Equations

For reciprocal circuits with symmetric $\mathcal{C}$, $\mathcal{L}$ and $\mathcal{G}$, we can further exploit the structure of the system matrices in (3.4) in order to reduce the computational complexity of Algorithms 2 and 3. In the sequel, we consider Algorithm 3 only. The other one can be modified analogously.

Consider the reciprocal system (3.3), (3.4), where $\mathcal{C}$, $\mathcal{L}$ and $\mathcal{G}$ are symmetric. Then

$$\hat{E}^T = S_{\rm int}\hat{E}S_{\rm int}, \quad \hat{A}^T = S_{\rm int}\hat{A}S_{\rm int}, \quad \hat{B}^T = S_{\rm ext}\hat{C}S_{\rm int}$$

with $S_{\rm int} = \mathrm{diag}(I_{n_\eta}, -I_{n_L}, -I_{n_V})$ and $S_{\rm ext} = \mathrm{diag}(I_{n_I}, -I_{n_V})$. Therefore, we have

$$\begin{aligned} \hat{P}_l &= S_{\rm int}\hat{P}_r^T S_{\rm int}, \\ Y_{br} &= S_{\rm int}X_{br}S_{\rm int} = S_{\rm int}\hat{R}\hat{R}^T S_{\rm int}^T = \hat{L}\hat{L}^T. \end{aligned} \tag{3.39}$$

Since $\hat{L}^T\hat{E}\hat{R} = \hat{R}^T S_{\rm int}\hat{E}\hat{R}$ and $(I - \hat{M}_0)S_{\rm ext}$ are both symmetric, the characteristic values γ_j and the matrices $\hat{B}_2$ and $\hat{C}_2$ can be determined from the eigenvalue decompositions of $\hat{R}^T S_{\rm int}\hat{E}\hat{R}$ and $(I - \hat{M}_0)S_{\rm ext}$ instead of a more expensive singular value decomposition. We summarize the resulting passivity-preserving balanced truncation method for reciprocal electrical circuits (PABTEC) in Algorithm 4. Note that this method also preserves reciprocity in the reduced-order model, see [64].

Algorithm 4 Passivity-preserving balanced truncation for electrical circuits (PABTEC).

Given passive $\mathbf{G} = (E, A, B, C, 0)$ with the system matrices as in (3.4), compute a reduced-order model $\tilde{\mathbf{G}} = (\tilde{E}, \tilde{A}, \tilde{B}, \tilde{C}, 0)$.

1. Compute the Cholesky factor $\hat{R}$ of the bounded real Gramian $X_{br} = \hat{R}\hat{R}^T$ that is the minimal solution of the projected Lur'e equation (3.32), where $\hat{E}$, $\hat{A}$, $\hat{B}$ and $\hat{C}$ are as in (3.15), the projectors $\hat{P}_r$ and $\hat{P}_l$ are given in (3.36) and (3.39), respectively, and $\hat{M}_0$ is as in (3.35).
2. Compute the eigenvalue decomposition $\hat{R}^T S_{\rm int}\hat{E}\hat{R} = [U_1, U_2]\mathrm{diag}(\Lambda_1, \Lambda_2)[U_1, U_2]^T$, where $[U_1, U_2]$ is orthogonal, $\Lambda_1 = \mathrm{diag}(\lambda_1, \ldots, \lambda_{\ell_f})$ and $\Lambda_2 = \mathrm{diag}(\lambda_{\ell_f+1}, \ldots, \lambda_{n_f})$.
3. Compute the eigenvalue decomposition $(I - M_0)S_{\rm ext} = U_0\Lambda_0 U_0^T$, where U_0 has orthonormal columns and $\Lambda_0 = \mathrm{diag}(\hat{\lambda}_1, \ldots, \hat{\lambda}_m)$ is nonsingular.
4. Compute the reduced-order system

$$\tilde{E} = \begin{bmatrix} I & 0 \\ 0 & 0 \end{bmatrix}, \quad \tilde{A} = \frac{1}{2}\begin{bmatrix} 2\hat{W}_1^T A\hat{T}_1 & \sqrt{2}\hat{W}_1^T B\hat{C}_2 \\ -\sqrt{2}\hat{B}_2 C\hat{T}_1 & 2I - \hat{B}_2\hat{C}_2 \end{bmatrix},$$
$$\tilde{B} = \begin{bmatrix} \hat{W}_1^T B \\ -\hat{B}_2/\sqrt{2} \end{bmatrix}, \quad \tilde{C}^T = \begin{bmatrix} (C\hat{T}_1)^T \\ \hat{C}_2^T/\sqrt{2} \end{bmatrix},$$

where

$$\hat{W}_1 = S_{\text{int}} R U_1 |\Lambda_1|^{-1/2}, \quad \hat{T}_1 = \hat{R} U_1 S_1 |\Lambda_1|^{-1/2},$$
$$\hat{B}_2 = S_0 |\Lambda_0|^{1/2} U_0^T S_{\text{ext}}, \quad \hat{C}_2 = U_0 |\Lambda_0|^{1/2},$$

with

$$|\Lambda_1| = \operatorname{diag}(|\lambda_1|, \ldots, |\lambda_{\ell_f}|), \quad S_1 = \operatorname{diag}(\operatorname{sign}(\lambda_1), \ldots, \operatorname{sign}(\lambda_{\ell_f})),$$
$$|\Lambda_0| = \operatorname{diag}(|\hat{\lambda}_1|, \ldots, |\hat{\lambda}_m|), \quad S_0 = \operatorname{diag}(\operatorname{sign}(\hat{\lambda}_1), \ldots, \operatorname{sign}(\hat{\lambda}_m)).$$

3.4 Numerical Methods for Matrix Equations

In this section, we consider numerical algorithms for the projected Lyapunov equations (3.19) and (3.20) and the projected Riccati equations (3.30), (3.31) and (3.37), (3.38) developed in [10, 75]. In practice, the numerical rank of the solutions of these equations is often much smaller than the dimension of the problem. Then such solutions can be well approximated by low-rank matrices. Moreover, these low-rank approximations can be determined directly in factored form. Replacing the Cholesky factors of the Gramians in Algorithms 1–4 by their low-rank factors reduces significantly the computational complexity and storage requirements in the balancing-related model reduction methods and makes these methods very suitable for large-scale DAE systems.

3.4.1 ADI Method for Projected Lyapunov Equations

First, we focus on solving the projected Lyapunov equation

$$EXA^T + AXE^T = -P_l BB^T P_l^T, \quad X = P_r X P_r^T, \tag{3.40}$$

using the *alternating direction implicit* (ADI) *method*. The dual equation can be treated analogously. The ADI method has been first proposed for standard Lyapunov equations [16, 49, 57, 80] and then extended in [75] to projected Lyapunov equations. The generalized ADI iteration for the projected Lyapunov equation (3.40) is given by

$$\begin{aligned}(E + \tau_k A) X_{k-1/2} A^T + A X_{k-1} (E - \tau_k A)^T &= -P_l BB^T P_l^T, \\ (E + \overline{\tau}_k A) X_k^T A^T + A X_{k-1/2}^T (E - \overline{\tau}_k A)^T &= -P_l BB^T P_l^T\end{aligned} \tag{3.41}$$

with an initial matrix $X_0 = 0$ and shift parameters $\tau_1, \ldots, \tau_k \in \mathbb{C}_-$. If the pencil $\lambda E - A$ is c-stable, then X_k converges towards the solution of the projected Lyapunov equation (3.40). The rate of convergence depends strongly on the choice of the shift parameters. The optimal shift parameters providing the superlinear convergence satisfy the generalized ADI minimax problem

$$\{\tau_1,\ldots,\tau_q\} = \operatorname*{arg\,min}_{\{\tau_1,\ldots,\tau_q\}\in\mathbb{C}_-} \max_{t\in Sp_f(E,A)} \frac{|(1-\overline{\tau}_1 t)\cdots(1-\overline{\tau}_q t)|}{|(1+\tau_1 t)\cdots(1+\tau_q t)|},$$

where $Sp_f(E, A)$ denotes the finite spectrum of the pencil $\lambda E - A$. Similarly to [57], the suboptimal ADI parameters can be obtained from a set of largest and smallest in modulus approximate finite eigenvalues of $\lambda E - A$ computed by an Arnoldi procedure. Other parameter selection techniques developed for standard Lyapunov equations [17, 69, 81] can also be used for the projected Lyapunov equation (3.40).

A low-rank approximation to the solution of the projected Lyapunov equation (3.40) can be computed in factored form $X \approx Z_k Z_k^T$ using a low-rank version of the ADI method (LR-ADI) as presented in Algorithm 5.

Algorithm 5 The generalized LR-ADI for the projected Lyapunov equation. Given $E, A \in \mathbb{R}^{n,n}$, $B \in \mathbb{R}^{n,m}$, projector P_l and shift parameters $\tau_1,\ldots,\tau_q \in \mathbb{C}_-$, compute a low-rank approximation $X \approx Z_k Z_k^T$ to the solution of the projected Lyapunov equation (3.40).

1. $Z^{(1)} = \sqrt{-2\mathrm{Re}(\tau_1)}(E+\tau_1 A)^{-1}P_l B, \quad Z_1 = Z^{(1)};$
2. FOR $k = 2, 3, \ldots$

$$Z^{(k)} = \sqrt{\frac{\mathrm{Re}(\tau_k)}{\mathrm{Re}(\tau_{k-1})}}\Big(I - (\overline{\tau}_{k-1}+\tau_k)(E+\tau_k A)^{-1}A\Big)Z^{(k-1)}, \quad Z_k = [Z_{k-1}, Z^{(k)}]$$

END FOR

In order to guarantee for the factors Z_k to be real in case of complex shift parameters, we take these parameters in complex conjugate pairs $\{\tau_k, \tau_{k+1} = \overline{\tau}_k\}$. At each iteration we have $Z_k = [Z^{(1)},\ldots,Z^{(k)}] \in \mathbb{R}^{n,mk}$. To keep the low-rank structure in Z_k for large mk, we can compress the columns of Z_k using the rank-revealing QR factorization [21] as described in [8].

Finally, note that the matrices $(E + \tau_k A)^{-1}$ in Algorithm 5 do not have to be computed explicitly. Instead, we solve linear systems of the form $(E + \tau_k A)x = P_l b$ either by computing (sparse) LU factorizations and forward/backward substitutions or by using iterative Krylov subspace methods [68].

3.4.2 Newton–Kleinman Method for Projected Riccati Equations

In this section, we consider the numerical solution of the projected Riccati equation

$$\mathscr{R}(X) \equiv EXF^T + FXE^T + EXH^THXE^T + \Pi_l QQ^T \Pi_l^T = 0, \quad X = \Pi_r X \Pi_r^T, \tag{3.42}$$

where

$$\begin{aligned} &F = A - P_l B(M_0 + M_0^T)^{-1} CP_r, \quad H = J_c^{-1} CP_r, \quad Q = BJ_c^{-T}, \\ &M_0 + M_0^T = J_c J_c^T, \quad \Pi_l = P_l, \quad \Pi_r = P_r \end{aligned} \tag{3.43}$$

in the positive real case and

$$\begin{aligned} F &= A - BC - 2\hat{P}_l B\hat{M}_0^T (I - \hat{M}_0 \hat{M}_0^T)^{-1} C\hat{P}_r, \\ H &= \sqrt{2} J_c^{-1} C\hat{P}_r, \quad Q = -\sqrt{2} BJ_o^{-1}, \\ J_o^T J_o &= I - \hat{M}_0^T \hat{M}_0, \quad J_c J_c^T = I - \hat{M}_0 \hat{M}_0^T, \quad \Pi_l = \hat{P}_l, \quad \Pi_r = \hat{P}_r \end{aligned} \tag{3.44}$$

in the bounded real case. The minimal solution $X_{\min}$ of (3.42) is at least *semi-stabilizing* in the sense that all the finite eigenvalues of $\lambda E - F - EX_{\min}H^TH$ are in the closed left half-plane. Such a solution can be computed via the *Newton–Kleinman method* [10] as presented in Algorithm 6.

Algorithm 6 The generalized Newton–Kleinman method for the projected Riccati equation.

Given E, $F \in \mathbb{R}^{n,n}, H \in \mathbb{R}^{m,n}$, $Q \in \mathbb{R}^{n,m}$, projectors Π_r, Π_l and a stabilizing initial guess X_0, compute an approximate solution of the projected Riccati equation (3.42).

FOR $j = 1, 2, \ldots,$

1. Compute $K_j = EX_{j-1}H^T$ and $F_j = F + K_jH$.
2. Solve the projected Lyapunov equation

$$EX_jF_j^T + F_jX_jE^T = -\Pi_l(QQ^T - K_jK_j^T)\Pi_l^T, \quad X_j = \Pi_r X_j \Pi_r^T.$$

END FOR

Similarly to the standard state space case [6, 78], one can show that if $\lambda E - F$ is c-stable, then for $X_0 = 0$, all $\lambda E - F_j$ are also c-stable and $\lim_{j\to\infty} X_j = X_{\min}$. The convergence is quadratic if the pencil $\lambda E - F - EX_{\min}H^TH$ is c-stable. Some difficulties may occur if the pencil $\lambda E - F$ has eigenvalues on the imaginary axis. For circuit equations, these eigenvalues are uncontrollable and unobservable [64]. In that case, similarly to [12], one could choose a special stabilizing initial guess X_0 that ensures the convergence of the Newton–Kleinman iteration. However, the computation of such a guess for large-scale problems remains an open problem.

A low-rank approximation to the minimal solution of the projected Riccati equation (3.42) can be computed in factored form $X_{\min} \approx \tilde{R}\tilde{R}^T$ with $\tilde{R} \in \mathbb{R}^{n,k}$, $k \ll n$ using the same approach as in [11]. Starting with $K_1 = EX_0H^T$ and $F_1 = F + K_1H$, in each Newton iteration we solve two projected Lyapunov equations

$$EX_{1,j}F_j^T + F_jX_{1,j}E^T = -\Pi_l QQ^T \Pi_l^T, \quad X_{1,j} = P_r X_{1,j} P_r^T, \tag{3.45}$$

$$EX_{2,j}F_j^T + F_jX_{2,j}E^T = -\Pi_l K_j K_j^T \Pi_l^T, \quad X_{2,j} = \Pi_l X_{2,j} \Pi_r^T, \tag{3.46}$$

for the low-rank approximations $X_{1,j} \approx R_{1,j}R_{1,j}^T$ and $X_{2,j} \approx R_{2,j}R_{2,j}^T$, respectively, and then compute $K_{j+1} = E(R_{1,j}R_{1,j}^T - R_{2,j}R_{2,j}^T)H^T$ and $F_{j+1} = F + K_{j+1}H$. If the convergence is observed after $j_{\max}$ iterations, then an approximate solution $X_{\min} \approx \tilde{R}\tilde{R}^T$ of the projected Riccati equation (3.42) can be computed in factored form by solving the projected Lyapunov equation

$$EXF^T + FXE^T = -\Pi_l Q_0 Q_0^T \Pi_l, \quad X = \Pi_r X \Pi_r^T \tag{3.47}$$

with $Q_0 = \left[Q, E(X_{1,j_{\max}} - X_{2,j_{\max}})H^T\right]$ provided $\lambda E - F$ is c-stable. For computing low-rank factors of the solutions of the projected Lyapunov equations (3.45)–(3.47), we can use the generalized LR-ADI method. Note that in this method we need to compute the products $(E + \tau F_j)^{-1}w$ with $\tau \in \mathbb{C}_-$ and $w \in \mathbb{R}^n$. For example, in the bounded real case we have $E + \tau F_j = E + \tau(A - BC) - \tau\hat{K}_jH$ with the low-rank matrices $H \in \mathbb{R}^{m,n}$ and $\hat{K}_j = \sqrt{2}\hat{P}_l B\hat{M}_0^T J_c^{-T} - K_j \in \mathbb{R}^{n,m}$. Then one can use the Sherman–Morrison–Woodbury formula [36, Section 2.1.3] to compute these products as

$$(E + \tau F_j)^{-1}w = w_1 + M_{\hat{K}_j}\left((I_m - HM_{\hat{K}_j})^{-1}Hw_1\right),$$

where $w_1 = (E + \tau(A - BC))^{-1}w$ and $M_{\hat{K}_j} = \tau(E + \tau(A - BC))^{-1}\hat{K}_j$ can be determined by solving linear systems with the sparse matrix $E + \tau(A - BC)$ either by computing sparse LU factorization or by using Krylov subspace methods [68].

3.5 Numerical Examples

In this section, we present some results of numerical experiments to demonstrate the efficiency of the passivity-preserving balancing-related model reduction methods for circuit equations described in Sect. 3.3.

Example 3 The first example is a three-port RC circuit provided by NEC Laboratories Europe. The passive reciprocal system of order $n = 2007$ was approximated by two models of order $\ell = 42$ computed by the positive real balanced truncation (PRBT) method and the bounded real balanced truncation (BRBT-M) method applied to the Moebius-transformed system. The minimal solutions of the projected Riccati equations (3.30) and (3.37) were approximated by the low rank matrices $X_{pr} \approx \tilde{R}_{pr}\tilde{R}_{pr}^T$ with $\tilde{R}_{pr} \in \mathbb{R}^{n,123}$ and $X_{br} \approx \tilde{R}_{br}\tilde{R}_{br}^T$ with $\tilde{R}_{br} \in \mathbb{R}^{n,125}$, respectively. Figure 3.3a shows the normalized residuals $\rho(X_j) = \|\mathscr{R}(X_j)\|_F / \|\Pi_l QQ^T \Pi_l^T\|_F$, where $\|\cdot\|_F$ denotes the Frobenius matrix

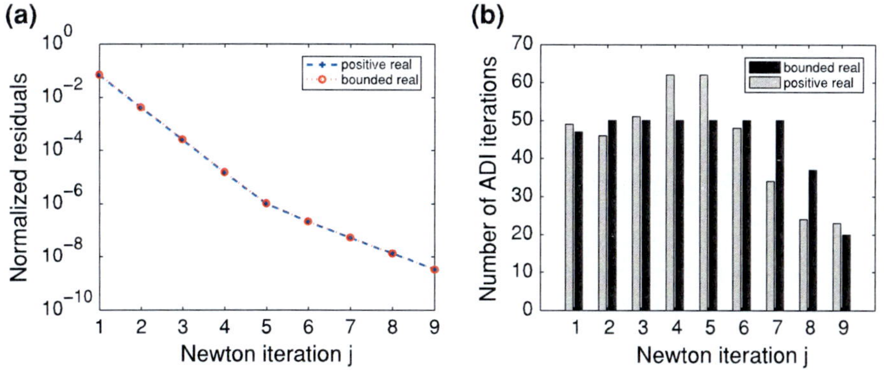

Fig. 3.3 RC circuit: (**a**) the convergence history of the low-rank Newton–Kleinman-ADI method; (**b**) the number of ADI iterations required for solving the projected Lyapunov equations at each Newton iteration

norm, $X_j = R_{1,j}R_{1,j}^T - R_{2,j}R_{2,j}^T$, Π_l and Q are given in (3.43) and (3.44) for (3.30) and (3.37), respectively. Figure 3.3b displays the number of ADI iterations required for solving the projected Lyapunov equations at each Newton iteration.

The spectral norms of the frequency responses $\|\mathbf{G}(\mathrm{i}\omega)\|$ and $\|\tilde{\mathbf{G}}(\mathrm{i}\omega)\|$ for a frequency range $\omega \in [1, 10^{15}]$ are presented in Fig. 3.4a. In Fig. 3.4b, we display the absolute errors $\|\tilde{\mathbf{G}}(i\omega) - \mathbf{G}(i\omega)\|$ for both reduced-order systems and also the error bounds (3.29) and (3.34). As expected, due to the properness of $\mathbf{G}$, the PRBT and BRBT-M methods are equivalent and provide similar results.

Example 4 The second example is a transmission line model [5] that consists of 20,000 RLC ladders. We approximate the DAE system of order $n = 60{,}000$ by a model of order $\ell = 32$ computed by the PABTEC method (Algorithm 4). The bounded real Gramian X_{br} was approximated by a low-rank matrix $X_{br} \approx \tilde{R}\tilde{R}^T$ with $\tilde{R} \in \mathbb{R}^{n,249}$. Figure 3.5a presents the bounded real characteristic values of the Moebius-transformed system computed as the absolute values of the eigenvalues of $\tilde{R}^T S_{\mathrm{int}} E \tilde{R}$. One can see that the characteristic values decay rapidly, so we can expect a good approximation by a reduced-order model. The frequency responses of the full-order and the reduced-order models are not presented, since they were impossible to distinguish. Figure 3.5b shows the absolute error $\|\tilde{\mathbf{G}}(i\omega) - \mathbf{G}(i\omega)\|$ for $\omega \in [1, 10^{20}]$ and the error bound (3.34).

3.6 Conclusions and Open Problems

In this paper, we have discussed balancing-related model reduction methods for linear DAEs arising in circuit simulation. The important property of these methods is the existence of computable error bounds that essentially distinguishes the

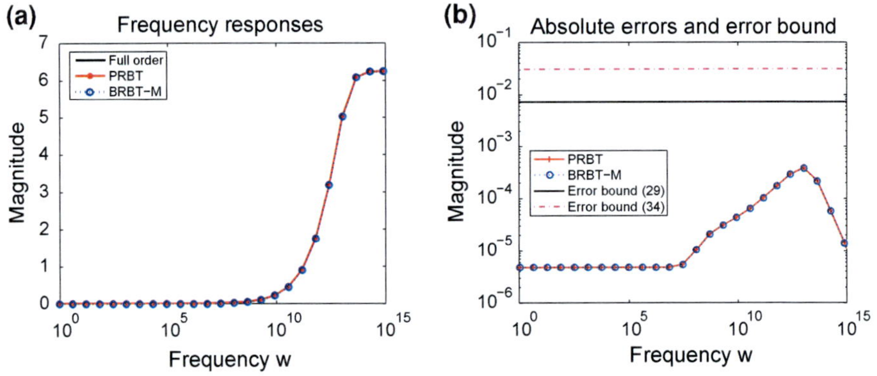

Fig. 3.4 RC circuit: (a) the frequency responses of the original and the reduced-order systems; (b) the absolute errors and error bounds

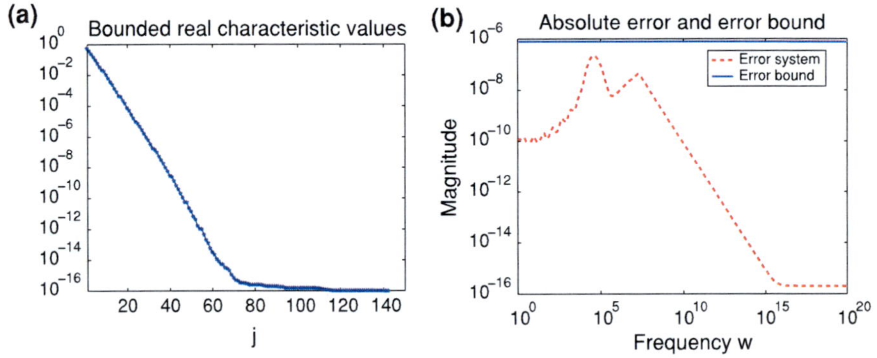

Fig. 3.5 Transmission line: (a) the bounded real characteristic values; (b) the absolute error and the error bound (3.34)

balanced truncation technique from other model reduction approaches. Moreover, positive real balanced truncation and bounded real balanced truncation applied to a Moebius-transformed system guarantee the preservation of passivity in a reduced-order model that makes these methods very suitable for circuit equations.

Balancing-related model reduction methods for DAEs involve solving projected Lyapunov, Lur'e and Riccati matrix equations. We have presented the efficient numerical algorithms for large-scale projected Lyapunov and Riccati equations based on the LR-ADI iteration and the Newton–Kleinman method, respectively. We have also shown that exploiting the topological structure of circuit equations reduces substantially the numerical complexity of balanced truncation.

Although considerable progress has recently been achieved in developing of balanced truncation methods for large-scale DAEs, some problems still remain open. For example, if $M_0 + M_0^T$ (or $I - \hat{M}_0\hat{M}_0^T$) is singular, then to compute the positive real (bounded real) Gramians we have to solve the projected Lur'e equations. Similarly to the standard state space case [82], for small to medium-sized DAE systems, these equations can be transformed to projected Riccati equations of smaller dimension. This approach becomes, however, prohibitive for large-scale problems due to the explicit use of state space transformations. Another problem, already mentioned in Sect. 3.4.2, is the computation of an appropriate stabilizing initial matrix in the Newton–Kleinman iteration in case when the pencil has pure imaginary eigenvalues. This problem could probably be solved for circuit equations by exploiting their special structure.

Finally, in some numerical experiments we observed a very slow convergence of the LR-ADI iteration caused by a poor choice of the shift parameters. The combination of the LR-ADI iteration with the Galerkin projection as proposed in [18, 43] for standard state space systems may improve the performance of the ADI method. Also, a generalization of an extended Krylov subspace method [71] to the projected Lyapunov equations remains for future work.

Acknowledgements This work was supported by the Research Network SyreNe—*System Reduction for Nanoscale IC Design* funded by the German Federal Ministry of Education and Science (BMBF), Grant No. 03STPAE3. Responsibility for the contents of this publication rests with the author. The author would like to thank Achim Basermann and Carsten Neff from NEC Laboratories Europe, IT Research Division for providing the circuit examples.

References

1. Anderson, B., Vongpanitlerd, S.: Network Analysis and Synthesis. Prentice Hall, Englewood Cliffs (1973)
2. Antoulas, A.: Approximation of Large-Scale Dynamical Systems. SIAM, Philadelphia (2005)
3. Antoulas, A.: A new result on passivity preserving model reduction. Syst. Control Lett. **54**(4), 361–374 (2005)
4. Bai, Z.: Krylov subspace techniques for reduced-order modeling of large-scale dynamical systems. Appl. Numer. Math. **43**, 9–44 (2002)
5. Bechtold, T., Verhoeven, A., ter Maten, E., Voss, T.: Model order reduction: an advanced, efficient and automated computational tool for microsystems. In: Cutello, V., Fotia, G., Puccio, L. (eds) Applied and Industrial Mathematics in Italy II, Selected contributions from the 8th SIMAI Conference, Series on Advances in Mathematics for Applied Sciences, vol. 75, pp. 113–124. World Scientific Publishing Co. Pte. Ltd, Singapore (2007)
6. Benner, P: Numerical solution of special algebraic Riccati equations via exact line search method. In: Proceedings of the European Control Conference (ECC97), Paper 786. BELWARE Information Technology, Waterloo, Belgium (1997)
7. Benner, P.: Advances in balancing-related model reduction for circuit simulation. In: Roos, J., Costa, L. (eds) Scientific Computing in Electrical Engineering SCEE 2008, Mathematics in Industry, vol. 14, pp. 469–482. Springer, Berlin (2010)
8. Benner, P., Quintana-Ortí, E.: Solving stable generalized Lyapunov equations with the matrix sign function. Numer. Algorithms **20**(1), 75–100 (1999)
9. Benner, P., Faßbender, H.: Numerische Methoden zur passivitätserhaltenden Modellreduktion (Numerical methods for passivity preserving model reduction). at-Automatisierungstechnik **54**(4), 153–160 (2006) (In German)

10. Benner, P., Stykel, T.: Numerical algorithms for projected generalized Riccati equations (in preparation)
11. Benner, P., Quintana-Ortí, E., Quintana-Ortí, G.: Efficient numerical algorithms for balanced stochastic truncation. Int. J. Appl. Math. Comput. Sci. **11**(5), 1123–1150 (2001)
12. Benner, P., Hernández, V., Pastor, A.: The Kleinman iteration for nonstabilizable systems. Math. Control Signals Syst. **16**, 76–93 (2003)
13. Benner, P., Quintana-Ortí, E., Quintana-Ortí, G.: Computing passive reduced-order models for circuit simulation. In: Proceedings of the International Conference on Parallel Computing in Electrical Engineering PARELEC 2004, pp. 146–151. IEEE Computer Society, Los Alamitos (2004)
14. Benner, P., Quintana-Ortí, E., Quintana-Ortí, G.: Parallel model reduction of large-scale linear descriptor systems via balanced truncation. In: Daydé, M., Dongarra, J., Hernández, V., Palma, J. (eds) High Performance Computing for Computational Science—VECPAR 2004. Lecture Notes in Computer Science, vol. 3402, pp. 340–353. Springer, Berlin (2004)
15. Benner, P., Mehrmann, V., Sorensen, D. (eds.): Dimension Reduction of Large-Scale Systems. Lecture Notes in Computational Science and Engineering, vol. 45. Springer, Berlin (2005)
16. Benner, P., Li, J.R., Penzl, T.: Numerical solution of large Lyapunov equations, Riccati equations, and linear-quadratic control problems. Numer. Linear Algebra Appl. **15**, 755–777 (2008)
17. Benner, P., Mena, H., Saak, J.: On the parameter selection problem in the Newton-ADI iteration for large-scale Riccati equations. Electron. Trans. Numer. Anal. **29**, 136–149 (2008)
18. Benner, P., Li, R.C., Truhar, N.: On the ADI method for Sylvester equations. J. Comput. Appl. Math. **233**(4), 1035–1045 (2009)
19. Brenan, K., Campbell, S., Petzold, L.: The Numerical Solution of Initial-Value Problems in Differential-Algebraic Equations Classics in Applied Mathematics. SIAM, Philadelphia (1996)
20. Campbell, S.: Singular Systems of Differential Equation I II. Pitman, San Francisco (1980)
21. Chan, T.: Rank revealing *QR* factorizations. Linear Algebra Appl. **88/89**, 67–82 (1987)
22. Chua, L., Desoer, C., Kuh, E.: Linear and Nonlinear Circuits. McGraw-Hill, New York (1987)
23. Clements, D., Anderson, B., Laub, A., Matson, L.: Spectral factorization with imaginary-axis zeros. Linear Algebra Appl. **250**, 225–252 (1997)
24. Dai, L.: Singular Control Systems. Lecture Notes in Control and Information Sciences, vol. 118. Springer, Berlin (1989)
25. Davison, E.: A method for simplifying linear dynamical systems. IEEE Trans. Automat. Contr. **11**, 93–101 (1966)
26. Enns, D.: Model reduction with balanced realization: an error bound and a frequency weighted generalization. In: Proceedings of the 23rd IEEE Conference on Decision and Control (Las Vegas, 1984), pp. 127–132. IEEE, New York (1984)
27. Estévez Schwarz, D., Tischendorf, C.: Structural analysis for electric circuits and consequences for MNA. Int. J. Circuit Theory Appl. **28**, 131–162 (2000)
28. Faßbender, H., Benner, P.: Passivity preserving model reduction via a structured Lanczos method. In: Proceedings of the IEEE International Symposium on Computer-Aided Control Systems Design (Munich, Germany, October 4–6, 2006), pp. 8–13. IEEE (2006)
29. Ferrante, A.: Positive real lemma: necessary and sufficient conditions for the existence of solutions under virtually no assumptions. IEEE Trans. Automat. Contr. **50**(5), 720–724 (2005)
30. Freund, R.: Model reduction methods based on Krylov subspaces. Acta Numerica **12**, 267–319 (2003)
31. Freund, R.: SPRIM: structure-preserving reduced-order interconnect macromodeling. In: Technical Digest of the 2004 IEEE/ACM International Conference on Computer-Aided Design, pp. 80–87. IEEE Computer Society Press, Los Alamos (2004)

32. Freund, R.: Structure-preserving model order reduction of RCL circuit equations. In: Schilders, W., van der Vorst, H., Rommes, (J. (eds) Model Order Reduction: Theory, Research Aspects and Applications. Mathematics in Industry, vol. 13, pp. 49–73. Springer, Berlin (2008)
33. Freund, R., Feldmann, P.: The SyMPVL algorithm and its applications in interconnect simulation. In: Proceedings of the 1997 International Conference on Simulation of Semiconductor Processes and Devices, pp. 113–116. IEEE, New York (1997)
34. Gantmacher, F.: Theory of Matrices. Chelsea Publishing Company, New York (1959)
35. Glover, K.: All optimal Hankel-norm aproximations of linear multivariable systems and their L^{∞}-error bounds. Int. J. Control **39**(6), 1115–1193 (1984)
36. Golub, G., Van Loan, C.: Matrix Computations. 3rd edn. The Johns Hopkins University Press, Baltimore (1996)
37. Griepentrog, E., März, R.: Differential-Algebraic Equations and Their Numerical Treatment. Teubner-Texte zur Mathematik, vol. 88. B.G. Teubner, Leipzig (1986)
38. Grimme, E.: Krylov projection methods for model reduction. Ph.D. thesis, University of Illinois, Urbana-Champaign (1997)
39. Gugercin, S., Antoulas, A.: A survey of model reduction by balanced truncation and some new results. Int. J. Control **77**(8), 748–766 (2004)
40. Ionescu, V., Oară, C., Weiss, M.: Generalized Riccati Theory and Robust Control: A Popov Function Approach. Wiley, Chichester (1999)
41. Ionutiu, R., Rommes, J., Antoulas, A.: Passivity-preserving model reduction using dominant spectral-zero interpolation. IEEE Trans. Comput. Aided Design Integr. Circuits Syst. **27**(12), 2250–2263 (2008)
42. Ishihara, J., Terra, M.: On the Lyapunov theorem for singular systems. IEEE Trans. Automat. Contr. **47**(11), 1926–1930 (2002)
43. Jbilou, K.: ADI preconditioned Krylov methods for large Lyapunov matrix equations. Linear Algebra Appl. **432**(10), 2473–2485 (2010)
44. Jungnickel, D.: Graphs, Network and Algorithms. Springer, Berlin (2005)
45. Knockaert, L., De Zutter, D.: Laguerre-SVD reduced-order modeling. IEEE Trans. Microw. Theory Tech. **48**(9), 1469–1475 (2000)
46. Kunkel, P., Mehrmann, V.: Differential-Algebraic Equations. Analysis and Numerical Solution. EMS Publishing House, Zürich (2006)
47. Lancaster, P., Rodman, L.: The Algebraic Riccati Equation. Oxford University Press, Oxford (1995)
48. Lewis, F.: A survey of linear singular systems. Circuits Syst. Signal Process **5**(1), 3–36 (1986)
49. Li, J.R., White, J.: Low rank solution of Lyapunov equations. SIAM J. Matrix Anal. Appl. **24**(1), 260–280 (2002)
50. Marschall, S.: An approximate method for reducing the order of a linear system. Contr. Eng. **10**, 642–648 (1966)
51. März, R.: Canonical projectors for linear differential algebraic equations. Comput. Math. Appl. **31**(4/5), 121–135 (1996)
52. Mehrmann, V., Stykel, T.: Balanced truncation model reduction for large-scale systems in descripter form. Chapter 3 (pp. 83–115) of [12]
53. Moore, B.: Principal component analysis in linear systems: controllability, observability, and model reduction. IEEE Trans. Automat. Contr. **26**(1), 17–32 (1981)
54. Mullis, T., Roberts, R.: Synthesis of minimum roundoff noise fixed point digital filters. IEEE Trans. Circuits Syst. **CAS-23**(9), 551–562 (1976)
55. Ober, R.: Balanced parametrization of classes of linear systems. SIAM J. Control Optim. **29**(6), 1251–1287 (1991)
56. Odabasioglu, A., Celik, M., Pileggi, L.: PRIMA: passive reduced-order interconnect macromodeling algorithm. IEEE Trans. Comput. Aided Design Integr. Circuits Syst. **17**(8), 645–654 (1998)

57. Penzl, T.: A cyclic low-rank Smith method for large sparse Lyapunov equations. SIAM J. Sci. Comput. **21**(4), 1401–1418 (1999/2000)
58. Penzl, T.: Algorithms for model reduction of large dynamical systems. Linear Algebra Appl. **415**(2–3), 322–343 (2006)
59. Perev, K., Shafai, B.: Balanced realization and model reduction of singular systems. Int. J. Syst. Sci. **25**(6), 1039–1052 (1994)
60. Phillips, J., Daniel, L., Silveira, L.: Guaranteed passive balancing transformations for model order reduction. IEEE Trans. Comput. Aided Design Integr. Circuits Syst. **22**(8), 1027–1041 (2003)
61. Reis, T.: Circuit synthesis of passive descriptor systems—a modified nodal approach. Int. J. Circuit Theory Appl. **38**(1), 44–68 (2010). doi:10.1002/cta.532
62. Reis, T.: Lur'e equations and even matrix pencils. Linear Algebra Appl. 434(4),152–173 (2011)
63. Reis, T., Stykel, T.: Positive real and bounded real balancing for model reduction of descriptor systems. Int. J. Control **83**(1), 74– 88 (2009) doi:10.1016/j.lao.2010.09.005
64. Reis, T., Stykel, T.: Passivity-preserving balanced truncation for electrical circuits. IEEE Trans. Comput. Aided Design Integr. Circuits Syst. **29**(9), 1354–1367 (2010)
65. Reis, T., Stykel, T.: Lyapunov balancing for passivity-preserving model reduction of RC circuits. SIAM J. Appl. Dyn. Syst. 10(1),1–34 (2011)
66. Riaza, R., Tischendorf, C.: Qualitative features of matrix pencils and DAEs arising in circuit dynamics. Dyn. Syst. **22**, 107–131 (2007)
67. Riaza, R.: Differential-Algebraic Systems: Analytical Aspects and Circuit Applications. World Scientific Publishing Co. Pte. Ltd, Hackensack (2008)
68. Saad, Y.: Iterative Methods for Sparse Linear Systems. PWS Publishing Company, Boston (1996)
69. Sabino, J.: Solution of large-scale Lyapunov equations via the block modified Smith method. Ph.D. thesis, Rice University, Houston (2006)
70. Schilders, W., van der Vorst, H., Rommes, J. (eds.): Model Order Reduction: Theory, Research Aspects and Applications. Mathematics in Industry, vol. 13. Springer, Berlin (2008)
71. Simoncini, V.: A new iterative method for solving large-scale Lyapunov matrix equations. SIAM J. Sci. Comput. **29**(3), 1268–1288 (2007)
72. Sorensen, D.: Passivity preserving model reduction via interpolation of spectral zeros. Syst. Control Lett. **54**(4), 347–360 (2005)
73. Stykel, T.: Stability and inertia theorems for generalized Lyapunov equations. Linear Algebra Appl. **355**, 297–314 (2002)
74. Stykel, T.: Gramian-based model reduction for descriptor systems. Math. Control Signals Syst. **16**, 297–319 (2004)
75. Stykel, T.: Low-rank iterative methods for projected generalized Lyapunov equations. Electron. Trans. Numer. Anal. **30**, 187–202 (2008)
76. Takaba, K., Morihira, N., Katayama, T.: A generalized Lyapunov theorem for descriptor system. Syst. Control Lett. **24**, 49–51 (1995)
77. Varga, A.: Efficient minimal realization procedure based on balancing. In: Moudni, A.E., Borne, P., Tzafestas, S. (eds.) Proceedings of IMACS/IFAC Symposium on Modelling and Control of Technological Systems (Lille, France, May 7–10, 1991), vol. 2, pp. 42–47 (1991)
78. Varga, A.: On computing high accuracy solutions of a class of Riccati equations. Control Theory Adv. Technol. **10**, 2005–2016 (1995)
79. Vlach, J., Singhal, K.: Computer Methods for Circuit Analysis and Design. Van Nostrand Reinhold, New York (1994)
80. Wachspress, E.: Iterative solution of the Lyapunov matrix equation. Appl. Math. Lett. **1**, 87–90 (1988)
81. Wachspress, E.: The ADI minimax problem for complex spectra. In: Kincaid, D., Hayes, L. (eds.) Iterative Methods for Large Linear Systems, pp. 251–271. Academic Press, San Diego (1990)

82. Weiss, H., Wang, Q., Speyer, J.: System characterization of positive real conditions. IEEE Trans. Automat. Contr. **39**(3), 540–544 (1994)
83. Yan, B., Tan, S.D., McGaughy, B.: Second-order balanced truncation for passive order reduction of RLCK circuits. IEEE Trans. Circuits Syst. II **55**(9), 942–946 (2008)
84. Yang, F., Zeng, X., Su, Y., Zhou, D.: RLC equivalent circuit synthesis method for structure-preserved reduced-order model of interconnect in VLSI. Commun. Comput. Phys. **3**(2), 376–396 (2008)

Chapter 4
Topics in Model Order Reduction with Applications to Circuit Simulation

Sanda Lefteriu and Athanasios C. Antoulas

Abstract When integrating several components on a chip, one may be faced with the task of combining parts which were not manufactured in house. Oftentimes, no accurate model is available for these constituents. In this scenario, one may perform measurements of each device in the frequency range of interest. This information can then be employed to construct a reduced order model which can be used in a simulator which integrates the behaviour of each component and analyzes the entire chip. This paper addresses the problem of modeling a system based on measurements of its frequency response using the concept of tangential interpolation, together with the tool of the Loewner matrix pencil. This concept is particularly useful in the context of modeling components with a large number of input and output ports. After a thorough theoretical analysis which motivates the approach and makes connections to system theory, we present two numerical results which show its applicability to different systems arising in circuit simulation.

Keywords Tangential interpolation · Lower marroit pencil · Multiport systems · System with delay

4.1 Introduction and Motivation

The current trend in VLSI circuits is decreasing the feature size while increasing the chip size, resulting in an ever-growing chip complexity. At the same time, the

S. Lefteriu (✉) · Athanasios C.Antoulas
Department of Electrical and Computer Engineering, Rice University, Houston, TX, USA
e-mail: sanda.lefteriu@rice.edu

Athanasios C.Antoulas
e-mail: aca@rice.edu

P. Benner et al. (eds.), *Model Reduction for Circuit Simulation*, Lecture Notes in Electrical Engineering, 74, DOI: 10.1007/978-94-007-0089-5_4, operating frequency is also increasing. Hence, an important step in the design process is to be able to simulate the desired chip layout in a reasonable amount of time. Model order reduction proves to be a very useful tool in this case, since it is capable of decreasing the simulation time of complex systems to a few seconds by constructing models of smaller dimensions which preserve the essential properties and characteristics of the original one.

Traditionally, model reduction has been applied to systems for which an accurate, but very complex model, is available, for example by discretizing Maxwell's equations. An alternative technique [2] is model reduction from time domain or frequency domain measurements of devices which have already been manufactured, for which no model is available. Constructing models from time or frequency domain data are essentially different research problems as each presents different challenges, but in this paper we focus on modeling from measurements of the frequency response. We do not investigate what is a good choice of sampling points, as this is a complex research question by itself. Assuming that the measurements capture the dynamic behaviour of interest, the next step is to construct a macromodel of low complexity which is consistent with the data. Reduced order modeling from frequency-domain measurements is related to the rational interpolation problem [2].

Several algorithms [6, 15, 24] are already available in the electronics community. The goal of most techniques is fitting the data, essentially interpreting it is an optimization problem. Thus, most procedures are based on least-squares approximations. Vector fitting (VF) [7, 13–17] is a robust technique which solves the problem in two stages: one for computing the poles of the transfer function and the other for finding the residues. While the first stage is an iterative process in which a set of better poles is determined by solving a least-squares problem at each step, finding the residues only needs one least-squares approximation. This algorithm has been extended to handle time domain data [12] and several variants were developed for parametric macromodeling [8, 26] which incorporate the dependence on parameters such as geometry or material properties.

The algorithm introduced in [6] uses Nevanlinna-Pick interpolation in the context of bounded-real interpolation of S-parameters. As shown in [3], passivity is enforced by interpolating at the original points $(s_i, \mathbf{H}(s_i))$, together with the associated *mirror-image* points $\left(-s_i^*, \frac{1}{\mathbf{H}(-s_i^*)}\right)$. Hence this method can not recover the original system, unless the latter satisfies the mirror-image interpolation constraints, which is generally not the case. Moreover, this technique uses matrix interpolation, by using the whole matrix transfer function for each sample. Algorithms which enforce passivity by construction are described in [11, 22].

Our approach has its foundations in system theory and employs the concept of tangential interpolation in the Loewner matrix pencil framework. The Loewner and shifted Loewner matrices can be expressed in terms of the generalized controllability and observability matrices, tools which were first used in the control community, but have found applications in many other areas, including circuit simulation. Tangential interpolation is of great importance in our context due to the fact that it allows to build reduced models for devices with a large number of input and output ports using a small computational time [20]. The issue of massive ports is problematic for currently available algorithms which are employed in macromodeling, as they are extremely CPU-demanding. Moreover, we show that if the number of available measurements is right, one can fully recover the underlying system. If more measurements are available than needed, plotting the singular values of the Loewner matrix pencil allows to identify the original system via an SVD truncation. Thus, in our approach, the data is revealing the order of the system, unlike vector fitting, which depends on a few parameters (the number of starting poles, the location of the starting poles, or the number of iterations to relocate the poles) to produce a good macromodel.

The numerical examples we investigate are systems arising in circuit simulation. First, we analyze a single input single output system, which, due to the delay, is infinite-dimensional. Our goal is to built a finite-dimensional model which is able to capture the low-frequency response. Second, we compare tangential interpolation to matrix interpolation on an RC circuit with a large number of ports.

This paper is organized as follows. Section 4.2 reviews some system-theoretic concepts related to model reduction from measurements. Next, we introduce the rational interpolation problem in the case of scalar data and matrix tangential data and propose a solution in the Loewner framework in Sect. 4.3. Moreover, this section provides guidelines on how to choose the sampling directions to ensure full recovery of the system. Section 4.4 briefly shows how to construct tangential data from given matrix data. Section 4.5 presents numerical examples which validate the proposed approach on systems arising in circuit simulation and we conclude in Sect. 4.6.

4.2 Background

An LTI system Σ with m inputs, p outputs and n internal variables in *descriptor-form representation* is given by a system of differential and algebraic equations:

$$\Sigma : \mathbf{E}\dot{\mathbf{x}}(t) = \mathbf{A}\mathbf{x}(t) + \mathbf{B}\mathbf{u}(t),\ \mathbf{y}(t) = \mathbf{C}\mathbf{x}(t) + \mathbf{D}\mathbf{u}(t), \tag{4.1}$$

where $\mathbf{x}(t)$ is an internal variable (the state, if $\mathbf{E}$ is invertible), $\mathbf{u}(t)$ is an input, $\mathbf{y}(t)$ is the corresponding output, while $\mathbf{E}, \mathbf{A} \in \mathbb{R}^{n\times n}$, $\mathbf{B} \in \mathbb{R}^{n\times m}$, $\mathbf{C} \in \mathbb{R}^{p\times n}$, $\mathbf{D} \in \mathbb{R}^{p\times m}$ are constant matrices with $\mathbf{E}$ possibly singular. The *transfer function* of Σ is $\mathbf{H}(s) = \mathbf{C}(s\mathbf{E} - \mathbf{A})^{-1}\mathbf{B} + \mathbf{D}$. It is *proper* if $\lim_{s\to\infty} \mathbf{H}(s) < \infty$ and *improper*, otherwise. Moreover, if $\lim_{s\to\infty} \mathbf{H}(s) = \mathbf{0}$, $\mathbf{H}(s)$ is *strictly proper*. The matrix pencil $(\mathbf{A}, \mathbf{E})$ is *regular* if the matrix $\mathbf{A} - \lambda\mathbf{E}$ is nonsingular for some finite $\lambda \in \mathbb{C}$. The set $[\mathbf{E}, \mathbf{A}, \mathbf{B}, \mathbf{C}, \mathbf{D}]$ is called a *realization* of $\mathbf{H}(s)$. The realization is not unique; the one of the smallest possible order n is called *a minimal realization* and the order n is referred to as the McMillan degree. A realization is minimal if it is *completely controllable* and *observable*. A controllable system implies that any realization is controllable, and the same holds for observability. A descriptor system with $(\mathbf{A}, \mathbf{E})$ regular is *completely controllable* [27] if $\text{rank}[\mathbf{A} - \lambda\mathbf{E}, \mathbf{B}] = n$, for all finite $\lambda \in \mathbb{C}$ and $\text{rank}[\mathbf{E}, \mathbf{B}] = n$ $\Leftrightarrow$ the matrix $\mathscr{R} := \left[(\lambda_1\mathbf{E} - \mathbf{A})^{-1}\mathbf{B} \ldots (\lambda_n\mathbf{E} - \mathbf{A})^{-1}\mathbf{B}\right]$ has rank n. *It is called completely observable* if $\text{rank}\left[\mathbf{A}^T - \lambda\mathbf{E}^T, \mathbf{C}^T\right] = n$, for all finite $\lambda \in \mathbb{C}$ and $\text{rank}\left[\mathbf{E}^T, \mathbf{C}^T\right] = n$, where $(\cdot)^T$ denotes transpose $\Leftrightarrow$ the matrix $\mathscr{O} := \begin{bmatrix} \mathbf{C}(\mu_1\mathbf{E} - \mathbf{A})^{-1} \\ \vdots \\ \mathbf{C}(\mu_n\mathbf{E} - \mathbf{A})^{-1} \end{bmatrix}$ has rank n. The *poles* are given by the eigenvalues of the matrix pencil $(\mathbf{A}, \mathbf{E})$. Σ is *stable* if all its finite poles are in the left-half plane: Σ stable $\Leftrightarrow$ $Re(\lambda(\mathbf{A}, \mathbf{E})) < 0$ for $|\lambda(\mathbf{A}, \mathbf{E})| < \infty$.

Model reduction from frequency-domain data can be formulated as a rational interpolation problem irrespective of the quantities which are measured. Assuming that our circuit is voltage-excited and the corresponding outputs are the port currents, we can measure the admittance parameters $\mathbf{Y}$. A system Σ models a data set with P measurements of the Y-parameters of a device with p outputs and m inputs

$$\left(f_i, \mathbf{Y}^{(i)} := \begin{bmatrix} Y_{11}^{(i)} & \ldots & Y_{1m}^{(i)} \\ \vdots & \vdots & \vdots \\ Y_{p1}^{(i)} & \ldots & Y_{pm}^{(i)} \end{bmatrix}\right), \quad i = 1, \ldots, P,$$

where f_i is the frequency at which the parameters $\mathbf{Y}^{(i)}$ are measured, if the associated transfer function evaluated at $j \cdot 2\pi f_i$ is close to $\mathbf{Y}^{(i)}$:

$$\mathbf{H}(j \cdot 2\pi f_i) \approx \mathbf{Y}^{(i)}, \quad i = 1, \ldots, P. \tag{4.2}$$

4.3 Theoretical Aspects

We start by analyzing the rational interpolation problem in the simple case of model reduction from scalar data (i.e., $m = p = 1$):

$$(s_i, z_i), \quad i = 1, \ldots, P, \; s_i \neq s_j, \; i \neq j,$$

where $s_i, z_i \in \mathbb{C}$. The sampling points s_i are not necessarily on the imaginary axis, as it was the case in (4.2). The rational interpolation problem is equivalent to finding $\mathbf{H}(s) = \frac{n(s)}{d(s)}$, with n, d coprime polynomials and $deg(n(s)) < deg(d(s))$, such that $\mathbf{H}(s_i) = z_i$, $i = 1, \ldots, P$. There always exists a solution in polynomial form, e.g., the Lagrange interpolating polynomial. Our approach is based on the Loewner matrix [4] which is constructed by partitioning the data in disjoint sets:

$$(\lambda_i, w_i), \quad i = 1, \ldots, k \quad \text{and} \quad (\mu_j, v_j), \quad j = 1, \ldots, h,$$

where $h, k \approx \left[\frac{P}{2}\right]$ such that $k + h = P$, using the formula:

$$\mathbb{L} = \begin{bmatrix} \frac{v_1 - w_1}{\mu_1 - \lambda_1} & \cdots & \frac{v_1 - w_k}{\mu_1 - \lambda_k} \\ \vdots & \ddots & \vdots \\ \frac{v_h - w_1}{\mu_h - \lambda_1} & \cdots & \frac{v_h - w_k}{\mu_h - \lambda_k} \end{bmatrix} \in \mathbb{C}^{h \times k}. \tag{4.3}$$

There are several reasons why this is the right tool to use. The rank of the Loewner matrices constructed using all possible partitions encodes the degree of the minimal interpolant of the data. This is due to the system theoretic interpretation [5] of the Loewner matrix as minus the product of the generalized observability matrix, $\mathcal{O}$, the underlying $\mathbf{E}$ matrix and the generalized controllability matrix, $\mathcal{R}$. In particular, suppose that the data is obtained by sampling a proper rational function $\mathbf{H}(s)$ with minimal representation $[\mathbf{E}, \mathbf{A}, \mathbf{B}, \mathbf{C}, \mathbf{0}]$ (i.e., $\mathbf{H}(s) = \mathbf{C}(s\mathbf{E} - \mathbf{A})^{-1}\mathbf{B}$). Then

$$\mathbb{L} = -\underbrace{\begin{bmatrix} \mathbf{C}(\mu_1\mathbf{E} - \mathbf{A})^{-1} \\ \vdots \\ \mathbf{C}(\mu_h\mathbf{E} - \mathbf{A})^{-1} \end{bmatrix}}_{\mathcal{O}} \cdot \mathbf{E} \cdot \underbrace{\left[(\lambda_1\mathbf{E} - \mathbf{A})^{-1}\mathbf{B} \ \ \ldots \ \ (\lambda_k\mathbf{E} - \mathbf{A})^{-1}\mathbf{B} \right]}_{\mathcal{R}}.$$

Using the fact that the system is completely controllable and observable, the rank of $\mathbb{L}$ is precisely the rank of $\mathbf{E}$ (assuming that $h, k \geq \text{rank}(E)$), which is the order of the minimal degree interpolant of the data. Last, in the special case of data consisting of a single point with multiplicity, namely: $(s_0; \phi_0, \phi_1, \ldots, \phi_{P-1})$, i.e. the value of a function at s_0 and that of a number of its derivatives are provided, it can be shown that the Loewner matrix has Hankel structure. Thus the Loewner matrix generalizes the Hankel matrix when interpolation at finite points is considered [1, 4, 5].

To help the reader understand this concept, we consider the following simple example. Suppose the underlying transfer function which generates the data is $\mathbf{H}(s) = \frac{1}{s+5}$ and we are given $P = 4$ samples of this function:

$$(s, z) = \left\{ \left(1, \frac{1}{6}\right), \left(2, \frac{1}{7}\right), \left(-1, \frac{1}{4}\right), \left(-2, \frac{1}{3}\right) \right\}. \tag{4.4}$$

We partition this set as follows:

$$k = 2, (\lambda, w) = \left\{ \left(1, \frac{1}{6}\right), \left(2, \frac{2}{7}\right) \right\}, \quad h = 2, (\mu, v) = \left\{ \left(-1, \frac{1}{4}\right), \left(-2, \frac{1}{3}\right) \right\}, \tag{4.5}$$

and build the Loewner matrix according to (4.3)

$$\mathbb{L} = \begin{bmatrix} -\frac{1}{24} & -\frac{1}{28} \\ -\frac{1}{18} & -\frac{1}{21} \end{bmatrix} \in \mathbb{C}^{2\times 2}. \tag{4.6}$$

It follows that the rank of $\mathbb{L}$ is 1, so the interpolant for these 4 points has minimal degree 1, which is indeed the case, since $\mathbf{H}(s) = \frac{1}{s+5}$ is of degree 1.

4.3.1 Tangential Interpolation

Tangential interpolation is a general form of the interpolation problem, where matrix data are sampled directionally on the left and on the right. Tangential data consist of the *right interpolation data*

$$\{(\lambda_i, \mathbf{r}_i, \mathbf{w}_i) | \lambda_i \in \mathbb{C}, \mathbf{r}_i \in \mathbb{C}^{m\times 1}, \mathbf{w}_i \in \mathbb{C}^{p\times 1}\}, \tag{4.7}$$

for $i = 1, \ldots, k$, or, more compactly,

$$\Lambda = \mathrm{diag}\,[\lambda_1, \ldots, \lambda_k] \in \mathbb{C}^{k\times k}, \tag{4.8}$$

$$\mathbf{R} = [\mathbf{r}_1, \ldots, \mathbf{r}_k] \in \mathbb{C}^{m\times k}, \tag{4.9}$$

$$\mathbf{W} = [\mathbf{w}_1, \ldots, \mathbf{w}_k] \in \mathbb{C}^{p\times k}, \tag{4.10}$$

and of the *left interpolation data*

$$\{(\mu_j, \ell_j, \mathbf{v}_j) | \mu_j \in \mathbb{C}, \ell_j \in \mathbb{C}^{1\times p}, \mathbf{v}_j \in \mathbb{C}^{1\times m}\}, \tag{4.11}$$

for $j = 1, \ldots, h$, or, more compactly,

$$M = \mathrm{diag}\,[\mu_1, \ldots, \mu_h] \in \mathbb{C}^{h\times h}, \tag{4.12}$$

$$\mathbf{L} = \begin{bmatrix} \ell_1 \\ \vdots \\ \ell_h \end{bmatrix} \in \mathbb{C}^{h\times p}, \mathbf{V} = \begin{bmatrix} \mathbf{v}_1 \\ \vdots \\ \mathbf{v}_h \end{bmatrix} \in \mathbb{C}^{h\times m}. \tag{4.13}$$

In this case, the rational interpolation problem consists of finding a realization in descriptor form $[\mathbf{E}, \mathbf{A}, \mathbf{B}, \mathbf{C}, \mathbf{D}]$, such that the associated transfer function $\mathbf{H}(s) = \mathbf{C}(s\mathbf{E} - \mathbf{A})^{-1}\mathbf{B} + \mathbf{D}$, satisfies the *right and left constraints*

$$\mathbf{H}(\lambda_i)\mathbf{r}_i = \mathbf{w}_i, \quad \ell_j\mathbf{H}(\mu_j) = \mathbf{v}_j. \tag{4.14}$$

The tools for addressing this problem are the *Loewner matrix*, variously called the *divided-difference matrix* and *null-pole coupling matrix* together with the *shifted Loewner matrix* associated with the data. For the material that follows, we refer to [21] for details on proofs and derivations.

4.3.2 The Loewner and the Shifted Loewner Matrices

Given $S = \{s_1, \ldots, s_P\}$ points in the complex plane and the evaluation of a rational matrix function $\mathbf{H}(s)$ at those points: $\{\mathbf{H}(s_1), \ldots, \mathbf{H}(s_P)\}$, we can partition S as:

$$S = \{\lambda_1, \ldots, \lambda_k\} \cup \{\mu_1, \ldots, \mu_h\},$$

where $k + h = P$. We construct the right and left data by appropriately selecting the sampling directions $\mathbf{r}_i$ and ℓ_j (see Sect. 4.3.4 for details).

The *Loewner matrix* and the *shifted Loewner matrix* are both of dimension $h \times k$, and are defined in terms of the data (4.7) and (4.11) as follows

$$\mathbb{L} = \begin{bmatrix} \frac{\mathbf{v}_1\mathbf{r}_1 - \ell_1\mathbf{w}_1}{\mu_1 - \lambda_1} & \cdots & \frac{\mathbf{v}_1\mathbf{r}_k - \ell_1\mathbf{w}_k}{\mu_1 - \lambda_k} \\ \vdots & \ddots & \vdots \\ \frac{\mathbf{v}_h\mathbf{r}_1 - \ell_h\mathbf{w}_1}{\mu_h - \lambda_1} & \cdots & \frac{\mathbf{v}_h\mathbf{r}_k - \ell_h\mathbf{w}_k}{\mu_h - \lambda_k} \end{bmatrix}, \tag{4.15}$$

$$\sigma\mathbb{L} = \begin{bmatrix} \frac{\mu_1\mathbf{v}_1\mathbf{r}_1 - \lambda_1\ell_1\mathbf{w}_1}{\mu_1 - \lambda_1} & \cdots & \frac{\mu_1\mathbf{v}_1\mathbf{r}_k - \lambda_k\ell_1\mathbf{w}_k}{\mu_1 - \lambda_k} \\ \vdots & \ddots & \vdots \\ \frac{\mu_h\mathbf{v}_h\mathbf{r}_1 - \lambda_1\ell_h\mathbf{w}_1}{\mu_h - \lambda_1} & \cdots & \frac{\mu_h\mathbf{v}_h\mathbf{r}_k - \lambda_k\ell_h\mathbf{w}_k}{\mu_h - \lambda_k} \end{bmatrix}. \tag{4.16}$$

Each entry in (4.15) and (4.16), for instance $\frac{\mathbf{v}_j\mathbf{r}_i - \ell_j\mathbf{w}_i}{\mu_j - \lambda_i}$, is a scalar quantity, and is obtained by taking inner products of the left data (row vectors) and right data (column vectors). Assuming that the rational matrix function $\mathbf{H}(s)$ generates the data, the shifted Loewner matrix is the Loewner matrix corresponding to $s\mathbf{H}(s)$. It is straightforward to see from (4.15) and (4.16) that these matrices satisfy the Sylvester equations

$$\mathbb{L}\Lambda - M\mathbb{L} = \mathbf{LW} - \mathbf{VR}, \quad \text{and} \quad \sigma\mathbb{L}\Lambda - M\sigma\mathbb{L} = \mathbf{LW}\Lambda - M\mathbf{VR}. \tag{4.17}$$

These expressions allow for an efficient MATLAB® implementation to initialize the matrices, by avoiding nested "for" loops to initialize entries one by one.

4.3.3 The Solution to the General Tangential Interpolation Problem in the Loewner Framework

We review the conditions for the solution of the general tangential interpolation problem by means of state-space matrices $[\mathbf{E}, \mathbf{A}, \mathbf{B}, \mathbf{C}, \mathbf{D}]$, as presented in [21].

Lemma 1 *Assume that* $k = h$ *and that* $\det(x\mathbb{L} - \sigma\mathbb{L}) \neq 0$, *for all* $x \in \{\lambda_i\} \cup \{\mu_j\}$ *(i.e., the matrix pencil* $(\sigma\mathbb{L}, \mathbb{L})$ *is regular and* $\mu_j, \lambda_i \notin \lambda(\sigma\mathbb{L}, \mathbb{L})$*). Then* $\mathbf{E} = -\mathbb{L}$, $\mathbf{A} = -\sigma\mathbb{L}$, $\mathbf{B} = \mathbf{V}$, $\mathbf{C} = \mathbf{W}$ *and* $\mathbf{D} = \mathbf{0}$ *is a minimal realization of an interpolant of the data. Thus, the associated transfer function*

$$\mathbf{H}(s) = \mathbf{W}(\sigma\mathbb{L} - s\mathbb{L})^{-1}\mathbf{V} \tag{4.18}$$

satisfies the left interpolation conditions $\ell_j\mathbf{H}(\mu_j) = \mathbf{v}_j$ *and the right interpolation conditions* $\mathbf{H}(\lambda_i)\mathbf{r}_i = \mathbf{w}_i$.

We revisit the previous example from the point of view of tangential interpolation. We consider the same transfer function, namely $\mathbf{H}(s) = \frac{1}{s+5}$, as well as right tangential data $(\lambda, \mathbf{r}, \mathbf{w}) = \left(1, 1, \frac{1}{6}\right)$ (i.e., $k = 1$) and left tangential data $(\mu, \ell, \mathbf{v}) = \left(-1, 1, \frac{1}{4}\right)$ (i.e., $h = 1$). Compactly, we write this as $\Lambda = [\,1\,]$, $\mathbf{R} = [\,1\,]$, $\mathbf{W} = \left[\frac{1}{6}\right]$ and $M = [\,-1\,]$, $\mathbf{L} = [\,1\,]$, $\mathbf{V} = \left[\frac{1}{4}\right]$. The Loewner matrix pencil is built according to (4.15) and (4.16): $\mathbb{L} = \left[-\frac{1}{24}\right] \in \mathbb{C}^{1\times 1}$, $\sigma\mathbb{L} = \left[\frac{5}{24}\right] \in \mathbb{C}^{1\times 1}$. Lemma 1's assumptions are satisfied ($k = h = 1$, $\det(x\mathbb{L} - \sigma\mathbb{L}) = \det(x\left(-\frac{1}{24}\right) - \frac{5}{24}) \neq 0$, for $x \in \{1, -1\}$ and $\mu_j = -1, \lambda_i = 1 \notin \lambda\left(\sigma = \frac{5}{24}, \mathbb{L} = -\frac{1}{24}\right)$), so we directly write the realization

$$\mathbf{E}_r = \frac{1}{24}, \quad \mathbf{A}_r = -\frac{5}{24}, \quad \mathbf{B}_r = \frac{1}{4}, \quad \mathbf{C}_r = \frac{1}{6}, \quad \mathbf{D} = \mathbf{0}$$

One can check that the transfer function associated to this realization $\mathbf{H}(s) = \frac{1}{6}\left(s\frac{1}{24} + \frac{5}{24}\right)^{-1}\frac{1}{4}$ is the same as the original $\frac{1}{s+5}$.

Next, we discuss another example. The transfer function generating the data is

$$\mathbf{H}(s) = \frac{1}{s^2 + 5s + 6}\begin{bmatrix} s & -6 \\ 1 & s+5 \end{bmatrix}. \tag{4.19}$$

As tangential interpolation data we have

$$k = 2, (\lambda, \mathbf{r}, \mathbf{w}) = \left\{ \left(i, \begin{bmatrix} 1 \\ 0 \end{bmatrix}, \frac{1}{5+5i}\begin{bmatrix} i \\ 1 \end{bmatrix} \right), \left(2i, \begin{bmatrix} 0 \\ 1 \end{bmatrix}, \frac{1}{2+10i}\begin{bmatrix} -6 \\ 5+2i \end{bmatrix} \right) \right\}, \tag{4.20}$$

$$h = 2, (\mu, \ell, \mathbf{v}) = \left\{ \left(-i, \begin{bmatrix} 1 \\ 0 \end{bmatrix}^T, \frac{-1}{5-5i}\begin{bmatrix} i \\ 6 \end{bmatrix}^T \right), \left(-2i, \begin{bmatrix} 0 \\ 1 \end{bmatrix}^T, \frac{1}{2-10i}\begin{bmatrix} 1 \\ 5-2i \end{bmatrix}^T \right) \right\}. \tag{4.21}$$

Tangential data leads to the following matrices

$$\Lambda = \begin{pmatrix} i & \\ & 2i \end{pmatrix}, \quad M = \begin{pmatrix} -i & \\ & -2i \end{pmatrix}, \quad \mathbf{R} = \mathbf{L} = \mathbf{I}_2,$$
$$\mathbf{W} = \begin{bmatrix} \frac{i}{5+5i} & \frac{-6}{2+10i} \\ \frac{1}{5+5i} & \frac{5+2i}{2+10i} \end{bmatrix}, \mathbf{V} = \begin{bmatrix} \frac{-i}{5-5i} & \frac{-6}{5-5i} \\ \frac{1}{2-10i} & \frac{5-2i}{2-10i} \end{bmatrix}$$

and the Loewner matrix pencil is given by

$$\mathbb{L} = \begin{bmatrix} \frac{1}{10} & \frac{51}{130} - \frac{21}{130}i \\ \frac{-17}{260} - \frac{7}{260}i & -\frac{23}{104} \end{bmatrix}, \quad \sigma\mathbb{L} = \begin{bmatrix} \frac{1}{10} & -\frac{18}{65} + \frac{12}{65}i \\ \frac{3}{65} + \frac{2}{65}i & \frac{15}{52} \end{bmatrix}. \tag{4.22}$$

Since the lemma's assumptions are satisfied, we can write a realization as

$$\mathbf{E}_r = -\begin{bmatrix} \frac{1}{10} & \frac{51}{130} - \frac{21}{130}i \\ \frac{-17}{260} - \frac{7}{260}i & -\frac{23}{104} \end{bmatrix}, \quad \mathbf{A}_r = -\begin{bmatrix} \frac{1}{10} & -\frac{18}{65} + \frac{12}{65}i \\ \frac{3}{65} + \frac{2}{65}i & \frac{15}{52} \end{bmatrix}, \tag{4.23}$$

$$\mathbf{B}_r = \begin{bmatrix} \frac{-i}{5-5i} & \frac{-6}{5-5i} \\ \frac{1}{2-10i} & \frac{5-2i}{2-10i} \end{bmatrix}, \quad \mathbf{C}_r = \begin{bmatrix} \frac{i}{5+5i} & \frac{-6}{2+10i} \\ \frac{1}{5+5i} & \frac{5+2i}{2+10i} \end{bmatrix}. \tag{4.24}$$

The transfer function is as expected:

$$\mathbf{H}_r(s) = \mathbf{C}_r(s\mathbf{E}_r - \mathbf{A}_r)^{-1}\mathbf{B}_r = \frac{1}{s^2 + 5s + 6}\begin{bmatrix} s & -6 \\ 1 & s+5 \end{bmatrix}. \tag{4.25}$$

Some remarks:

- Instead of the matrix obtained by evaluating the transfer function at $s = i$,

$$\mathbf{H}(i) = \begin{bmatrix} \frac{i}{i^2+5i+6} & \frac{-6}{i^2+5i+6} \\ \frac{1}{i^2+5i+6} & \frac{i+5}{i^2+5i+6} \end{bmatrix} = \begin{bmatrix} \frac{i}{5i+5} & \frac{-6}{5i+5} \\ \frac{1}{5i+5} & \frac{i+5}{5i+5} \end{bmatrix},$$

  which one would use in matrix interpolation, tangential data consists only of the vector $\mathbf{w}_1 = \mathbf{H}(i)\mathbf{r}_1$, which in our case is the first column of $\mathbf{H}(i)$, since $\mathbf{r}_1$ is the first identity vector. A similar observation holds for $\mathbf{w}_2$ and the left data $\mathbf{v}_1$, $\mathbf{v}_2$.
- In the realization $[\mathbf{E}_r, \mathbf{A}_r, \mathbf{B}_r, \mathbf{C}_r, \mathbf{D}_r]$, the matrices have complex entries, so using the same matrix data and considering the tangential data as

$$k = 2, \quad (\lambda, \mathbf{r}, \mathbf{w}) = \left\{ \left(i, \begin{bmatrix} 1 \\ 0 \end{bmatrix}, \frac{1}{5+5i}\begin{bmatrix} i \\ 1 \end{bmatrix} \right), \left(-i, \begin{bmatrix} 1 \\ 0 \end{bmatrix}, \frac{1}{5-5i}\begin{bmatrix} -i \\ 1 \end{bmatrix} \right) \right\}, \tag{4.27}$$

$$h = 2, (\mu, \ell, \mathbf{v}) = \left\{ \left(2i, \begin{bmatrix} 1 \\ 0 \end{bmatrix}^T, \frac{1}{2+10i}\begin{bmatrix} 2i \\ -6 \end{bmatrix}^T \right), \left(-2i, \begin{bmatrix} 1 \\ 0 \end{bmatrix}^T, \frac{1}{2-10i}\begin{bmatrix} -2i \\ -6 \end{bmatrix}^T \right) \right\}, \tag{4.28}$$

(i.e., points and their conjugates on the same side) leads to the realization

$$\mathbf{E}_r = -\mathbb{L} = -\begin{bmatrix} -\frac{4}{65}-\frac{6}{65}i & \frac{3}{65}-\frac{2}{65}i \\ \frac{3}{65}+\frac{2}{65}i & -\frac{4}{65}+\frac{6}{65}i \end{bmatrix}, \quad \mathbf{A}_r = -\sigma\mathbb{L} = -\begin{bmatrix} \frac{37}{130}-\frac{3}{130}i & \frac{21}{130}-\frac{1}{130}i \\ \frac{21}{130}+\frac{1}{130}i & \frac{37}{130}+\frac{3}{130}i \end{bmatrix},$$

$$\mathbf{B}_r = \mathbf{V} = \begin{bmatrix} \frac{5}{26}+\frac{1}{26}i & -\frac{3}{26}+\frac{15}{26}i \\ \frac{5}{26}-\frac{1}{26}i & -\frac{3}{26}-\frac{15}{26}i \end{bmatrix}, \quad \mathbf{C}_r = \mathbf{W} = \begin{bmatrix} \frac{1}{10}+\frac{1}{10}i & \frac{1}{10}-\frac{1}{10}i \\ \frac{1}{10}-\frac{1}{10}i & \frac{1}{10}+\frac{1}{10}i \end{bmatrix}, \quad \mathbf{D}_r = \mathbf{0}_2.$$

After applying a coordinate change (see Sect. 4.4 for details): $\Pi = \frac{1}{\sqrt{2}}\begin{bmatrix} 1 & -j \\ 1 & j \end{bmatrix}$, a realization with real entries is obtained as:

$$\mathbf{E}_r = -\mathbb{L} = -\begin{bmatrix} -\frac{1}{65} & -\frac{4}{65} \\ \frac{8}{65} & -\frac{7}{65} \end{bmatrix}, \quad \mathbf{A}_r = -\sigma\mathbb{L} = -\begin{bmatrix} \frac{29}{65} & -\frac{1}{65} \\ \frac{2}{65} & \frac{8}{65} \end{bmatrix},$$

$$\mathbf{B}_r = \mathbf{V} = \begin{bmatrix} \frac{486}{1787} & -\frac{460}{2819} \\ -\frac{473}{8696} & -\frac{1458}{1787} \end{bmatrix}, \quad \mathbf{C}_r = \mathbf{W} = \begin{bmatrix} \frac{197}{1393} & \frac{197}{1393} \\ \frac{197}{1393} & -\frac{197}{1393} \end{bmatrix}, \quad \mathbf{D}_r = \mathbf{0}_2.$$

The associated transfer function is as in (4.25), which is what we expected.

- The tangential directions in (4.27) and (4.28) are equal, i.e., $\mathbf{r}_1 = \mathbf{r}_2$, $\ell_1 = \ell_2 = \mathbf{r}_1{}^T$, but this still allows us to identify the system (see Sect. 4.3.4 for the theoretical reasoning).

4.3.4 System Identification

In this section, we show that one is able to identify the underlying system given the right number of measurements. Assume that our system has McMillan degree n, that the number of inputs and outputs is equal ($m = p$) and that the underlying $\mathbf{D}$ matrix is full rank. Thus, we are looking for a realization of the system of order n which is minimal (both controllable and observable).

Let us begin our argument by noting that, even though our realizations have $\mathbf{D} = \mathbf{0}$, we can still identify systems with a non-zero $\mathbf{D}$. This is due to the fact that the $\mathbf{D}$-term can be incorporated either in the $\mathbf{B}$ or the $\mathbf{C}$ matrices. Assume a realization of the system as $[\mathbf{E}, \mathbf{A}, \mathbf{B}, \mathbf{C}, \mathbf{D}]$ with $\mathbf{E}$, $\mathbf{A}$ invertible. The following realization

$$\widetilde{\mathbf{E}} = \begin{bmatrix} \mathbf{E} & \mathbf{0} \\ \mathbf{0} & \mathbf{0} \end{bmatrix}, \quad \widetilde{\mathbf{A}} = \begin{bmatrix} \mathbf{A} & \mathbf{0} \\ \mathbf{0} & \mathbf{I} \end{bmatrix}, \quad \widetilde{\mathbf{B}} = \begin{bmatrix} \mathbf{B} \\ -\mathbf{I} \end{bmatrix}, \quad \widetilde{\mathbf{C}} = [\mathbf{C} \;\; \mathbf{D}], \quad \widetilde{\mathbf{D}} = \mathbf{0} \tag{4.29}$$

corresponds to a system of order $n + p$ (i.e., both $\mathbf{A}$ and $\mathbf{E}$ are of dimension $(n + p) \times (n + p)$), but a zero $\mathbf{D}$-term. One can easily check that the transfer function of the system in the new realization is as before.

Next, we show that when samples at $n+p$ left tangential data and $n+p$ right tangential data are provided, and the conditions in Lemma 1 are satisfied, we have recovered the underlying system.

The number of free parameters for a rational matrix function $\mathbf{H}(s) \in \mathbb{C}^{p\times p}$ with $\mathbf{D} \neq \mathbf{0}$ and McMillan degree n is $2np + 2p^2$, which is obtained by considering the control canonical form [18] of the system. Thus, we can freely choose n entries on p rows of $\mathbf{A}$, p entries on p rows of $\mathbf{B}$, np entries of $\mathbf{C}$, as well as the p^2 entries of $\mathbf{D}$. On the other hand, the number of free parameters in tangential interpolation are $2(n+p)$ for $\lambda_i, \mu_j, i, j = 1, \ldots, n+p$, as well as the tangential directions $\mathbf{r}_i \in \mathbb{C}^{p\times 1}, i = 1, \ldots, n+p$ and $\ell_j \in \mathbb{C}^{1\times p}, j = 1, \ldots, n+p$. Note that once the sampling points λ_i and μ_j and the sampling directions $\mathbf{r}_i$ and ℓ_j were chosen, the left and right data $\mathbf{w}_i$ and $\mathbf{v}_j$ are obtained as $\mathbf{H}(\lambda_i)\mathbf{r}_i = \mathbf{w}_i$ and $\ell_j\mathbf{H}(\mu_j) = \mathbf{v}_j$, respectively. This gives a total number of free parameters of $2(n+p) + p(n+p) + (n+p)p = 2n + 2p + 2np + 2p^2$. Thus, the number of the free parameters needed to uniquely define the transfer function matrix is smaller than the number of unknown parameters when $n+p$ tangential measurements are chosen on the left and on the right.

In the following, we show that the Loewner matrix constructed from tangential data still has a system theoretic interpretation in terms of tangential controllability and observability matrices. A similar expression exists for the shifted Loewner matrix. Given $n+p$ left interpolation data $\mathbf{v}_j = \ell_j\mathbf{H}(\mu_j), j = 1, \ldots, n+p$, and the same number of right interpolation data $\mathbf{w}_i = \mathbf{H}(\lambda_i)\mathbf{r}_i, i = 1, \ldots, n+p$, we have that

$$\begin{aligned}
\mathbb{L}_{j,i} &= \frac{\mathbf{v}_j\mathbf{r}_i - \ell_j\mathbf{w}_i}{\mu_j - \lambda_i} = \frac{\ell_j\mathbf{H}(\mu_j)\mathbf{r}_i - \ell_j\mathbf{H}(\lambda_i)\mathbf{r}_i}{\mu_j - \lambda_i} \\
&= \ell_j\frac{\mathbf{C}(\mu_j\mathbf{E} - \mathbf{A})^{-1}\mathbf{B} + \mathbf{D} - \mathbf{C}(\lambda_i\mathbf{E} - \mathbf{A})^{-1}\mathbf{B} - \mathbf{D}}{\mu_j - \lambda_i}\mathbf{r}_i \\
&= \ell_j\mathbf{C}\frac{(\mu_j\mathbf{E} - \mathbf{A})^{-1} - (\lambda_i\mathbf{E} - \mathbf{A})^{-1}}{\mu_j - \lambda_i}\mathbf{B}\mathbf{r}_i \\
&= -\ell_j\mathbf{C}(\mu_j\mathbf{E} - \mathbf{A})^{-1}\mathbf{E}(\lambda_i\mathbf{E} - \mathbf{A})^{-1}\mathbf{B}\mathbf{r}_i,\ j, i = 1, \ldots, n+p.
\end{aligned}$$

Similarly,

$$\begin{aligned}
\sigma\mathbb{L}_{j,i} &= \frac{\mu_j\mathbf{v}_i\mathbf{r}_i - \lambda_i\ell_j\mathbf{w}_i}{\mu_j - \lambda_i} = \ell_j\mathbf{C}\frac{\mu_j(\mu_j\mathbf{E} - \mathbf{A})^{-1} - \lambda_i(\lambda_i\mathbf{E} - \mathbf{A})^{-1}}{\mu_j - \lambda_i}\mathbf{B}\mathbf{r}_i + \ell_j\mathbf{D}\mathbf{r}_i \\
&= \ell_j\frac{\mu_j\mathbf{C}(\mu_j\mathbf{E} - \mathbf{A})^{-1}\mathbf{B} + \mu_j\mathbf{D} - \lambda_i\mathbf{C}(\lambda_i\mathbf{E} - \mathbf{A})^{-1}\mathbf{B} - \lambda_i\mathbf{D}}{\mu_j - \lambda_i}\mathbf{r}_i \\
&= -\ell_j\mathbf{C}(\mu_j\mathbf{E} - \mathbf{A})^{-1}\mathbf{A}(\lambda_i\mathbf{E} - \mathbf{A})^{-1}\mathbf{B}\mathbf{r}_i + \ell_j\mathbf{D}\mathbf{r}_i,\ j, i = 1, \ldots, n+p.
\end{aligned}$$

Thus

$$\mathbb{L} = -\underbrace{\begin{bmatrix} \ell_1 \mathbf{C}(\mu_1 \mathbf{E} - \mathbf{A})^{-1} \\ \vdots \\ \ell_{n+p} \mathbf{C}(\mu_{n+p} \mathbf{E} - \mathbf{A})^{-1} \end{bmatrix}}_{\mathscr{O}} \mathbf{E} \underbrace{\begin{bmatrix} (\lambda_1 \mathbf{E} - \mathbf{A})^{-1} \mathbf{B} \mathbf{r}_1 & \dots & (\lambda_{n+p} \mathbf{E} - \mathbf{A})^{-1} \mathbf{B} \mathbf{r}_{n+p} \end{bmatrix}}_{\mathscr{R}} \tag{4.30}$$

$$\sigma\mathbb{L} = -\underbrace{\begin{bmatrix} \ell_1 \mathbf{C}(\mu_1 \mathbf{E} - \mathbf{A})^{-1} \\ \vdots \\ \ell_{n+p} \mathbf{C}(\mu_{n+p} \mathbf{E} - \mathbf{A})^{-1} \end{bmatrix}}_{\mathscr{O}} \mathbf{A} \underbrace{\begin{bmatrix} (\lambda_1 \mathbf{E} - \mathbf{A})^{-1} \mathbf{B} \mathbf{r}_1 & \dots & (\lambda_{n+p} \mathbf{E} - \mathbf{A})^{-1} \mathbf{B} \mathbf{r}_{n+p} \end{bmatrix}}_{\mathscr{R}} + \mathbf{LDR}. \tag{4.31}$$

The pairs $(\mathbf{E}, \mathbf{A}, \mathbf{B})$ and $(\mathbf{C}, \mathbf{E}, \mathbf{A})$ are controllable and observable, respectively, so provided that the sampling directions are chosen appropriately, the rank of the Loewner matrix is precisely the rank of the underlying $\mathbf{E}$ matrix, while the rank of the shifted Loewner matrix is p more than the rank of the underlying $\mathbf{A}$ matrix, which is as expected from (4.29). Next we provide an outline on how to choose the sampling tangential directions:

- If $\mathbf{D}$=0, then $\mathbb{L} = -\mathscr{O}\mathbf{E}\mathscr{R}$ and $\sigma\mathbb{L} = -\mathscr{O}\mathbf{A}\mathscr{R}$ and we can choose the same sampling direction for all sampling points: $\ell_1 = \ell_2 = \ldots = \ell_{n+p}$ and $\mathbf{r}_1 = \mathbf{r}_2 = \ldots = \mathbf{r}_{n+p}$. This is because $(\mathbf{E}, \mathbf{A}, \mathbf{B})$ is controllable, so by using a linear combination of the columns of $\mathbf{B}$, one can still control the system. Similarly, as $(\mathbf{C}, \mathbf{E}, \mathbf{A})$ is observable, choosing a linear combination of the rows of $\mathbf{C}$ still makes the system observable. Denoting $\ell_1\mathbf{C}$ by $\mathbf{c}$ and $\mathbf{B}\mathbf{r}_1$ by $\mathbf{b}$, we have that

$$\mathbb{L} = -\underbrace{\begin{bmatrix} \mathbf{c}(\mu_1 \mathbf{E} - \mathbf{A})^{-1} \\ \vdots \\ \mathbf{c}(\mu_{n+p} \mathbf{E} - \mathbf{A})^{-1} \end{bmatrix}}_{\mathscr{O}} \mathbf{E} \underbrace{\begin{bmatrix} (\lambda_1 \mathbf{E} - \mathbf{A})^{-1} \mathbf{b} & \dots & (\lambda_{n+p} \mathbf{E} - \mathbf{A})^{-1} \mathbf{b} \end{bmatrix}}_{\mathscr{R}} \tag{4.32}$$

  and similarly for $\sigma\mathbb{L}$.
- The statement above does not hold when $(\mathbf{A}, \mathbf{E})$ has at least one multiple eigenvalue with the same algebraic and geometric multiplicity (e.g., $\mathbf{E} = \mathbf{A} = \mathbf{I}$), as one would need more than one vector $\mathbf{b}$ to make $(\mathbf{E}, \mathbf{A}, \mathbf{b})$ controllable. Similarly, more than one vector $\mathbf{c}$ is required to make $(\mathbf{c}, \mathbf{E}, \mathbf{A})$ observable. Hence, the sampling directions for each evaluating point should be different from the others.
- If $\mathbf{D} \neq \mathbf{0}$, (4.31) shows that the $\mathbf{D}$-term cannot be recovered unless we choose at least p linearly independent vectors $\mathbf{r}_i$ and ℓ_j. In particular, one can choose the sampling directions as unit vectors of dimension p (see Sect. 4.4).

Lemma 1 provides a realization of the system in terms of the Loewner matrix pencil together with the **V** and **W** matrices. Therefore, this method constructs a descriptor-form representation of an underlying dynamical system exclusively from the data, just by arranging it in a convenient form, with no further manipulations involved.

Consider the following transfer function

$$\mathbf{H}(s) = \frac{1}{s^2+5s+6}\begin{bmatrix} s & -6 \\ 1 & s+5 \end{bmatrix} + \begin{bmatrix} 1 & 2 \\ 3 & 4 \end{bmatrix}. \tag{4.33}$$

Note that this time, $\mathbf{D} \neq \mathbf{0}$, so the purpose of this example is to show how the **D** term affects the realization. Suppose that we are given the matrix data at i, $2i$, $3i$, $4i$ and their complex conjugates (i.e., $\mathbf{H}(i), \ldots, \mathbf{H}(4i)$ and $\mathbf{H}(-i) = \overline{\mathbf{H}(i)}, \ldots,$ $\mathbf{H}(-4i) = \overline{\mathbf{H}(4i)}$). We construct tangential data as

$$\begin{aligned} k = 4, (\lambda, \mathbf{r}, \mathbf{w}) &= \{(i, \mathbf{e}_1, \mathbf{H}(i)\mathbf{e}_1), (-i, \mathbf{e}_1, \mathbf{H}(-i)\mathbf{e}_1), (3i, \mathbf{e}_2, \mathbf{H}(3i)\mathbf{e}_2), (-3i, \mathbf{e}_2, \mathbf{H}(-3i)\mathbf{e}_2)\}, \\ h = 4, (\mu, \ell, \mathbf{v}) &= \{(2i, \mathbf{e}_1{}^T, \mathbf{e}_1{}^T\mathbf{H}(2i)), (-2i, \mathbf{e}_1{}^T, \mathbf{e}_1{}^T\mathbf{H}(-2i)), (4i, \mathbf{e}_2{}^T, \mathbf{e}_2{}^T\mathbf{H}(4i)), \\ &\qquad (-4i, \mathbf{e2}^T, \mathbf{e}_2{}^T\mathbf{H}(-4i))\}, \end{aligned}$$

and obtain the following realization, after performing the appropriate basis change:

$$\mathbf{E}_r = -\mathbb{L} = -\begin{bmatrix} -\frac{1}{130} & -\frac{2}{65} & 0 & -\frac{3}{26} \\ \frac{4}{65} & -\frac{7}{130} & \frac{1}{13} & -\frac{5}{26} \\ -\frac{1}{250} & \frac{1}{125} & -\frac{2}{325} & \frac{19}{650} \\ -\frac{4}{125} & \frac{3}{125} & -\frac{38}{975} & \frac{83}{975} \end{bmatrix},$$

$$\mathbf{A}_r = -\sigma\mathbf{L} = -\begin{bmatrix} \frac{159}{130} & -\frac{1}{130} & \frac{29}{13} & 0 \\ \frac{1}{65} & \frac{4}{65} & 0 & \frac{3}{13} \\ \frac{743}{250} & -\frac{1}{250} & \frac{1291}{325} & -\frac{6}{325} \\ \frac{2}{125} & -\frac{4}{125} & \frac{8}{325} & -\frac{38}{325} \end{bmatrix},$$

$$\mathbf{B}_r = \mathbf{V} = \begin{bmatrix} \frac{31}{26} & \frac{49}{26} \\ -\frac{1}{26} & -\frac{15}{26} \\ \frac{149}{50} & \frac{203}{50} \\ \frac{1}{25} & \frac{7}{25} \end{bmatrix}, \quad \mathbf{C}_r = \mathbf{W} = \begin{bmatrix} \frac{11}{10} & \frac{1}{10} & \frac{27}{13} & \frac{5}{13} \\ \frac{31}{10} & -\frac{1}{10} & \frac{161}{39} & -\frac{14}{39} \end{bmatrix}, \quad \mathbf{D}_r = \mathbf{0}_2.$$

The singular values of $\mathbf{E}_r$, which is the Loewner matrix up to a sign change, are $\frac{487}{1811}, \frac{345}{6374}, 0, 0$, while those of $\mathbf{A}_r$ are $\frac{2349}{422}, \frac{397}{1246}, \frac{364}{1359}, \frac{44}{55193}$, so indeed our realization has order $4 = 2 + 2 = n + p$ with $\mathbf{A}_r$ full rank and $\mathbf{E}_r$ of rank $2 = n$. Moreover, the computed eigenvalues of the matrix pencil are $-5 \cdot 10^{14}, 2 \cdot 10^{14}, -3, -2$. The first two correspond to poles at infinity, while the last two are precisely the poles of our original system (given by the roots of the equation $s^2 + 5s + 6 = 0$).

4.4 Tangential Interpolation for Modeling Y-Parameters

The data sets are assumed to contain P samples of the multi-port Y-parameters, $\mathbf{Y}^{(i)}$, at frequency points $j\omega_i$, for $i = 1,\ldots,P$. To obtain a real system, the condition $\overline{\mathbf{H}(s)} = \mathbf{H}(\overline{s})$ needs to be satisfied. Therefore, the Y-parameters at the complex conjugate values of the sample points $-j\omega_i$ should equal the complex conjugates of the measurements at $j\omega_i$, namely $\overline{\mathbf{Y}}^{(i)}$.

The right interpolation data can be chosen as

$$\left(\lambda_i = j\omega_i, \lambda_{i+1} = -j\omega_i; \mathbf{r}_i, \mathbf{r}_{i+1} = \mathbf{r}_i; \mathbf{w}_i = \mathbf{Y}^{(i)}\mathbf{r}_i, \mathbf{w}_{i+1} = \overline{\mathbf{w}_i} = \overline{\mathbf{Y}}^{(i)}\mathbf{r}_i\right), \quad i = 1,3,\ldots,P, \tag{4.34}$$

where $\omega_i = 2\pi f_i \in \mathbb{R}$. The right sampling directions may be chosen as $\mathbf{r}_1 = \mathbf{r}_2 = \mathbf{e}_1 \in \mathbb{R}^{p\times 1}$, $\mathbf{r}_3 = \mathbf{r}_4 = \mathbf{e}_2,\ldots$, $\mathbf{r}_{2p-1} = \mathbf{r}_{2p} = \mathbf{e}_p$, $\mathbf{r}_{2p+1} = \mathbf{r}_{2p+2} = \mathbf{e}_1$, $\mathbf{r}_{2p+3} = \mathbf{r}_{2p+4} = \mathbf{e}_2$, $\ldots$, where $\mathbf{e}_m$ denotes the m-th column of the identity matrix $\mathbf{I}_p$, $m = 1,\ldots,p$. Consequently, the right data are columns of $\mathbf{Y}^{(i)}$ and $\overline{\mathbf{Y}}^{(i)}$, e.g. $\mathbf{w}_1 = \mathbf{Y}^{(1)}_{:,1} \in \mathbf{C}^{p\times 1}$, $\mathbf{w}_2 = \overline{\mathbf{Y}}^{(1)}_{:,1}$, $\mathbf{w}_3 = \mathbf{Y}^{(2)}_{:,2}$, $\mathbf{w}_4 = \overline{\mathbf{Y}}^{(2)}_{:,2}$, etc. More compactly,

$$\Lambda = \mathrm{diag}\left[j\omega_1, -j\omega_1, j\omega_3, -j\omega_3, \ldots, j\omega_{P-1}, -j\omega_{P-1}\right], \tag{4.35}$$

$$\mathbf{R} = \left[\mathbf{r}_1, \mathbf{r}_2 = \mathbf{r}_1, \mathbf{r}_3, \mathbf{r}_4 = \mathbf{r}_3, \ldots, \mathbf{r}_{P-1}, \mathbf{r}_P = \mathbf{r}_{P-1}\right] \in \mathbb{C}^{p\times P}, \tag{4.36}$$

$$\mathbf{W} = \left[\mathbf{w}_1, \mathbf{w}_2 = \overline{\mathbf{w}_1}, \mathbf{w}_3, \mathbf{w}_4 = \overline{\mathbf{w}_3}, \ldots, \mathbf{w}_{P-1}, \mathbf{w}_P = \overline{\mathbf{w}_{P-1}}\right] \in \mathbb{C}^{p\times P}, \tag{4.37}$$

The left interpolation data is constructed as

$$\left(\mu_i = j\omega_{i+1}, \mu_{i+1} = -j\omega_{i+1}; \ell_i, \ell_{i+1} = \ell_i; \mathbf{v}_i = \ell_i\mathbf{Y}^{(i+1)}, \mathbf{v}_{i+1} = \overline{\mathbf{v}_i} = \ell_i\overline{\mathbf{Y}}^{(i+1)}\right), \quad i = 1,3,\ldots,P. \tag{4.38}$$

The left sampling directions ℓ_i may be chosen as rows of the identity matrix $\mathbf{I}_p$, so the left data are rows of the admittance parameters $\mathbf{Y}^{(i)}$ and $\overline{\mathbf{Y}}^{(i)}$. Compactly,

$$M = \mathrm{diag}\left[j\omega_2, -j\omega_2, j\omega_4, -j\omega_4, \ldots, j\omega_P, -j\omega_P\right], \tag{4.39}$$

$$\mathbf{L}^* = \left[\ell_1^* \quad \ell_1^* \quad \ell_3^* \quad \ell_3^* \quad \cdots \quad \ell_P^* \quad \ell_P^*\right], \quad \mathbf{L} \in \mathbb{C}^{P\times p}, \tag{4.40}$$

$$\mathbf{V}^* = \left[\mathbf{v}_1{}^* \quad \overline{\mathbf{v}_1}^* \quad \mathbf{v}_3{}^* \quad \overline{\mathbf{v}_3}^* \quad \cdots \quad \mathbf{v}_P{}^* \quad \overline{\mathbf{v}_P}^*\right], \quad \mathbf{V} \in \mathbb{C}^{P\times p}. \tag{4.41}$$

Without loss of generality, we assumed that we have an even number of samples. After tangential data are constructed, the Loewner and shifted Loewner matrices are built according to formulas (4.15) and (4.16). The following change of basis needs to be performed to ensure that the matrix entries are real: $\hat{\Lambda} = \hat{\Pi}^*\Lambda\hat{\Pi}$, $\hat{M} = \hat{\Pi}^*M\hat{\Pi}$, $\hat{\mathbf{L}} = \hat{\Pi}^*\mathbf{L}$, $\hat{\mathbf{V}} = \hat{\Pi}^*\mathbf{V}$, $\hat{\mathbf{R}} = \mathbf{R}\hat{\Pi}$, $\hat{\mathbf{W}} = \mathbf{W}\hat{\Pi}$ $\hat{\mathbb{L}} = \hat{\Pi}^*\mathbb{L}\hat{\Pi}$, $\sigma\hat{\mathbb{L}} = \hat{\Pi}^*\sigma\mathbb{L}\hat{\Pi}$, where

$$\hat{\Pi} = \text{diag}[\Pi, \ldots, \Pi] \in \mathbb{C}^{P \times P}, \quad \Pi = \frac{1}{\sqrt{2}} \begin{bmatrix} 1 & -j \\ 1 & j \end{bmatrix}.$$

Note that $\Pi^* \Pi = \Pi \Pi^* = \mathbf{I}_2$.

When too many samples are available, the pencil obtained after using all measurements is singular. Hence, the assumption that the Loewner pencil is regular does not hold since the realization given in Lemma 1 is not minimal. The underlying system can still be identified by plotting the singular values of the Loewner and shifted Loewner matrices and noticing where the largest drop in singular values occurs. Thus, the singular part can be projected out via a singular value decomposition:

$$x\mathbb{L} - \sigma\mathbb{L} = \mathbf{Y\Sigma X}, \quad x \in \{j\omega_i, -j\omega_i\}, \quad i = 1, \cdots, k,$$

where rank $(x\mathbb{L} - \sigma\mathbb{L}) =: N$ (the *dimension of the regular part*), $\mathbf{Y} \in \mathbb{C}^{k \times N}$ and $\mathbf{X} \in \mathbb{C}^{N \times k}$. For strictly proper systems, $N = n$, while for proper systems $N = n + \text{rank}(\mathbf{D})$. Using the singular vectors as projectors, the realization is given as

$$\mathbf{E} = -\mathbf{Y}^*\mathbb{L}\mathbf{X}, \mathbf{A} = -\mathbf{Y}^*\sigma\mathbb{L}\mathbf{X}, \mathbf{B} = \mathbf{Y}^*\mathbf{V}, \mathbf{C} = \mathbf{WX}, \mathbf{D} = 0. \tag{4.43}$$

This approach is costly for sets with a large number of samples k, as the complexity of the SVD of $x\mathbb{L} - \sigma\mathbb{L}$ is $O(k^3)$. The adaptive and recursive procedures presented in [20] are cheaper, but still identify when the pencil becomes singular.

Remark Our algorithms are able to recover the underlying system, therefore, for data sets coming from real-world systems, the resulting models are stable (after extracting the necessary **D**-term).

4.5 Numerical Results

We investigate two examples arising in circuit simulation. The first is an infinite-dimensional delay system, for which we build a finite-dimensional, low-order model which captures the low-frequency behaviour. The second is an RC circuit with 70 ports for which we compare the tangential and the matrix interpolation approaches.

4.5.1 Delay System

Electrical circuits with transmission lines can be described by differential equations containing delays. These expressions are obtained from the d'Alembert solution to the telegrapher's equations [23]. The delay factor turns out to be $\sqrt{LC}$.

In the following, we consider the system:

$$\mathbf{E}\dot{\mathbf{x}}(t) = \mathbf{A}_0\mathbf{x}(t) + \mathbf{A}_1\mathbf{x}(t-\tau) + \mathbf{b}u(t), \quad \mathbf{y}(t) = \mathbf{c}\mathbf{x}(t), \tag{4.44}$$

with $\mathbf{E}, \mathbf{A}_0, \mathbf{A}_1 \in \mathbb{R}^{500\times 500}$ and $\tau = 1$. The associated transfer function is given by

$$\mathbf{H}(s) = \mathbf{c}(s\mathbf{E} - \mathbf{A}_0 - \mathbf{A}_1 e^{-\tau s})^{-1}\mathbf{b}. \tag{4.45}$$

The matrices $\mathbf{E}$, $\mathbf{A}_0$ and $\mathbf{A}_1$ are symmetric. Moreover, $\mathbf{E}$ is positive definite, while $\mathbf{A}_0$ and $\mathbf{A}_1$ are negative definite. Due to the time delay, this is an infinite-dimensional system. However, as the matrices $\mathbf{E}$, $\mathbf{A}_0$ and $\mathbf{A}_1$ are diagonalized by the same transformation, the system can be decoupled into 500 independent equations and the poles of the system can be found by solving nonlinear equations of the type:

$$s\lambda_{\mathbf{E}} - \lambda_{\mathbf{A}_0} - \lambda_{\mathbf{A}_1}e^{-\tau s} = 0, \tag{4.46}$$

where $\lambda_{\mathbf{E}}$ is an eigenvalue of $\mathbf{E}$, and $\lambda_{\mathbf{A}_0}$ and $\lambda_{\mathbf{A}_1}$ are eigenvalues of $\mathbf{A}_0$ and $\mathbf{A}_1$, respectively. Setting s as $x + j \cdot y$ allows us to write (4.46) as a system of 2 nonlinear equations corresponding to the real and imaginary part:

$$x\lambda_{\mathbf{E}} - \lambda_{\mathbf{A}_0} - \lambda_{\mathbf{A}_1}\cos(\tau x) = 0, \quad y\lambda_{\mathbf{E}} + \lambda_{\mathbf{A}_1}\sin(\tau y) = 0. \tag{4.47}$$

Figure 4.1a shows poles with imaginary part between ± 130. Each of the 500 systems of equations of the form (4.47) were solved using the MATLAB function `fsolve` with initial guess 0 for x and different guesses for y, between -130 and 130. Thus, for each value on the imaginary axis, there is one pole obtained from solving one of the systems of nonlinear equations. Even though Fig. 4.1a shows only poles with imaginary part between -130 and 130, they extend well above that value.

We are looking for a low-order finite-dimensional model as in (4.1) to approximate this infinite-dimensional system. Our approach consists of taking 1000 measurements of the frequency response logarithmically distributed between

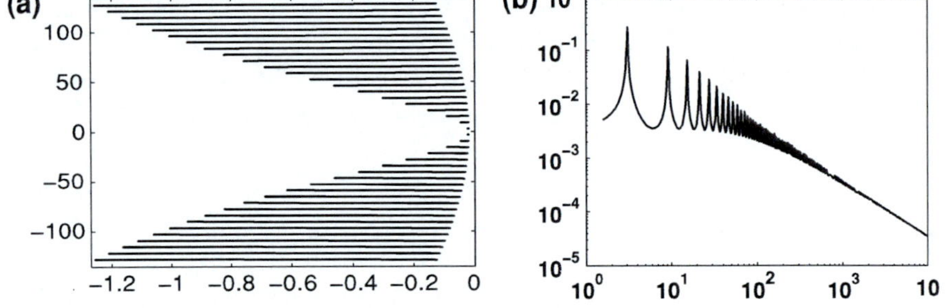

Fig. 4.1 Poles and frequency response of the delay system: **a** Poles **b** Frequency response

$10^{-0.6}$ and $10^{3.2}$ and using this information together with the theoretical concepts described in Sect. 4.3 and the numerical implementation presented in Sect. 4.4 to construct such a reduced model. We also employ state-of-the-art *vector fitting* in our comparison. The accuracy was assessed using two error measures: the normalized $\mathscr{H}_\infty$-norm,

$$\mathscr{H}_\infty \text{ error} = \frac{\max_{i=1,\dots,k}\left|Y^{(i)} - \mathbf{H}_{\text{model}}(j \cdot 2\pi f_i)\right|}{\max_{i=1,\dots,k}\left|Y^{(i)}\right|},$$

which evaluates the maximum deviation in the singular values and the normalized $\mathscr{H}_2$-norm of the error system,

$$\mathscr{H}_2 \text{ error} = \sqrt{\frac{\sum_{i=1}^{k}\left|Y^{(i)} - \mathbf{H}_{\text{model}}(j \cdot 2\pi f_i)\right|^2}{\sum_{i=1}^{k}\left|Y^{(i)}\right|^2}},$$

which evaluates the error in the magnitude of all entries, proving to be a good estimate of the overall performance.

Let us begin our discussion with the analysis of the singular value drop of the matrix pencil $(\sigma\mathbb{L}, \mathbb{L})$ constructed using all measurements. This is shown in Fig. 4.2. We notice that the rank of the pencil $(\sigma\mathbb{L}, \mathbb{L})$ constructed using these measurements is about 700. Since the system is infinite-dimensional, one may expect the matrix pencil to be full rank. The fact that it has a numerical rank of 700, instead of 1000, is because of the small magnitude of the response and small oscillation peaks at high frequencies, which can essentially be approximated by a line.

We build models of order $n = 41$. Results of the comparison of our approach against vector fitting are shown in Table 4.1, while Fig. 4.3 presents plots of the errors in absolute value between the original system and the different reduced models.

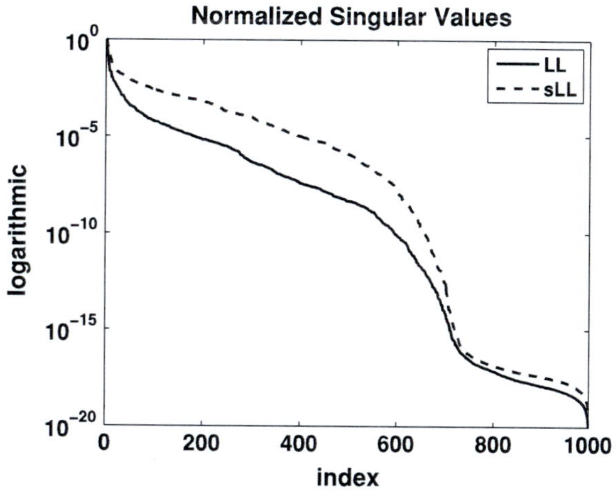

Fig. 4.2 Drop of the singular values of the matrix pencil $(\sigma\mathbb{L}, \mathbb{L})$

Table 4.1 Results for $n = 41$ order models of an infinite-dimensional delay system

Algorithm	Iterations	$\mathscr{H}_\infty$ error	$\mathscr{H}_2$ error
SVD Approach	N/A	1.4995e-002	3.5498e-002
Vector Fitting	50	3.4159e-002	5.4146e-002

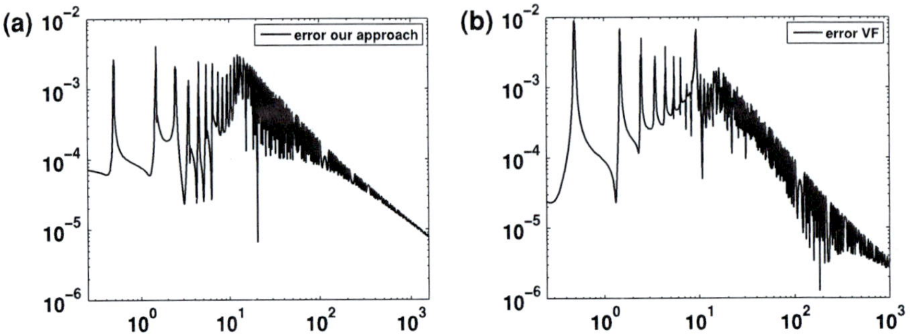

Fig. 4.3 Models obtained with our approach and with vector fitting: **a** SVD approach **b** VF (50 iterations)

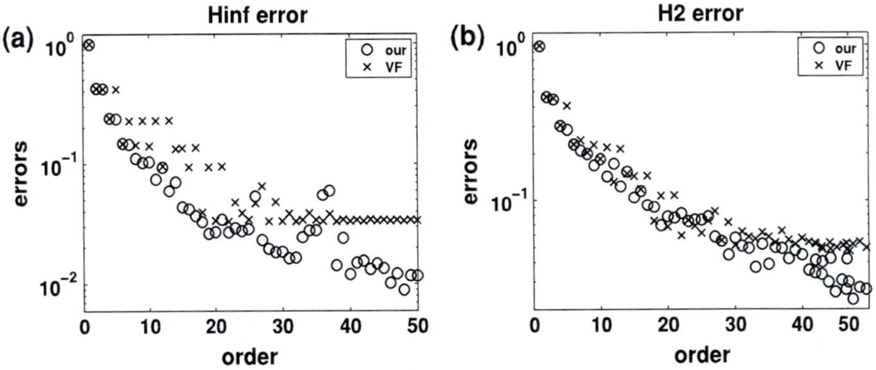

Fig. 4.4 $\mathscr{H}_\infty$ and $\mathscr{H}_2$ errors for our approach compared to vector fitting: **a** $\mathscr{H}_\infty$ error **b** $\mathscr{H}_2$ error

Figure 4.4a compares the $\mathscr{H}_\infty$ error between all our models with orders between $n = 1$ and $n = 50$ and the same order model built with vector fitting (in 50 iterations). Similarly, Fig. 4.4b shows the corresponding $\mathscr{H}_2$ errors. We notice that, in most cases, the errors for our models are lower than the VF models. Moreover, in Fig. 4.4a, we observe that VF reaches a plateau, in the sense that the errors stay almost constant with increasing the model order. The opposite

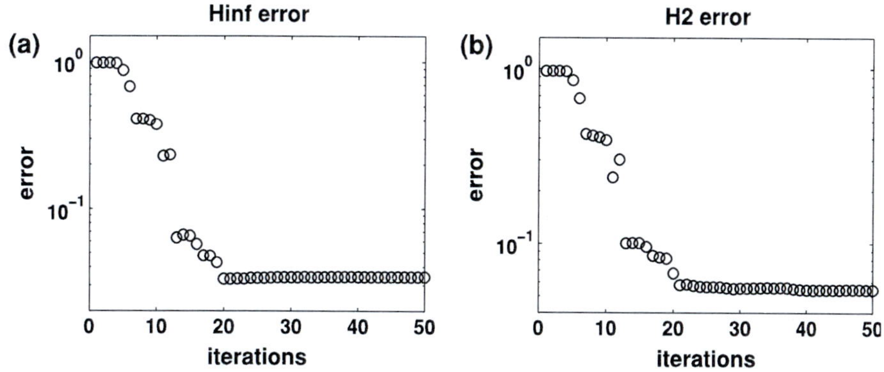

Fig. 4.5 $\mathscr{H}_\infty$ and $\mathscr{H}_2$ errors for VF as a function of the number of iterations: **a** $\mathscr{H}_\infty$ error **b** $\mathscr{H}_2$ error

behaviour happens for our models, as they become better when we increase the desired dimension.

Since vector fitting is solving an optimization problem, its performance greatly depends on the number of iterations. Figure 4.5 shows the decay of the $\mathscr{H}_\infty$ and $\mathscr{H}_2$ errors as a function of the number of iterations performed (between 1 and 50) when the procedure starts with 41 starting poles. The errors stagnate after about 20 iterations.

4.5.2 Circuit with $p = 70$ Ports

We analyze a dynamical system which describes a linear subcircuit of a large post-layout circuit (i.e. with parasitics due to layout). It was provided by Qimonda AG under the name *rc549* and has previously been investigated in [25] from the point of view of terminal reduction. One of the approaches currently used for model order reduction of systems with a large number of terminals is terminal reduction [9, 10]. This system has 141 poles: 84 infinite, 56 real between $-1.9 \cdot 10^{15}$ and $-1.1 \cdot 10^{13}$ and one at $-2 \cdot 10^{-1}$. The poles are depicted in Fig. 4.6b and the frequency response of this system is shown in Fig. 4.6a.

Our goal is to compare the tangential approach to the matrix interpolation approach on this example arising in circuit simulation. For tangential interpolation, we take $P = 400$ measurements of the transfer function logarithmically distributed between 10^9 and 10^{20} and, for each measurement, we select one sampling direction as one of the unit vectors of dimension p. Thus, for each frequency measurement, we use either a column (as right data) or a row (as left data) of the matrix measurement. The resulting Loewner and shifted Loewner matrices are of dimension 400 and a plot of their normalized singular values is shown in Fig. 4.7a.

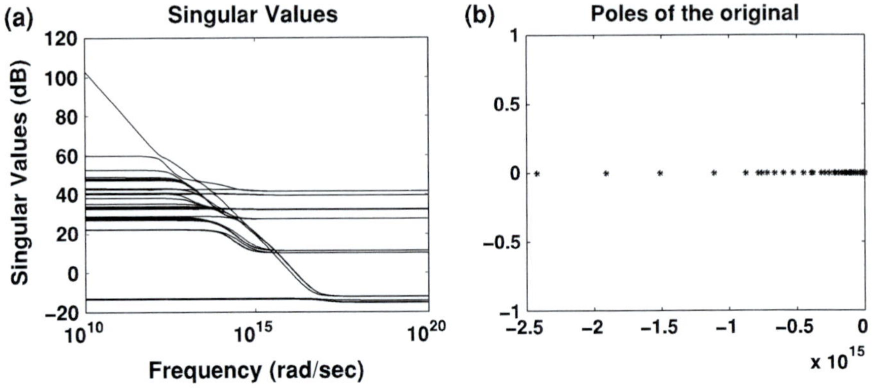

Fig. 4.6 Frequency response and finite poles of the original system: **a** Frequency response **b** Finite poles

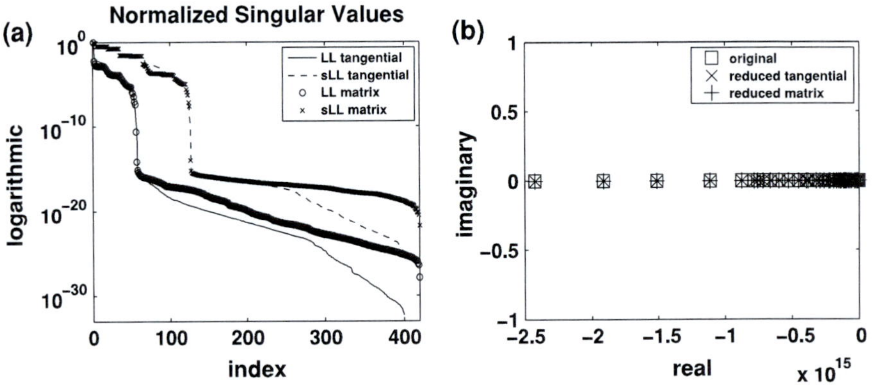

Fig. 4.7 Singular values of the Loewner matrix pencil and poles of the original and reduced systems: **a** Drop of the singular values of the Loewner matrix pencil for tangential and matrix interpolation **b** Poles of the original and reduced systems

For matrix interpolation, we take $P = 6$ measurements of the transfer function logarithmically distributed between 10^9 and 10^{20} and, for each measurement, we select p sampling directions as p unit vectors (all columns and rows of the identity matrix of dimension $p \times p$ are used as right directions and left directions, respectively). Thus, for each frequency measurement, we use the entire $p \times p$ matrix measurement. The resulting Loewner and shifted Loewner matrices are of dimension $6 \times 70 = 420$ and a plot of their normalized singular values is also shown in Fig. 4.7a. We notice that both methods yield Loewner and shifted Loewner matrices of the same rank, which suggest that the underlying system can be reduced to order $n = 127$.

Table 4.2 Results for the tangential and matrix interpolation approaches

Approach	$\mathcal{H}_\infty$ error	$\mathcal{H}_2$ error
Tangential interpolation	3.3422e-010	3.3589e-010
Matrix interpolation	6.9040e-009	6.9056e-009

Figure 4.7b shows the finite poles of the original system, together with the poles of the reduced systems obtained with the tangential and the matrix interpolation approaches. We notice that all poles were recovered. In particular, we notice that the reduced systems of order $n = 127$ with $\mathbf{D} = \mathbf{0}$ contain 57 finite poles (the same as the original one), and 70 infinite poles which hide an underlying non-zero $\mathbf{D}$-term, as in (4.29). Hence, all redundant poles have been eliminated..

Similarly to the scalar case (the delay system), we use the following error measures to asses the accuracy of the reduced models with respect to the measurements:

$$\mathcal{H}_\infty \text{ error} = \frac{\max_{i=1\ldots P} \sigma_1\left(\mathbf{H}_{\text{model}}(j \cdot 2\pi f_i) - \mathbf{Y}^{(i)}\right)}{\max_{i=1\ldots P} \sigma_1\left(\mathbf{Y}^{(i)}\right)},$$

$$\mathcal{H}_2 \text{ error} = \sqrt{\frac{\sum_{i=1}^{k} \left\|\mathbf{H}_{\text{model}}(j \cdot 2\pi f_i) - \mathbf{Y}^{(i)}\right\|_F^2}{\sum_{i=1}^{k} \left\|\mathbf{Y}^{(i)}\right\|_F^2}},$$

where $\sigma_1(\cdot)$ stands for the largest singular value of the matrix $(\cdot)$ and $\|\cdot\|_F^2$ stands for the Frobenius-norm, which computes the sum in the magnitude squared of all entries. For devices with many ports, computing the error in each entry is unfeasible, as for $p = 70$ ports, the errors in all $70^2 = 4900$ entries would have to be assessed. Instead, these error measures give a good indication of the model's quality.

Table 4.2 shows that both approaches yield small errors, with the tangential interpolation approach yielding slightly smaller errors than matrix interpolation.

Figure 4.8 shows the frequency response (sigma plot) of the top left and bottom right 10×10 entries of the original system, together with the reduced systems obtained via the two approaches: tangential and matrix interpolation.

The purpose of this example is to show that tangential interpolation yields comparable results to matrix interpolation provided that the sampling directions are chosen as outlined in Sect. 4.3.4. Moreover, both methods are able to eliminate the redundancy and produce reduced models which are very close to the original.

4.6 Conclusion

This paper discusses topics in model order reduction related to accurate macromodeling from frequency-domain data. In particular, the system theoretic concept of the Loewner matrix pencil constructed in the framework of tangential

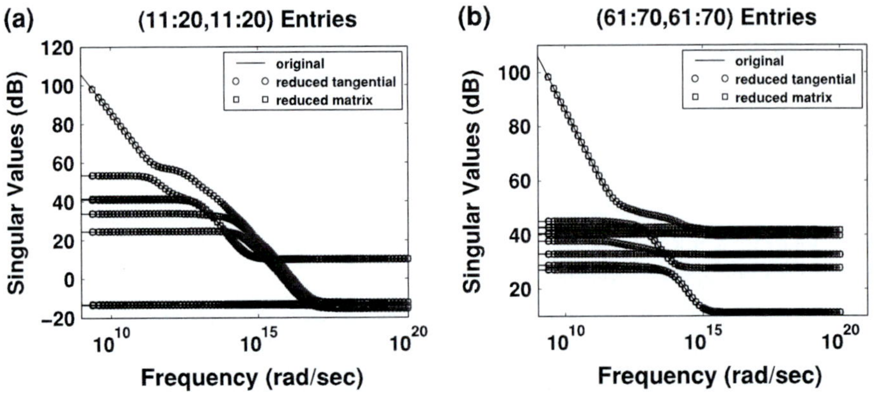

Fig. 4.8 Entries of the original transfer function plotted against those of the reduced systems: **a** (11:20,11:20) entries of the transfer function **b** Bottom right 10×10 entries of the transfer function

interpolation is analyzed from a theoretical point of view, by showing that a system can be recovered from tangential interpolation data given that the right amount of measurements are available and the sampling directions are chosen appropriately. We applied this concept to two numerical examples arising in circuit simulation: a delay system and a subcircuit with a large number of ports. Several other numerical examples which support our approach are presented in [19, 20].

References

1. Anderson, B.D.O., Antoulas, A.C.: Rational interpolation and state-variable realizations. Linear Algebra Appl. **137/138**, 479–509 (1990)
2. Antoulas, A.C.: Approximation of Large-Scale Dynamical Systems. SIAM, Philadelphia (2005)
3. Antoulas, A.C.: On the construction of passive models from frequency response data. Automatisierungstechnik **56**, 447–452 (2008)
4. Antoulas, A.C., Anderson, B.D.O.: On the scalar rational interpolation problem. IMA J. Math. Control Inf. **3**(2–3), 61–88 (1986)
5. Antoulas, A.C., Anderson, B.D.O.: State space and polynomial approaches to rational interpolation. In: Kaashoek, M., van Schuppen, J., Ran, A. (eds.) Progress in Systems and Control Theory III: Realization and Modeling in System Theory, pp. 73–82. Birkhäuser (1990)
6. Coelho, C., Silveira, L., Phillips, J.R.: Passive constrained rational approximation algorithm using Nevanlinna-Pick interpolation. In: DATE '02: Proceedings of the Conference on Design, Automation and Test in Europe, pp. 923–930. Washington, DC (2002)
7. Deschrijver, D., Dhaene, T.: Passivity-based sample selection and adaptive vector fitting algorithm for pole-residue modeling of sparse frequency-domain data. In: Proceedings of IEEE International Conference on Behavioral Modeling and Simulation (BMAS'04), pp. 68–73 (2004)

8. Dhaene, T., Deschrijver, D.: Stable parametric macromodeling using a recursive implementation of the vector fitting algorithm. IEEE Microw. Wirel. Compon. Lett. **19**(2), 59–61 (2009)
9. Feldmann, P.: Model order reduction techniques for linear systems with large numbers of terminals. In: Proceedings on Design, Automation and Test Conference in Europe, pp. 944–947 (2004)
10. Feldmann, P., Liu, F.: Sparse and efficient reduced order modeling of linear subcircuits with large number of terminals. In: Proceedings of IEEE/ACM International Conference on Computer-Aided Design(ICCAD'04), pp. 88–92 (2004)
11. Gao, R., Mekonnen, Y., Beyene, W., Schutt-Aine, J.: Black-box modeling of passive systems by rational function approximation. IEEE Trans. Adv. Packag. **28**(2), 209–215 (2005)
12. Grivet-Talocia, S.: Package macromodeling via time-domain vector fitting. IEEE Microw. Wirel. Compon. Lett. **13**(11), 472–474 (2003)
13. Grivet-Talocia, S., Bandinu, M.: Improving the convergence of vector fitting for equivalent circuit extraction from noisy frequency responses. IEEE Trans. Electromagn. Compat. **48**, 104–120 (2006)
14. Gustavsen, B.: Improving the pole relocation properties of vector fitting. IEEE Trans. Power Del. **21**(3), 1587–1592 (2006)
15. Gustavsen, B., Semlyen, A.: Rational approximation of frequency domain responses by vector fitting. IEEE Trans. Power Del. **14**, 1052–1061 (1999)
16. Gustavsen, B., Semlyen, A.: Enforcing passivity for admittance matrices approximated by rational functions. IEEE Trans. Power Syst. **16**(1), 97–104 (2001)
17. Gustavsen, B., Semlyen, A.: A robust approach for system identification in the frequency domain. IEEE Trans. Power Del. **19**, 1167–1173 (2004)
18. Kailath, T.: Linear Systems. Prentice-Hall (1979)
19. Lefteriu, S.: New approaches to modeling multi-port scattering parameters. Master's thesis, Rice University, Houston (2008)
20. Lefteriu, S., Antoulas, A.C.: A new approach to modeling multi-port systems from frequency-domain data. IEEE Trans. CAD **29**(1), 14–27 (2010)
21. Mayo, A.J., Antoulas, A.C.: A framework for the solution of the generalized realization problem. Linear Algebra Appl. **405**, 634–662 (2007)
22. Min, S.H., Swaminathan, M.: Construction of broadband passive macromodels from frequency data for simulation of distributed interconnect networks. IEEE Trans. Electromagn. Compat. **46**(4), 544–558 (2004)
23. Reis, T.: Systems theoretic aspects of PDAEs and applications to electrical circuits. Ph.D. thesis, Technische Universität Kaiserslautern (2006)
24. Saraswat, D., Achar, R., Nakhla, M.: A fast algorithm and practical considerations for passive macromodeling of measured/simulated data. IEEE Transactions on Advanced Packaging **27**(1), 57–70 (2004)
25. Schneider, A.: Matrix decomposition based approaches for model order reduction of linear systems with a large number of terminals. Diplomarbeit, Technische Universität Chemnitz (2008)
26. Triverio, P., Grivet-Talocia, S., Nakhla, M.S.: A parameterized macromodeling strategy with uniform stability test. IEEE Trans. Adv. Packag. **32**(1), 205–215 (2009)
27. Yip, E., Sincovec, R.: Solvability, controllability, and observability of continuous descriptor systems **26**(3), 702–707 (1981)

Part II
Contributed Papers

Chapter 5
Forward and Reverse Modeling of Low Noise Amplifiers Based on Circuit Simulations

Luciano De Tommasi, Joost Rommes, Theo Beelen, Marcel Sevat, Jeroen A. Croon and Tom Dhaene

Abstract Forward and reverse modeling of RF circuit blocks are useful approaches in design space exploration. The underlying idea of forward modeling is the creation of accurate surrogate models, which can be used to predict the circuit performances replacing (expensive) circuit simulations. On the other hand, reverse modeling concerns multiobjective optimization to explore relevant trade-offs between performances. This paper provides a discussion of application of surrogate models and multiobjective optimization to narrow-band low noise amplifier design. We discuss numerical difficulties encountered when the forward model is derived by using surrogate models of low noise amplifier admittances to compute performance figures via analytical equations. Afterward, we provide an example

L. D. Tommasi (✉)
University of Antwerp and NXP Semiconductors, High Tech Campus 37, Post Box WY4-01, 5656 AE, Eindhoven, The Netherlands
e-mail: luciano.de.tommasi@gmail.com

J. Rommes · T. Beelen · M. Sevat · J. A. Croon
NXP Semiconductors, High Tech Campus 37, Post Box WY4-01, 5656 AE, Eindhoven, The Netherlands
e-mail: joost.rommes@nxp.com

T. Beelen
e-mail: theo.beelen@nxp.com

M. Sevat
e-mail: marcel.sevat@nxp.com

J. A. Croon
e-mail: jeroen.croon@nxp.com

T. Dhaene
Department of Information Technology (INTEC), Ghent University-IBBT, Gaston Crommenlaan 8 Bus 201, 9050, Ghent, Belgium
e-mail: tom.dhaene@ugent.be

P. Benner et al. (eds.), *Model Reduction for Circuit Simulation*,
Lecture Notes in Electrical Engineering, 74, DOI: 10.1007/978-94-007-0089-5_5,

where direct performace modeling leads to a more accurate result even when the simplest surrogate model type (a lookup table) is used. Finally, a detailed tutorial of the normal boundary intersection optimization method is provided.

Keywords Response surface modelling · Surrogate modelling · Forward modelling · Reverse modelling · Low noise amplifier · Design space exploration · Multiobjective optimization · Pareto front

5.1 Introduction

An RF circuit block, such as a Low Noise Amplifier (LNA), shown in Fig. 5.1, is characterized by performance figures (e.g., voltage gain, linearity, noise figure, etc.) which are functions of the design parameters (e.g., width and length of transistors, bias conditions, values of passive components) [10]. The goal of the design process is to figure out one or more sets of design parameters resulting in a circuit which fulfills the specifications, i.e., constraints given on the performances.

Nowadays most of the RF integrated circuit design methods are still based on experience. Typically, circuit designers start guessing one or more solutions of the design problem, possibly relying on simplified mathematical models reflecting coarse approximations of the physics describing the circuit. Afterwards, they verify the circuit behavior through circuit simulations, which instead exploit highly accurate models of electronic components. Design parameters are then adjusted until circuit specifications are achieved. Hence, there is a demand from industry for the development of systematic methods and tools, which fill the gap between the generality of several existing modeling and optimization software and the concrete applications at hand.

The design problem can be formulated as a single objective optimization problem, where a single performance or a weighted sum of performances is minimized. For example, typical LNA design strategies try to minimize noise

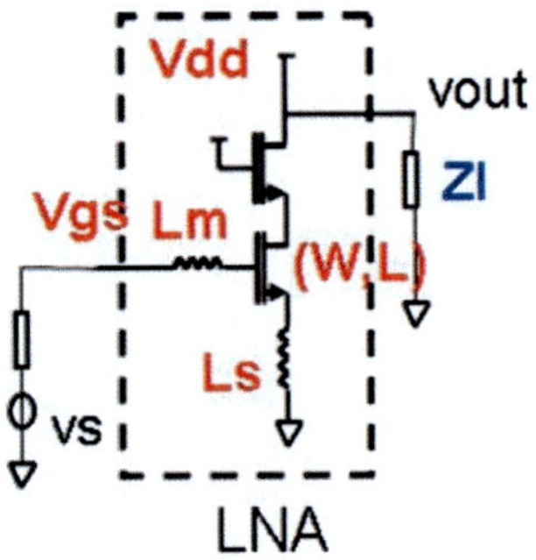

Fig. 5.1 Schematic of a (weakly-nonlinear) narrowband low noise amplifier. Most important design parameters, on which performances depend on, are: transistor width W, inductances L_s and L_m, frequency f, and bias condition V_{GS} and transistor length L

figure (NF) (e.g. [12]) or power dissipation or to maximize linearity, while enforcing proper constraints on the remaining performances.

A more insightful and systematic approach than single objective optimization, even though computationally more expensive, is multi-objective optimization. Circuit performances normally are in conflict with each other: for instance, to improve the voltage gain of an LNA, the linearity has to be reduced. Hence, a multiobjective optimization method (e.g. the normal boundary intersection algorithm (NBI) [3, 16] or the strength pareto evolutionary algorithm 2 (SPEA2) [15, 20]) can be used to figure out and graphically display such trade-offs as Pareto fronts. Since the sample points belonging to a Pareto front are obtained by optimizing performances with respect to design parameters, the result of such multi-objective optimization is a 'reverse model' of the circuit block, which determines a correspondence between samples of the Pareto front and samples belonging to the design space.

Design tools based on optimization methods can be classified in two categories: simulation-based tools and stand-alone tools. To compute circuit performances in the optimization loop, the former category of tools calls a commercially available circuit simulator, such as Spectre or Hspice. The second category includes a built-in performance evaluation module based on a symbolic performance model of the circuit [11, 14, 18]. Although symbolic models enable a faster evaluation of performances than circuit simulations, their accuracy is lower. For this reason, the authors of [18] propose to use their tool as an "auxiliary tool generating an initial design for a commercial design environment". On the other hand, an extensive exploration of the design space with simulation-based tools is limited by the high computational cost of numerical simulations.

In this work, a simulation-based methodology, aiming to combine accuracy of circuit simulations with computational efficiency of symbolic models, is described. Circuit simulations are replaced by a cheap-to-evaluate surrogate model, derived from a proper set of precomputed simulations [6, 7, 14, 19]. An approximation of the circuit performance figures based on one or more surrogate models is referred to as 'forward model' of the circuit. A forward model can be either obtained via direct modeling of circuit performances or by using surrogate models of a convenient set of behavioural parameters (e.g. admittances and noise functions) to compute performances via analytical equations. The former approach involves the direct approximation of the function:

$$\mathbf{p} = \mathbf{p}(\mathbf{x}, \mathbf{y}), \tag{5.1}$$

where $\mathbf{p} = (p_1 \cdots p_j)$ is the vector of circuit performances, $\mathbf{x} = (x_1 \cdots x_k)$ is the vector of design parameters and $\mathbf{y} = (y_1 \cdots y_l)$ is a vector holding extra-circuit parameters (e.g. load impedances). Surrogate modeling involves the identification of a vector of functions:

$$\tilde{\mathbf{p}} = \tilde{\mathbf{p}}(\mathbf{x}, \mathbf{y}), \tag{5.2}$$

such that $\tilde{\mathbf{p}} \approx \mathbf{p}$.

Circuit performances can also be written as functions of the behavioural parameters $\mathbf{b} = \mathbf{b}(\mathbf{x})$ and the extra-parameters $\mathbf{y}$:

$$\mathbf{p} = \mathbf{p}(\mathbf{b}(\mathbf{x}), \mathbf{y}). \tag{5.3}$$

The latter modeling approach consists in the approximation of the behavioural parameters $\mathbf{b} = \mathbf{b}(\mathbf{x})$ through the surrogate models $\tilde{\mathbf{b}} = \tilde{\mathbf{b}}(\mathbf{x})$, such that $\tilde{\mathbf{b}} \approx \mathbf{b}$. The forward model is then given by:

$$\tilde{\mathbf{p}} = \mathbf{p}(\tilde{\mathbf{b}}(\mathbf{x}), \mathbf{y}). \tag{5.4}$$

Modeling of behavioural parameters is attractive because they do not depend on the extra-parameters $\mathbf{y}$, therefore the approximation problem is simplified. However, in this paper we show that forward modeling from behavioural parameters is likely to introduce an unacceptable amplification of the numerical error, i.e. $\tilde{\mathbf{b}} \approx \mathbf{b}$ does not necessarily imply that $\tilde{\mathbf{p}} \approx \mathbf{p}$. Therefore, direct approximation of performances has to be currently preferred, even though the inclusion of all circuit parameters in the model is still an open problem.

Methods described in this paper have been implemented into a prototype of an auxiliary design tool. Currently, it can only handle narrowband LNA design and does disregard circuit layout optimization. Further extensions can enable wideband and multiband LNAs design [5, 12], and layout-level optimization [13].

The paper is organized as follows. In Sect. 5.2, forward and reverse modeling problem descriptions are introduced. Section 5.3 focuses on forward modeling and the numerical error amplification problem. Section 5.4 describes reverse modeling via NBI optimization technique. Section 5.5 explains how a reverse model can exploit a forward model to speed-up the computation of a Pareto-front. Section 5.6 discusses presented results and concludes the paper.

5.2 Forward and Reverse Modeling: Problem Descriptions

A forward model of a circuit determines a correspondence between design parameters and its performances. It involves the generation of a set of surrogate models or response surface models (RSMs), which are cheaper to evaluate than the set of circuit simulations needed for evaluating the same performances. On the other hand, a reverse model links Pareto-optimal points of the performance space with the corresponding design parameters.

A block diagram showing the connection between circuit simulations, surrogate models, design parameters and performances is depicted in Fig. 5.2. Computational times are indicative. In particular, the computational time needed to generate the forward model depends on the specific set of functions that are approximated via surrogate modeling.

A typical setup used to build a forward model of an RF circuit block consists of modeling software (e.g. the SUMO Toolbox [17]) linked to a circuit simulator. The

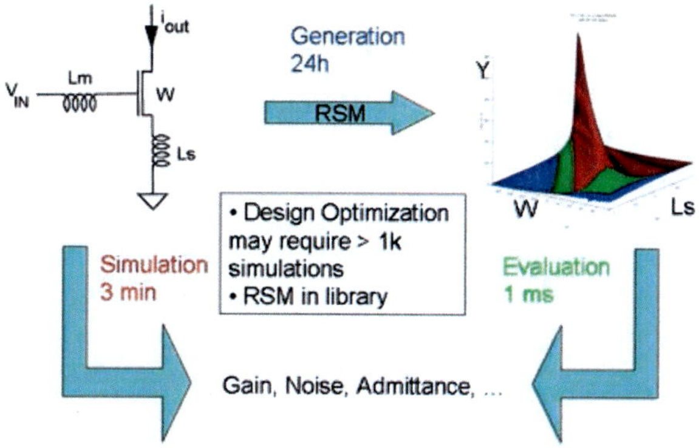

Fig. 5.2 Block diagram showing the connection between circuit simulations, surrogate models, design parameters and performances

modeling software directly drives the circuit simulator, requesting new simulations in order to accurately cover the entire design space. Additional data samples may be selected according to adaptive sampling techniques, to allocate more samples in those regions of the design space which are more difficult to model and/or where the error is the largest [1]. This way, the obtained model is intended to be valid throughout the entire design space. This approach was adopted in [6, 7].

Several design problems can be formulated as *multiobjective optimization* problems, involving the simultaneous optimization of two or more conflicting objectives subject to constraints. For instance, for an LNA one wants to maximize

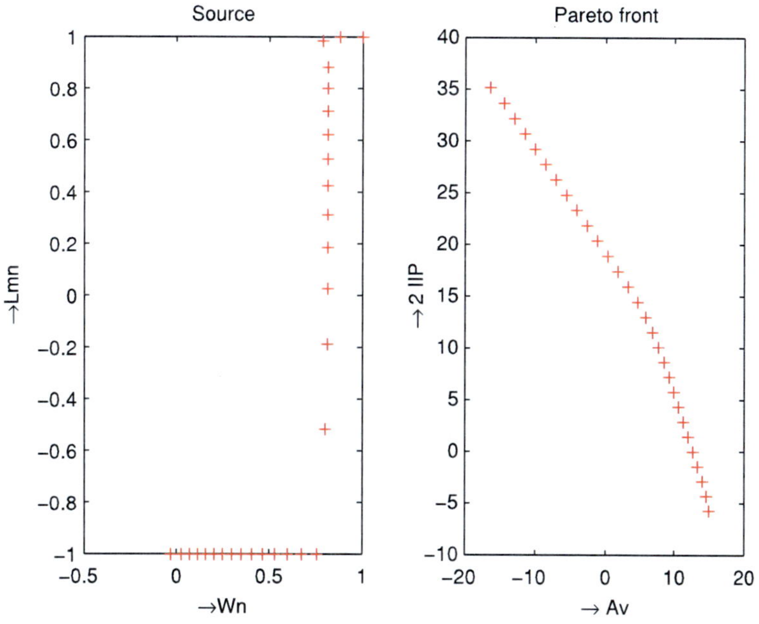

Fig. 5.3 Pareto-front representing the trade-off between voltage gain A_v and linearity *IIP*2 of an LNA. *Left*: design space. *Right*: performance space

linearity $IIP2$ but at the same time, maximize the voltage gain A_v as well. Since these are two conflicting objectives, the trade-off has to be studied. One way to do this is to compute a Pareto-front. An example is given in Fig. 5.3.

The final result of such optimization process is a reverse model, which provides the design parameters corresponding to each Pareto-optimal solution. In the example shown in Fig. 5.3, the left side plot shows the points of the design space (W_n, L_{mn})[1] which correspond to the Pareto-optimal points of the right side plot.

Since the construction of a reverse model requires a lot of performance evaluations, it could exploit a forward model in order to make the computation of Pareto-front fast. This will be shown in Sect. 5.4.

5.3 Forward Modeling

Forward modeling of RF circuit blocks is a challenging research area, since it demands to obtain accurate models with a small number of data samples distributed in high dimensional design spaces. For instance, a forward model of an LNA should provide the following performance figures: power consumption P, voltage gain A_v, input reflection Γ_a, second-order linearity $IIP2$, third-order linearity $IIP3$ and noise figure NF, as functions of the most important design parameters.

In [7] the authors used the SUMO Toolbox [17] to generate surrogate models of LNA behavioral parameters (admittances and noise functions). Such behavioral parameters were given by a simplified analytic model, in order to allow a fast model identification. It was tried to achieve a Root Relative Square Error (RRSE) ≤ 0.05 with a maximum number of samples of 1500. Results indicated that this is possible only including the first four parameters as 'inputs' of the model W, L_s, f, L (when V_{GS} and L_m are kept constant).

A study reported in [4] considers a complete characterization of an LNA provided by circuit simulations. This includes the complete set of admittances needed to describe the weakly non-linear behavior [2], which were not taken into account in [7], but the number of inputs was limited to two (W and L_s). Results indicate that a greater number of samples is needed to achieve a similar level of accuracy as using the analytic model.

In order to build a forward model, surrogate models of admittances and noise functions have to be used to compute performances (P, A_v, Γ_a, $IIP2$, $IIP3$ and NF), which are expressed as analytical functions of admittances and noise functions.[2] This means that errors in the surrogate models may be amplified and hence result in inaccurate performance models (a general discussion of this issue is provided in

[1] The subscript 'n' indicates that design variables W and L_m have been normalized such that they vary between -1 and 1.

[2] Admittances and noise functions are the elements of the vector **b** introducted in Sect. 5.1.

[3] Z_s, Z_l and L_m are elements of the vector **y** introduced in Sect. 5.1.

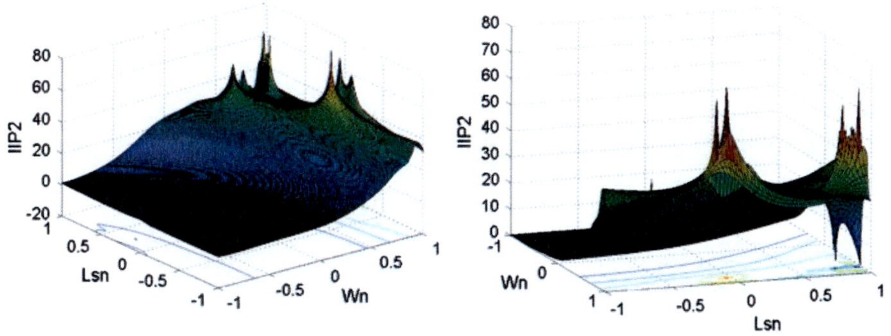

Fig. 5.4 *IIP*2 as function of *W* and L_s (computed by means of surrogate models)

the introduction). On the other hand, direct modeling of performances would require the inclusion of the source and load impedances, respectively Z_s and Z_l, as additional inputs (Fig. 5.1).[3] This would make the modeling more difficult. With the computation of performances from admittances and noise functions, the load conditions enter into the analytical expression and do not need to be modeled.

5.3.1 Performance Figures via Surrogate Models

Forward modeling from behavioural parameters is prone to introduce an unacceptable amplification of the surrogate model numerical error when computing circuit performances from behavioural parameter models. This is demonstrated by the following example.

Performance figures are expected to be smooth functions, 'local' spikes are not expected. We computed *IIP*2 with respect to the design parameters *W* and L_s, by exploiting the models computed in [9] (Fig. 5.4). It is seen that several spikes appear. *IIP*2 can be expressed as function of the ratio of two voltages v_{21}/v_{2p}:

$$IIP2 = 10\log_{10}\left[\left|\left(\frac{1000}{8R_S}\right)\left(\frac{v_{21}}{v_{2p}}\right)^2\right|\right], \tag{5.5}$$

where v_{21} is the voltage amplitude of the 'wanted' signal at output and v_{2p} is the voltage amplitude of 'unwanted' second harmonic at output [10]. It can be verified that all spikes are due to v_{2p} and $1/v_{2p}$, being v_{21} a smooth function.

Furthermore, by decomposing v_{2p} into numerator and denominator $v_{2p} = \frac{num}{den}$ it can be verified that, being *den* a smooth function, all the peaks are due to *num* and 1/*num* (Fig. 5.5).

[4] *IIP*2 is expressed in dBm, i.e. logarithmic scale.

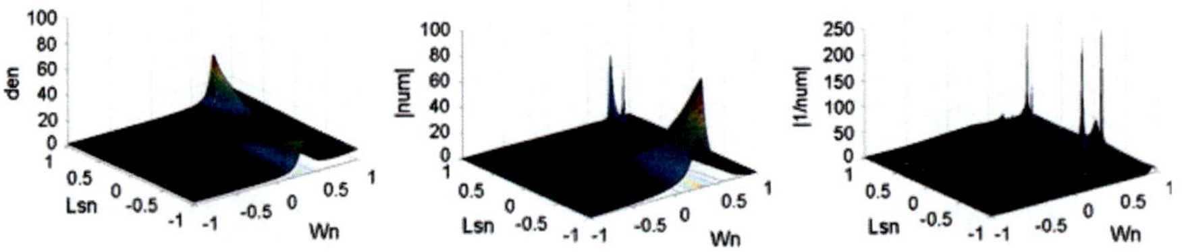

Fig. 5.5 Magnitude of v_{2p} denominator, numerator and inverse of numerator

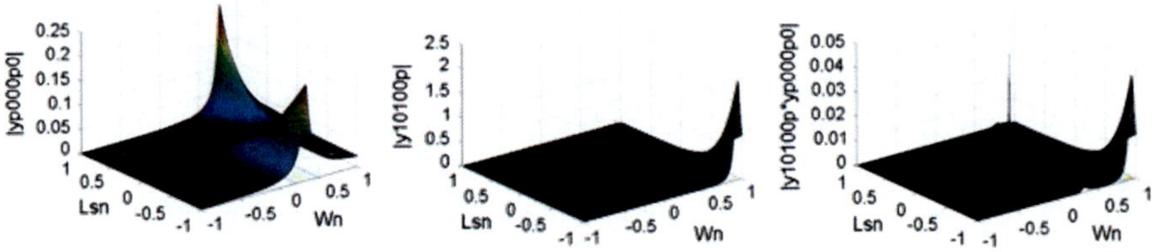

Fig. 5.6 Magnitude of y_{p000p0}, y_{10100p} and $y_{p000p0}y_{10100p}$

In particular, 'positive' peaks of $IIP2$ are due to peaks of $\frac{1}{num}$. They appear when num assumes values close to zero. On the other hand, negative peaks of $IIP2$ are due to peaks of num.[4]

The origin of 'negative' peaks is easy to investigate, since the function num may be readily decomposed into a sum of several terms, which are functions of surrogate models. A complete analysis is out of scope of this paper, we just provide a single example. A subpart of num, num', is proportional to the product of two admittance surrogate models: $num' \propto y_{10100p}y_{p000p0}$. The model of y_{10100p} has a small peak due to modeling errors, which is amplified when multiplied by y_{p000p0}, since it is located in the region of the main resonance of y_{p000p0} (Fig. 5.6). This peak is reflected also onto $IIP2$.

5.4 Reverse Modeling with the NBI Method

As indicated in the introduction, a natural way to deal with trade-offs between performances is to use a suitable multiobjective optimization technique.

Multiobjective optimization algorithms can be categorized globally into deterministic and evolutionary methods. Examples of such methods are NBI method and SPEA2, respectively. In this paper, we focus on the NBI method [3].

To outline this method we have to introduce a few notions in a more formal mathematical setting. The design parameters and performance parameters are assumed to be in the design space $\mathscr{D}$ and performance space $\mathscr{P}$, respectively. We assume that the region $\mathscr{D}$ is *feasible*, i.e., all design parameters in $\mathscr{D}$ satisfy the imposed constraints:

$$\mathscr{D} = \{\mathbf{x} \in \mathscr{R}^m | \ \mathbf{c}(\mathbf{x}) \leq 0\} \quad \text{with } \mathbf{c}(\mathbf{x}) \in \mathscr{R}^q. \tag{5.6}$$

The performance space $\mathscr{P}$ contains all performance values resulting from feasible design values in the region $\mathscr{D}$ after applying a performance function **f**. Furthermore, the performance values can also be constrained, i.e.,

$$\mathscr{P} = \{\mathbf{f} \in \mathscr{R}^n | \ \exists_{\mathbf{x} \in \mathscr{D}} \mathbf{f} = \mathbf{f}(\mathbf{x}), \mathbf{g}(\mathbf{f}) \leq 0\}. \tag{5.7}$$

The general problem at hand is the minimization of several objectives *simultaneously*,[5] i.e.,

$$\underset{\mathbf{x} \in \mathscr{D}}{\text{minimize}} \, \mathbf{f}(\mathbf{x}) = \begin{pmatrix} f_1(\mathbf{x}) \\ \vdots \\ f_n(\mathbf{x}) \end{pmatrix} \quad \text{such that } \mathbf{g}(\mathbf{f}) \leq 0. \tag{5.8}$$

However, in general there is no single $\mathbf{x} \in \mathscr{D}$ that minimizes all components $f_i, i = 1, \ldots, n$ simultaneously. Consequently, the multi-objective optimization (5.8) leads to trade-off situations where it is only possible to improve one performance at the cost of another. A trade-off situation needs to be based mathematically on a so-called *dominance* relation between vectors (see for example [16]).

Definition 1 Let **a** and **b** be vectors in $\mathscr{R}^n$. Vector $\mathbf{a} = (a_1, \ldots, a_n)$ is said to *dominate* vector $\mathbf{b} = (b_1, \ldots, b_n)$ if and only if

$$\mathbf{a} \prec \mathbf{b} :\Leftrightarrow \forall_{i \in \{1,\ldots,n\}} (a_i \leq b_i) \wedge \exists_{i \in \{1,\ldots,n\}} (a_i < b_i). \tag{5.9}$$

With this definition, a performance vector f^* f* is said to be *Pareto-optimal* if it is *nondominated* within $\mathscr{P}$, i.e.,

$$\neg \exists_{\mathbf{f} \in \mathscr{P}} [\mathbf{f} \prec \mathbf{f}^*]. \tag{5.10}$$

The set of all Pareto-optimal points is called the *Pareto front* of $\mathscr{P}$. The set of points in the design space corresponding to the Pareto front are called the *source of the Pareto front*.

In literature there are several methods available to calculate the Pareto front. In this paper we will describe the commonly used Normal-Boundary Intersection method (NBI method [3, 16]). We will closely follow [16].

For simplicity reasons we describe the NBI method for the 2D case only. Thus, we assume the design space $\mathscr{D}$ and performance space $\mathscr{P}$ as given above (with $m = n = 2$). However, it can easily be extended to higher dimensions.

Description of the NBI algorithm

[5] Maximization of a performance p is equivalent to minimization of $-p$.

1. Start with individually minimizing f_1 and f_2 as a function of $\mathbf{x}$, i.e., compute $\mathbf{f}^{*k} = \mathbf{f}(\mathbf{x}^{*k}) = \min\{\mathrm{f_k}(\mathbf{x})|\ \mathbf{x} \in \mathscr{D}\}, \mathrm{k} = 1, 2.$
2. Determine the straight line $\mathscr{L}$ in $\mathscr{P}$ between these points $\mathbf{f}^{*1}$ and $\mathbf{f}^{*2}$.
3. Determine the vector $\mathbf{n}$ normal to this straight line, pointing into the direction of decreasing $\mathbf{f}$.
4. Select N points $\mathbf{f}_k$ on the line $\mathscr{L}$ between $\mathbf{f}^{*1}$ and $\mathbf{f}^{*2}$: $\mathbf{f}_k = \lambda_k \mathbf{f}^{*1} + (1 - \lambda_k)\mathbf{f}^{*2}$ for some $0 \leq \lambda_k \leq 1$.
5. For each point $\mathbf{f}_k$, find the point $\mathbf{p}_k$ that maximizes the distance along $\mathbf{n}$ in the direction of decreasing $\mathbf{f}$, and starting in $\mathbf{f}_k$.
6. The points $\mathbf{p}_k, 1 \leq k \leq N$ are on the Pareto front.

Notice that in fact the NBI method consists of N subproblems in which a 1D line search optimization problem has to be solved. Clearly, several modifications to the above approach can be made to speed up the calculations. However, we will not go into details here.

5.5 Reverse Modeling Using Transistor Level Simulations

The NBI algorithm described in Sect. 5.4 was tested by using a simplified analytical model of an LNA [8]. However, to obtain an accurate representation of trade-offs between competing performances, a transistor level simulator has to be used. In principle, NBI can directly drive a circuit simulator by requiring data samples needed in the optimization process. However, this is nearly infeasible in practice, because of the large amount of simulations needed (a typical order of magnitude is 10^4, with a cost of about 2–3 minutes per simulation). A more applicable idea is the usage of a forward model as described in Sect. 5.3. However, the final accuracy of the Pareto-front may be poor even when surrogate models of behavioral parameters are fairly accurate, due to the error amplification process described in Sect. 5.3.1.

A better solution relies on the direct approximation of circuit performances, as explained in the introduction. With this approach, no error propagation from the behavioral parameter space to the performance space occurs. A simple precomputed lookup table, which is linearly interpolated when NBI requests a new performance evaluation, will be used to demonstrate the validity of this approach.

Despite of its simplicity, linear interpolation is a robust way to approximate a function, which is doable even in high dimensional spaces. Moreover, a lookup table is generated at the only cost of circuit simulations. There is not additional cost due to optimization of surrogate model parameters (e.g. training of a neural network and number of layers and neurons per layer optimization [4, 6, 7, 8]). Linear interpolation occurs when a new sample of the performance has to be used in the Pareto front computation and not in an earlier stage.

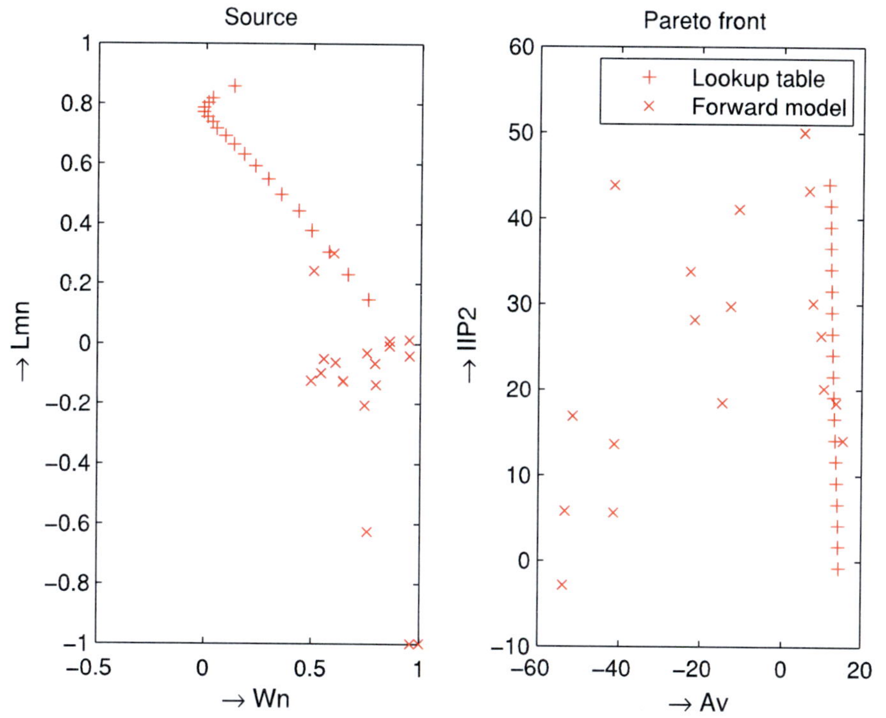

Fig. 5.7 Comparison between Pareto-fronts obtained with a lookup table and with a forward model (trade-off between voltage gain A_v and linearity $IIP2$ of an LNA). *Left*: design space. *Right*: performance space

5.5.1 Numerical Results

In this section we show an example comparing forward model and lookup table methods to generate a Pareto-front (Fig. 5.7) with the NBI method. Similarly to Fig. 5.3, we show the trade-off between voltage gain A_v and linearity $IIP2$ with respect to the (normalized) design parameters W_n and L_{mn} (other design parameters are kept constant to convenient values). However, we remark that results in Fig. 5.3 were generated by relying on a simplified LNA analytical model [7], whereas here both forward model and lookup table are based on transistor-level circuit simulations. Details about surrogate models exploited in performance computations are given in [4]. The lookup table is a regular grid of 100 × 100

[6] It is worthwhile to note that only 100 circuit simulations corresponding to different values of W need to be performed to generate the table, since the variability of L_m is taken into account in the analytical performance equations, being L_m an element of the vector $\mathbf{y}$ introduced in Sect. 5.1. Specific LNA performance equations are not included in this paper, because they can be found in any relevant text-book.

points.[6] However, it is evident that the number of circuit simulations for each design parameter taken into account, has to be reduced as the dimensionality of the considered design space is increased.

It is readily seen from Fig. 5.3 that the forward model based on surrogate models of admittance function, returns a 'sparse' set of points which fail to approximate the Pareto-front. Indeed such inaccurate results could have been expected from the discussion developed in Sect. 5.3.1. On the other hand, the forward model based on the performance lookup table provides a reasonable front.

Finally, we note that in principle the error amplification effect discussed in Sect. 5.3.1, could be negligible if surrogate models had an error sufficiently lower than ones in [4]. However, in practice this is not easy to achieve. Moreover, how much accurate the models should be, and in which part of design space, are still open issues.

5.6 Discussion and Conclusions

This paper has reviewed forward and reverse modeling concepts of RF circuit blocks, referring to a narrowband LNA.

A number of conclusions can be drawn:

- In principle forward models can be exploited to speed-up the computation of Pareto-fronts (i.e. reverse modeling), but the accuracy can be limited by error amplification from the surrogate models of behavioral parameters (admittances and noise functions) to circuit performances.
  Direct modeling of circuit performances solves the problem of error amplification from behavioural parameters to performances (Sect. 5.3.1).
- NBI optimization is proven as a promising approach to investigate LNA performance trade-offs.
- Forward models based on performance lookup tables (and linear interpolation) are likely to be sufficient for obtaining accurate Pareto-fronts when only two design parameters are taken into account in the forward model.

However, an LNA design problem may have ten or more design variables. Therefore, possible limitations of the uniformly filled lookup tables are:

- Uniform sampling of performances would require a too large number of samples because of the oversampling of the smooth regions, where sensitivity of performances as functions of design parameters is modest. It is expected that adaptive sampling techniques are needed when more design parameters are considered [1, 4, 17].
- Interpolation of the resulting large lookup tables can be computationally less efficient than modeling such a large number of design variables (which on the other hand is technically much more challenging).

Some important points have to be investigated further:

- How does the lookup table approach scale to more design parameters?
- How can the lookup table be filled and processed in an efficient and accurate way?
- How accurate should a forward model be in order to be successfully used in reverse modeling?

Finally, the surrogate modeling approach implemented in [17] will be directly applied to performances, comparing the accuracy with the table-based method.

Acknowledgements This work was supported by the European Commission through the Marie Curie Actions of its Sixth Framework Program under the contract number MTKI-CT-2006-042477.

References

1. Crombecq, K.: A gradient based approach to adaptive metamodeling. Technical report, University of Antwerp. (2007)
2. Croon, J.A., Leenaerts, D.M.W., Klaassen, D.B.M.: Accurate modeling of RF circuit blocks: weakly-nonlinear narrowband LNAs. In: Proceedings of the IEEE Custom Integrated Circuits Conference 2007, CICC 2007, pp. 865–868. 16–19 Sept 2007
3. Das, I., Dennis, J.E.: Normal-boundary intersection: a new method for generating Pareto optimal points in multicriteria optimization problems. SIAM J. Optim. **8**, Nr. 3, 631–657 (1998)
4. De Tommasi, L., Gorissen, D., Croon, J.A., Dhaene, T.: Surrogate modeling of low noise amplifiers based on transistor level simulations. In: Roos J., Costa L.R.J. (eds.) Scientific Computing in Electrical Engineering SCEE 2008, Mathematics in Industry, Springer, Berlin (2009)
5. Eeckelaert, T., McConaghy, T., Gielen, G.: Efficient multiobjective synthesis of analog circuits using hierarchical pareto-optimal performance hypersurfaces. In: Proceedings of the Conference on Design, Automation and Test in Europe, DATE 2005, **2**, pp. 1070–1075 (2005).
6. Gorissen, D., De Tommasi, L., Croon, J., Dhaene, T.: Automatic model type selection with heterogeneous evolution: an application to RF circuit block modeling. In: Proceedings of IEEE World Congress on Computational Intelligence, WCCI 2008, pp. 989–996. Hong Kong, June (2008).
7. Gorissen, D., De Tommasi, L., Crombecq, K., Dhaene, T.: Sequential modeling of a low noise amplifier with neural networks and active learning. Neural Comput. Appl. **18**(5), 485–494 (2009)
8. Hendrickx, W., Gorissen, D., Dhaene, T.: Grid enabled sequential design and adaptive metamodeling. In: Proceedings of the 2006 Winter Simulation Conference, WSC 2006, pp. 872–881 (2006).
9. Karer, E.: Design Space Exploration of RF-Circuit Blocks, Master's thesis, TU Eindhoven, July 2007.
10. Lee, T.H.: The Design of CMOS Radio-Frequency Integrated Circuits, 2nd edn. Cambridge University Press (2003).
11. Nieuwoudt, A., Ragheb, T., Massoud, Y.: SOC-NLNA: Synthesis and optimization for fully integrated narrow-band CMOS low noise amplifiers. In: Proceedings of IEEE/ACM Design Automation Conference, pp. 879–884 (2006).
12. Nieuwoudt, A., Ragheb, T., Nejati, H., Massoud, Y.: Numerical design optimization methodology for wideband and multi-band inductively degenerated cascode CMOS low noise amplifiers. IEEE Trans. circuits syst. **I**, 1088–1101 (2009).

13. Park, J., Choi, K., Allstot, D.: Parasitic-aware design and optimization of a fully integrated CMOS wideband amplifier. In: Proceedings of Asia and South Pacific Design Automation Conference, pp. 904–907 (2003).
14. Ranjan, M., Bhaduri, A., Verhaegen, W., Mukherjee, B., Vemuri, R., Gielen, G., Pacelli, A.: Use of symbolic performance models in layout-inclusive RF low noise amplifier synthesis. In: Proceedings of the 2004 IEEE International Behavioral Modeling and Simulation Conference, BMAS 2004, pp. 130–134 (2004).
15. SPEA2—Source files in C, http://www.tik.ee.ethz.ch/pisa/selectors/spea2/spea2_c_source.html accessed on 23/01/2011
16. Stehr, G., Graeb, H.E., Antreich, K.J.: Analog performance space exploration by normal-boundary intersection and by Fourier-Motzkin elimination. IEEE Trans. Comp. Aided Des. Integr. Circuits Syst. **26**, Nr. 10 (2007)
17. The SUrrogate MOdeling Toolbox. Wiki Page. URL http://www.sumowiki.intec.ugent.be/.
18. Tulunay, G., Balkir, S.: A synthesis tool for CMOS RF low-noise amplifiers. IEEE Trans. Comput. Aided Des. Integr. Circuits Syst. **27**(5), pp. 977–982 (2008).
19. Zhang, Q.J., Gupta, K.C.: Neural Networks for RF and Microwave Design. Artech House, Boston (2000).
20. Zitzler, E., Laumanns, M., Thiele, L.: SPEA2: improving the strength pareto evolutionary algorithm, Technical Report TIK Report 103, ETH Zürich (2001).

Chapter 6
Recycling Krylov Subspaces for Solving Linear Systems with Successively Changing Right-hand Sides Arising in Model Reduction

Peter Benner and Lihong Feng

Abstract We discuss the numerical solution of successive linear systems of equations $Ax = b_i,\ i = 1, 2, \ldots, m$, by iterative methods based on recycling Krylov subspaces. We propose various recycling algorithms which are based on the generalized conjugate residual (GCR) method. The recycling algorithms reuse the descent vectors computed while solving the previous linear systems $Ax = b_j$, $j = 1, 2, \ldots, i-1$, such that a lot of computational work can be saved when solving the current system $Ax = b_i$. The proposed algorithms are robust for solving sequences of linear systems arising in circuit simulation. Sequences of linear systems need to be solved, e.g., in model order reduction (MOR) for systems with many terminals. Numerical experiments illustrate the efficiency and robustness of the proposed method.

Keywords Linear systems of equations · GCR · GMRES · Recycling Krylov subspace

6.1 Introduction

In some fields of circuit design, such as macromodeling of linear networks with many terminals [2], noise analysis of RF circuits [19] etc., successive linear systems of equations $A_i x = b_i, i = 1, 2, \ldots, m$, have to be solved. Here the dimension

P. Benner
Professur Mathematik in Industrie und Technik, Fakultät für Mathematik, Technische Universität Chemnitz, Reichenhainer Str. 41, 01926 Chemnitz, Germany
e-mail: benner@mathematik.tu-chemnitz.de

L. Feng (✉)
Institut für Mikrosystemtechnik, Universität Freiburg, Freiburg, Germany
e-mail: lihong.feng@mathematik.tu-chemnitz.de

P. Benner et al. (eds.), *Model Reduction for Circuit Simulation*,
Lecture Notes in Electrical Engineering, 74, DOI: 10.1007/978-94-007-0089-5_6,

n of the unknown vector $x \in \mathbb{R}^n$ is very large. In general, the A_i, $i = 1, 2, \ldots, m$, form a set of different, nonsymmetric matrices. In some applications [2], only the right-hand sides b_i vary while $A_i \equiv A$ is fixed, which can and should be exploited. In this paper we focus on the latter case, i.e.

$$Ax = b_i, \quad i = 1, 2, \ldots, m. \tag{6.1}$$

If a direct method like a sparse LU or Cholesky decomposition is applicable, then the solution of the linear systems (6.1) can be obtained by sharing the LU/Cholesky decomposition of A and using successive forward/backward solves. Whether or not a sparse direct solver is applicable depends on the size of the matrix, its sparsity pattern and adjacency graph which is highly application-dependent. For details see [3]. Here, we are interested in cases where the coefficient matrix A does not allow the efficient application of a sparse direct solver. In this situation, iterative methods are to be used. For this purpose, any standard solver can be applied, e.g., depending on the structure of A, the generalized conjugate residual (GCR) method, the generalized minimized residual (GMRES) method, the quasi minimal residual method (QMR), the preconditioned conjugate gradient (PCG) method are possible choices (see, e.g., [17] for hints on the choice of method). Doing this in a naive way, the chosen method is started anew for each right-hand side vector b_i. Since the linear systems of equations in (6.1) share the same coefficient matrix A, we try to exploit the information derived from the previous system's solution to speed up the solution of the next system. Recycling is one possible choice here which turns out to be a suitable approach for the applications in MOR we have in mind. Due to a quite easy mechanism for recycling, we focus here on GCR. That is, the recycling algorithms proposed in this paper reuse the descent vectors in GCR for the solution of the previous systems as the descent vectors for the solution of the current system. This way, the computation of many matrix–vector products can be saved as compared to restarting GCR for each subsystem. The algorithm has the same convergence property as the standard GCR method, that is: for each system, in exact arithmetic the approximate solution computed at each iteration step minimizes the residual in the affine subspace comprised of initial guess and descent vectors.

Note that very recently, other Krylov subspace methods have been suggested for recycling. We comment on GMRES-based recycling in Sect. 6.4 and have further investigated this in [9]. CGS- and BiCGStab-recycling is suggested in [1], but the investigation of these new variants for our purposes is beyond the scope of this paper.

However, the number of the recycled descent vectors may become too large if there are many systems in the sequence. If the size of A is very large, then storing and computing all the descent vectors requires very large memory and also slows down the speed of computation due to orthogonalization needs. Therefore, we propose a restarting technique which limits the number of recycled descent vectors so that the required workspace can be fixed a priori.

Conventional block iterative methods (see [12, 18] and references therein) cannot be applied to (6.1) if the b_i are supplied successively (like in the MOR applications to be discussed), since these methods require that the right-hand sides are available simultaneously. A new method for solving systems $A_i x_i = b_i$ is proposed in [16], where the focus is on varying A_i. This approach turns out to be quite involved for systems with the same coefficient matrix A as in (6.1). The proposed algorithms are simpler and easier to implement.

In the next section, we propose various recycling algorithms based on standard GCR. We discuss the convergence properties of these algorithms. In Sect. 6.3, we review the MOR method from [2] for circuits with massive inputs and outputs. We show where sequences of linear systems as in (6.1) arise in this context and how the recycling algorithms are applied. In Sect. 6.4, we show simulation results of the proposed algorithms and the standard GCR method when applied to MOR of a circuit with numerous inputs and outputs. The recycling algorithms are compared with GCR by the number of matrix–vector products needed for solving each system. We also compare recycled GCR with GMRES-based recycling as suggested in [20]. Conclusions are given in the end.

6.2 Methods Based on Recycling Krylov Subspaces

In this section, we first introduce standard GCR and discuss some of its properties. Based on this method, we propose several recycling algorithms for solving sequences of linear systems as in (6.1). In addition, we present a convergence analysis of one of the recycling algorithms in exact arithmetic. (A finite precision analysis is beyond the scope of this paper.)

6.2.1 The Standard Generalized Conjugate Residual Method

The standard GCR method is proposed to solve a single system of linear equations $Ax = b$ in [5]. GCR belongs to a kind of descent methods for nonsymmetric systems, which have the general form as below:

Algorithm 1 Descent method

given x_1 (initial guess), compute $r_1 = b - Ax_1$;
set $p_1 = r_1$; $k = 1$;
`while` $\|r_k\| > \text{tol}$ `do`
  $\alpha_k = \frac{(r_k, Ap_k)}{(Ap_k, Ap_k)}$;
  $x_{k+1} = x_k + \alpha_k p_k$;
  $r_{k+1} = r_k - \alpha_k A p_k$;
  compute p_{k+1};
  $k = k + 1$;
`end while`

The parameter α_k is chosen to minimize $\|r_{k+1}\|_2$ as a function of α, so that the residual is reduced step by step. The descent methods differ in computing the new descent vector p_{k+1}. For GCR, the descent vector p_{k+1} is computed so that it satisfies $(Ap_{k+1}, Ap_j) = 0$ for $1 \leq j \leq k$ (see [5] for details; $(y, z) = z^T y$ denotes the Euclidian inner product in $\mathbb{R}_n$). For completeness, we present a version of standard GCR in the following algorithm.

Algorithm 2 GCR

given x_1 (initial guess), compute $r_1 = b - Ax_1$; set $k = 1$;
`while` $\|r_k\|_2 >$ tol do
  $\hat{p}_k = r_k; \hat{q}_k = A\hat{p}_k;$
  `for` $j = 1$ `to` $k - 1$
    $\beta_{kj} = (\hat{q}_k, q_j);$
    $\hat{q}_k = \hat{q}_k - \beta_{kj} q_j;$
    $\hat{p}_k = \hat{p}_k - \beta_{kj} p_j;$
  end `for`
  $q_k = \hat{q}_k / \|\hat{q}_k\|_2, p_k = \hat{p}_k / \|\hat{q}_k\|_2;$
  $\alpha_k = (r_k, q_k);$
  $x_{k+1} = x_k + \alpha_k p_k;$
  $r_{k+1} = r_k - \alpha_k q_k;$
  $k = k + 1;$
end `while`

It is not difficult to prove that the descent vectors p_k in Algorithm 2 satisfy $(Ap_{k+1}, Ap_j) = 0$ for $j = 1, \ldots, k$—this is the effect of the `for`-loop which serves the purpose of $A^T A$-conjugation of the descent vectors. It is also proven in [5] that x_{k+1} minimizes $R(w) = \|b - Aw\|_2$over the affine space $x_1 + \text{span}\{p_1, p_2, \ldots, p_k\}$, where $\text{span}\{p_1, p_2, \ldots, p_k\}$ constitutes a Krylov subspace, i.e. $\text{span}\{p_1, p_2, \ldots, p_k\} = \text{span}\{p_1, Ap_1, \ldots, A^{k-1} p_1\}$.

In the next subsection, we will show how to employ the descent vectors from the previous systems for solving the current system. This way, the standard GCR method can be accelerated and many matrix–vector products can be saved when solving all linear systems in a sequence as in (6.1).

6.2.2 Recycling Methods

In this subsection, we propose three variants of recycling algorithms based on the standard GCR method. The idea of recycling descent vectors in GCR is motivated by the recycling algorithm in [19], where sequences of linear systems $(I + \sigma_i A) = b_i$, $i = 1, 2, \ldots, m$, are solved. Unfortunately, the algorithm in [19] is not convergent for some cases, because the computed x_k does not minimize $R(w) = \|b - \tilde{A}_i w\|_2$ $(\tilde{A}_i = I + \sigma_i A)$ at each iteration step. In the following, we present a convergent recycling algorithm: Algorithm 3. We will further show that for each system $Ax = b_i$, x_k^i generated by Algorithm 3 minimizes $R(w) = \|b_i - Aw\|_2$ at step k.

Algorithm 3 GCR with recycling for $Ax = b_i, i = 1, 2, \ldots m$.

```
NumDirecs = 0; (number of descent vectors p_j, j = 1,2,...)
for i = 1 to m
    given x_1^i (initial guess for the ith linear system), compute r_1^i = b_i - A x_1^i;
    set k = 1;
    while ||r_k^i||_2 > tol do
        if k > NumDirecs then
            p̂_k = r_k^i;
            q̂_k = A p̂_k;
            for j = 1 to k - 1
                β_kj = (q̂_k, q_j);
                q̂_k = q̂_k - β_kj q_j;
                p̂_k = p̂_k - β_kj p_j;
            end for
            q_k = q̂_k/||q̂_k||_2, p_k = p̂_k/||q̂_k||_2;
            NumDirecs = k;
        end if
        α_k^i = (r_k^i, q_k);
        x_{k+1}^i = x_k^i + α_k^i p_k;
        r_{k+1}^i = r_k^i - α_k^i q_k;
        k = k + 1;
    end while
end for
```

In Algorithm 3, every system $Ax = b_i, i = 2, 3, \ldots$, reuses the descent vectors p_j computed from the previous systems. For example, in the beginning of solving system $Ax = b_k$, *NumDirecs* denotes the number of descent vectors computed when solving $Ax = b_i, i = 1, 2, \ldots, k-1$. These descent vectors are reused when computing the approximate solution of $Ax = b_k$. If these descent vectors cannot produce an accurate enough solution, then new descent vectors are computed, followed by the matrix–vector multiplication $q_k = Ap_k$ in the *if* loop. This way, a considerable number of matrix–vector products together with the corresponding conjugations are saved when solving the current system. Altogether the amount of computations can be significantly reduced as compared to starting an iterative solver from scratch for each new system in the sequence. Furthermore, Algorithm 3 has similar convergence properties as standard GCR as is shown in the following. We first state the results and then comment on the proofs.

Corollary 1 *The descent vectors generated by Algorithm 3 are A^TA-orthogonal, that is,* $(Ap_i, Ap_j) = 0, i \neq j$.

Corollary 2 *For each linear system* $Ax = b_i$ in (6.1), x^i_{k+1} *generated by Algorithm 3 minimizes* $R(x) = \|b_i - Ax\|_2$ *over the subspace* $x^i_1 + \mathrm{span}\{p_1, p_2, \ldots, p_k\}$.

Following the corresponding proof of Theorem 3.1 in [5], the proof of Corollary 1 is straightforward. (Essentially, the proof consists of observing that the `for`-loop in Algorithm 3 is modified Gram-Schmidt A^TA-orthogonalization.) Because x^i_{k+1} in Algorithm 3 is constructed analogously to x_{k+1} in Algorithm 2, and using the proof of Theorem 3.1 in [5], Corollary 2 can also be easily proven if the descent vectors $p_j, j = 1, 2, \ldots k$ are A^TA-orthogonal, which is guaranteed by Corollary 1. Due to space limitation, a complete proof is not provided here.

In Algorithm 3, the search space is actually composed of several Krylov subspaces. For example, when solving the second system, the recycled vectors constitute the Krylov subspace

$$K_{n_1}(A, r^1_1) = \{r^1_1, Ar^1_1, \ldots, A^{n_1-1}r^1_1\}.$$

Furthermore, if new descent vectors are computed to obtain the solution of the second system, they constitute a second Krylov subspace

$$K_{n_2}(A, r^2_{n_1+1}) = \{r^2_{n_1+1}, Ar^2_{n_1+1}, \ldots, A^{n_2-1}r^2_{n_1+1}\}.$$

Here, n_i, $i = 1, 2$ are the number of vectors in the Krylov subspaces. That is, the approximation x^2 of $Ax = b_2$ minimizes the residual over

$$x^2_1 + K_{n_1}(A, r^1_1) + K_{n_2}(A, r^2_{n_1+1}),$$

which, if analyzed further, is a nested Krylov subspace since $r^2_{n_1+1}$ is generated using $K_{n_1}(A, r^1_1)$, for details see [4].

Generally, for the ith system, the recycled vectors are all the linearly independent vectors from the previous Krylov spaces $K_{n_1}(A, r^1_1), K_{n_2}(A, r^2_{n_1+1})$, $K_{n_3}(A, r^3_{n_1+n_2+1}), \ldots$. Thus, in general, the number of recycled Krylov vectors increases with the number of linear systems to be solved. We will see in Sect. 6.4 that this causes a problem as often, for each new system, some new vectors have to be added to attain the required accuracy. Thus, the major drawback of Algorithm 3 is: if there are a large number of systems in (6.1), the number of recycled descent vectors generated may become simply too large. Since all the descent vectors are reused for the next system, they all have to be stored till the final system is solved. For very large-scale systems, the required memory resources will be massive. Moreover, the conjugation part of the algorithm will slow down the algorithm. Therefore, we propose Algorithm 4 in which a restarting technique is used. This keeps the number of recycled descent vectors within a maximum limit which can be based upon the available memory and the dimension of the linear systems.

Algorithm 4 Algorithm 3 with restarts.

1. *NumDirecs* = 0; (number of descent vectors p_j, $j = 1, 2, \ldots$)
2. `for` $i = 1$ `to` m

```
3.     if NumDirecs > MaxDirecs
4.         NumDirecs = 0;
5.     end if
6.     perform steps 3.–21. of Algorithm 3;
7. end for
```

Algorithm 4 is equivalent to Algorithm 3, except for the first `if`-branch in the beginning of the algorithm. In the algorithm, *MaxDirecs* is the maximum number of recycled descent vectors, and the recycling is restarted whenever *MaxDirecs* is reached. When recycling is restarted, the algorithm is identical with standard GCR. This restart technique is somewhat heuristic, because the optimal *MaxDirecs* is difficult to determine. Here "optimal" is meant in the sense that the overall number of matrix–vector products is minimized. However, from experimental experience, the algorithm requires the least matrix–vector products if *MaxDirecs* is set to be a little larger than the actual number of iteration steps needed by standard GCR for solving a single system.

Since Algorithm 4 reduces to the standard GCR method whenever it restarts, no vector is recycled for the current system. As information is wasted this way, this can be considered as a drawback of the algorithm. In the following Algorithm 5, we propose another possibility of controlling the number of recycled vectors in Algorithm 3. Defining M to be the number of recycled vectors used in each (except for the first) system, we then use the M descent vectors which are computed when solving the first system as the recycled vectors for each of the following systems. Thus, each system is guaranteed to be able to recycle certain vectors. The second advantage of Algorithm 5 over Algorithm 4 is that M can be set much smaller than *MaxDirecs*; therefore it can outperform Algorithm 4 by saving computations in the orthogonalization process (Step 10–14 in Algorithm 3).

Algorithm 5 Algorithm 3 with fixed number of recycled vectors.

```
1. for i = 1 to m
2.     if i = 1
3.         NumDirecs = 0;
4.     else
5.         NumDirecs = M; (number of fixed recycled vectors)
6.     end if
7.     given x_1^i (initial guess for the ith linear system), compute r_1^i = b_i − A x_1^i;
8.     k = 1;
9.     while ||r_k^i||_2 > tol do
10.        if k > NumDirecs
11.            perform steps 7.–14. of Algorithm 3;
12.        end if
13.        perform steps 17.–20. of Algorithm 3;
14.    end while
15. end for
```

Algorithm 4 is the restarted version of Algorithm 3, therefore the recycled vectors also constitute a union of several Krylov subspaces. Once the algorithm restarts, new recycled Krylov subspaces will be generated. Algorithm 5 always recycles the same search space, which is generated when solving the first linear system(s) (in case M or more vectors are needed for solving $Ax = b_1$, these are taken from $K_{n_1}(A, r_1^1)$, otherwise some Krylov vectors from the next Krylov subspaces are also included).

It is known that GCR may break down in some cases. However, if the coefficient matrix is positive real, i.e., $A + A^T$ is positive definite, then it can be shown that no breakdown occurs [5]. In the following section, we will discuss this issue for those linear systems of equations that we are interested in, that is, system matrices A arising in the context of model reduction algorithms for circuit simulation.

6.3 Application to Model Order Reduction

In this section, we review the model order reduction (MOR) problem for linear systems. Based on this, the method SPRIMA [2] for MOR of circuits with massive inputs and outputs is briefly described. We show that in SPRIMA, successive systems of linear equations as in (6.1) need to be solved.

MOR [7, 15] has been widely applied to the fast numerical simulation of large-scale systems in circuit design. Using MOR, the original large dimensional system is approximated by a system of (much) smaller dimension. If the approximation error is within a given tolerance, only the small system needs to be simulated, which will in general take much less time and computer memory than the original large system would do. Here, we consider linear, time-invariant descriptor systems of the form

$$C\dot{x}(t) + Gx(t) = Bu(t), \quad y(t) = L^T x(t) \tag{6.2}$$

with system matrices $C, G \in \mathbb{R}^{n \times n}, B \in \mathbb{R}^{n \times \ell}$, and $L \in \mathbb{R}^{n \times r}$. Here, $x(t) \in R^n$ denotes the unknown descriptor vector of the system, while $u(t)$ and $y(t)$ stand for ℓ inputs and r outputs, n is the order of the system. Conventional moment-matching MOR methods, as, e.g., PRIMA [15] derive the reduced-order model by computing a projection matrix V having orthonormal columns and satisfying

$$\begin{aligned} \text{range}\{V\} = \text{span}\{&(s_0C + G)^{-1}B, -(s_0C + G)^{-1}C(s_0C + G)^{-1}B, \\ &\ldots, (-(s_0C + G)^{-1}C)^p(s_0C + G)^{-1}B\}, \quad p \ll n. \end{aligned} \tag{6.3}$$

The reduced-order model is obtained by using the approximation $x \approx Vz$, leading to the projected system

$$\hat{C}\dot{z}(t) + \hat{G}z(t) = \hat{B}u(t), \quad \hat{y}(t) = \hat{L}^T z(t), \tag{6.4}$$

where $\hat{C} = V^T CV, \hat{G} = V^T GV \in \mathbb{R}^{q\times q}, \hat{B} = V^T B \in \mathbb{R}^{q\times \ell}, \hat{L} = V^T L \in \mathbb{R}^{q\times r}$. The matrix $V \in \mathbb{R}^{n\times q}$ with $q \ll n$ depending on p and ℓ satisfies $V^T V = I$. The new descriptor vector is $z(t) \in \mathbb{R}^q$.

If there are many inputs and outputs in (6.2), conventional MOR methods as the one described above are inefficient in obtaining a reduced-order model. It is not difficult to see from (6.3) that each term in the Krylov subspace is given as a matrix with l columns, where l is the number of inputs as well as the number of columns in B. If the number of linearly independent columns in B is very large, then the number q of columns of the projection matrix V increases very quickly. This may cause that a system fails to be reduced, the dimension of the reduced-order model can become as large as that of the original system.

Several methods have been proposed for MOR of systems with massive inputs/outputs, including the ones discussed in [2, 8, 13, 14]. One of the newly proposed methods is SPRIMA [2]. When there is a large number of inputs, SPRIMA reduces the original system (6.2) by decoupling it into l single-input systems. The basic idea is described in the following. Based on the superposition principle for linear systems, the output response of the multi-input multi-output system (6.2) can be obtained by the sum of the output responses of the following single-input systems:

$$\begin{aligned} C\dot{x}(t) + Gx(t) &= b_i u_i(t), \\ y_i(t) &= L^T x(t), \end{aligned} \quad \text{for } i = 1, 2, \ldots, \ell, \tag{6.5}$$

i.e., $y(t) = y_1(t) + y_2(t) + \cdots + y_\ell(t)$, with b_i being the ith column of B.

Since PRIMA [15] is efficient for MOR of single-input systems, by applying PRIMA to each single-input system in (6.5), we can get approximate output responses $\hat{y}_1, \hat{y}_2, \ldots, \hat{y}_l$ of $y_1, y_2, \ldots, y_\ell$ from the reduced-order models of the single-input systems

$$\begin{aligned} \hat{C}\dot{z}(t) + \hat{G}z(t) &= \hat{b}_i u(t), \\ \hat{y}_i(t) &= \hat{L}^T z(t), \end{aligned} \quad \text{for } i = 1, 2, \ldots, \ell. \tag{6.6}$$

The output response of system (6.2) is thus approximated by $y(t) \approx \hat{y}_1(t) + \hat{y}_2(t) + \cdots + \hat{y}_\ell(t)$.

Based on the above considerations, we can see that by decoupling the original system with numerous inputs into systems with a single input, SPRIMA [2] overcomes the difficulties of PRIMA. As can be seen from (6.6), we actually create several reduced-order models, but the final output of the original system can be computed by the superposition of all the reduced-order models. Moreover, the final transfer function of the original system can also be represented by the transfer functions of the reduced models [2].

As has been introduced in (6.3), for the ith single-input system in (6.6), PRIMA requires the matrices V_i to be computed as

$$\begin{aligned} \text{range}\{V_i\} = \text{span}\{&(s_0C + G)^{-1}b_i, -(s_0C + G)^{-1}C(s_0C + G)^{-1}b_i, \\ &\ldots, (-(s_0C + G)^{-1}C)^{p_i}(s_0C + G)^{-1}b_i\}. \end{aligned} \tag{6.7}$$

It is clear from (6.7) that in order to compute V_i, $i = 1, \ldots, \ell$, we have to solve $\sum_{i=1}^{l} p_i + \ell$ systems of linear equations with identical coefficient matrix $s_0C + G$, but different right-hand sides. In [2], it is proposed to use the (sparse) LU decomposition $s_0C + G$ to solve systems of medium dimension. For very large-scale systems, standard iterative solvers are employed to solve them one by one. For circuit systems with numerous inputs/outputs, thousands of such systems are produced during the MOR process, the computational complexity in solving them increases much too fast. Therefore, efficient computational algorithms for solving successive systems of linear equations are very important to improve the whole MOR process. Thus, we propose to use the recycling algorithms proposed in Sect. 6.2 to solve the sequence of systems arising from (6.7).

Please note that for each single-input system in (6.5), MOR methods other than PRIMA can also be employed to derive the reduced single-input systems in (6.6). For example, one-sided rational Krylov methods [11] (which allow multiple expansion points) are possible here.

It remains to discuss the breakdown issue in GCR for the linear systems of equations encountered here. If RLC circuits are modeled by modified nodal analysis (MNA), the matrices in (6.5) have the structure

$$C = \begin{bmatrix} C_1 & 0 \\ 0 & C_2 \end{bmatrix}, \quad G = \begin{bmatrix} G_1 & G_2 \\ -G_2^T & 0 \end{bmatrix},$$

where G_1, C_1 are symmetric positive semidefinite and C_2 is symmetric positive definite [10]. As a consequence, both G, C and thus $A = s_0C + G$ have a positive semidefinite symmetric part for any real s_0. (Similar conclusions can be drawn for complex s_0 using complex conjugate transposes.) Unfortunately, the no-breakdown condition in [5] requires A to be positive real, i.e., its symmetric part to be positive definite. But from the structure, we can only infer that the symmetric part of A is positive semidefinite. Although we have not encountered breakdown in any of the numerical examples performed for RLC circuit equations, the possibility of breakdown cannot be excluded. As we are fairly close to the necessary positive realness condition for A, there is hope that breakdown will only be encountered in pathological cases. Moreover, in many situations (if the circuit neither contains voltage sources nor a cutset consisting of inductances and/or current sources only, see [6]), it can be shown that $s_0C_1 + G_1$ is positive definite; in these cases, $A = s_0\, C + G$ will be positive real and thus, breakdown cannot occur.

6.4 Simulation Results

In this section, we illustrate the efficiency of the proposed recycling algorithms when used within SPRIMA. As a test case, we employ an RLC circuit with $\ell = 854$ inputs and $r = 854$ outputs. The dimension of the system is fairly small,

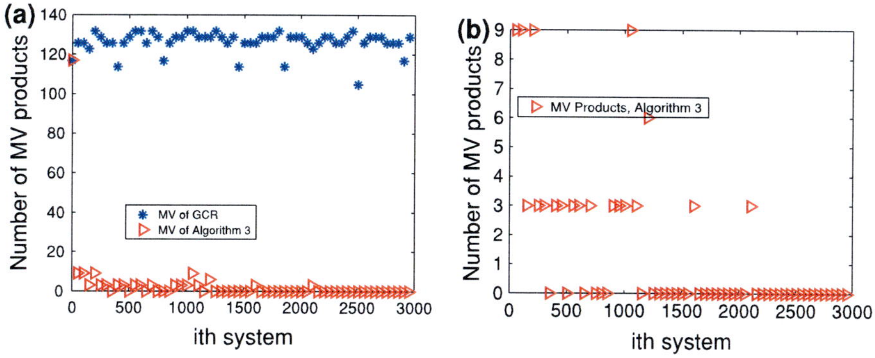

Fig. 6.1 Algorithm 3: ♯ of matrix–vector products. **a** All systems and **b** all but 1st system

$n = 1{,}726$, but it allows to demonstrate the effects of recycling. The model was generated with the circuit simulator TITAN.[1]

We compare Algorithms 3–5 with the standard GCR method (Algorithm 2). If we set $p_i \equiv 19$ in (6.7), then $m = (p_i + 1)\ell = 20 \times 854 = 17{,}080$ systems of linear equations have to be solved in order to obtain the projection matrices V_i, $i = 1, 2, \ldots, \ell$. In this case, $p_i = 19$ is necessary to obtain the desired accuracy of the reduced-order model. As for a preconditioner, we employ an incomplete LU factorization with threshold 0.01 (in MATLAB® notation: `ilu(A,0.01)`) within all algorithms.

We first compare standard GCR with Algorithm 3 by solving the first 3,000 systems. (Similar results for all the systems are also available.) The number of matrix–vector (MV) products for solving each system is plotted in Fig. 6.1. The number of MV products required by Algorithm 3 is much less than that of GCR. Only very few MV products are required after solving the first several systems. Starting from the 1,000 th system, there are almost no MV products necessary.

Figure 6.1 suggests that Algorithm 3 is much more efficient than GCR. However, this would only be true if n would be significantly larger than the number of systems to be solved (which is not the case in the example considered here). Figure 6.2 illustrates that in this example, the number of recycled vectors reaches n before all systems are solved so that storing all recycled vectors need a full $n \times n$ array which is not feasible for larger problems. In order to solve this problem, we propose to use Algorithm 4 or Algorithm 5.

Figures 6.3 and 6.4a compare the MV products required by GCR with the corresponding numbers resulting from Algorithm 4 and Algorithm 5, respectively. Figure 6.3 shows that although Algorithm 4 needs more MV products than Algorithm 3, it is applicable to much larger systems as the number of recycled vectors remains limited. Figure 6.4a illustrates that Algorithm 5 can save MV

[1] TITAN was then developed by Qimonda AG, Neubiberg (Germany); after insolvency of Qimonda in 2009, TITAN is now owned by Infineon Technologies AG, Neubiberg.

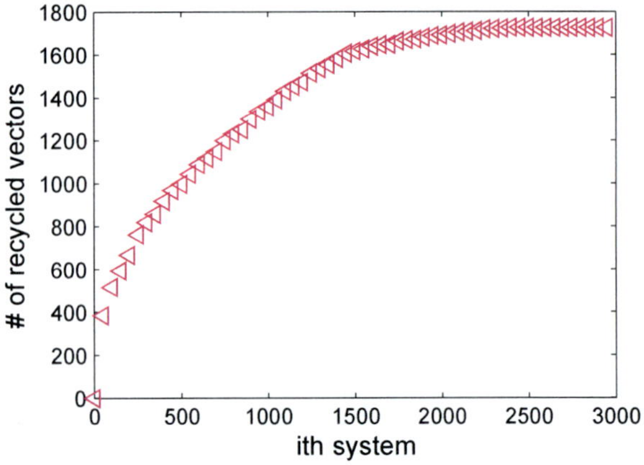

Fig. 6.2 Algorithm 3: ♯ of recycled vectors

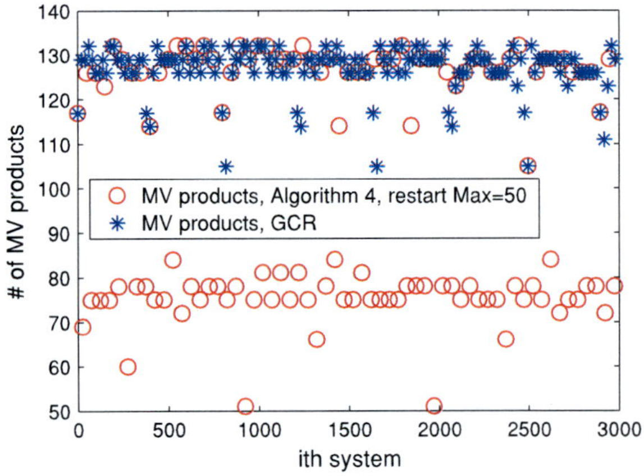

Fig. 6.3 Algorithm 4: ♯ of matrix–vector products

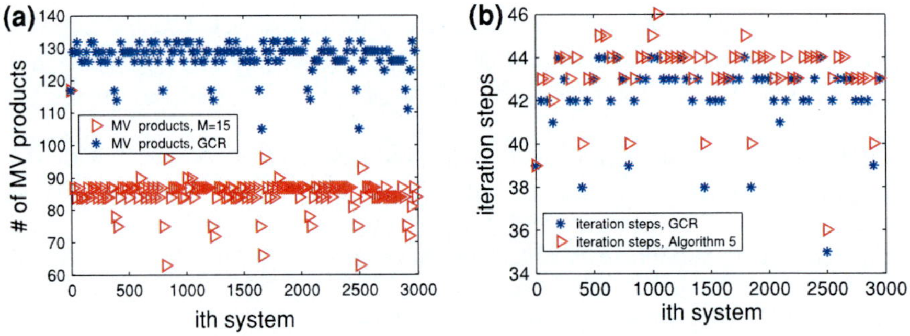

Fig. 6.4 Results for Algorithm 5. **a** ♯ of matrix–vector products **b** ♯ of iteration steps

products for each system, whereas Algorithm 4 cannot save MV products for all the systems as is shown in Fig. 6.3, because Algorithm 4 falls back on GCR whenever it restarts. One can also conclude from Fig. 6.4b that the number of iteration steps used in Algorithm 5 is almost the same as that of GCR for solving each system. Notice that Algorithm 5 uses $M = 15$ recycled vectors for each system; therefore, it actually saves $M = 15$ MV products for each system in the standard form as 1 MV product is computed per iteration. However, if incomplete LU preconditioning is used, there will be three MV products at each iteration step in GCR: one is due to the coefficient matrix itself, the other two are due to the backward/forward solves employed by the preconditioner. Therefore, $M = 15$ iteration steps actually lead to 45 MV products.

In Table 6.1, we list the MV products saved by Algorithms 4 and 5 for solving several groups of systems of linear equations. It is clear that the recycling algorithms can save many MV products when solving long sequences of systems, and therefore can improve the overall efficiency of MOR.

In order to show the accuracy of the reduced-order model obtained by applying the recycling algorithms, we compare the simulation results of the reduced-order model with those of the original model in Fig. 6.5a, b. We use step functions at all the terminals as the input signals and observe the output response at the first terminal. The reduced model is derived by using Algorithm 5. The output responses of the reduced model and the original system at the first terminal are plotted in Fig. 6.5a and the maximum relative error between them is 4.8×10^{-6}.

Table 6.1 Number of MV products saved by Algorithms 4 and 5

No. of systems solved	MV saved by Algorithm 4	MV saved by Algorithm 5
500	13,320	21,288
1,000	26,337	42,381
2,000	52,530	84,354
3,000	78,630	126,039

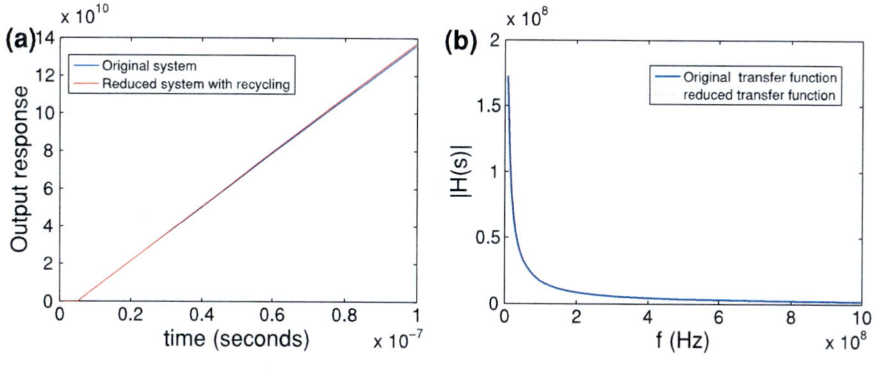

Fig. 6.5 Accuracy of the reduced model. **a** Comparison of the output response and **b** comparison of the transfer function

We have similar accuracy for the output signals at the other terminals. The magnitudes of the transfer functions for both systems are shown in Fig. 6.5b. The transfer function shown here is the mapping from the input signal at the first terminal to the output response at the same terminal. The relative error (computed point-wise at 201 frequency samples) of the magnitude of the reduced transfer function is around 10^{-10}, therefore the two waveforms are indistinguishable in the figure. The accuracy of the transfer functions at any other two terminals is also acceptable, with errors usually smaller than 10^{-4}.

A new recycling method based on GMRES is proposed in [20]. The basic idea of recycling in this method, called MKR_GMRES, is quite similar to Algorithm 3. The main difference is that MKR_GMRES uses GMRES as the basic algorithm for recycling, the recycled vectors are the orthogonal basis vectors generate by the underlying Arnoldi process. When solving the ith system, the Arnoldi vectors generated by all the previous linear systems are reused to generate a "better" initial solution. If the initial solution is not accurate, new Arnoldi vectors are computed and the solution is generated in the augmented Krylov subspace spanned by all the Arnoldi vectors. Because all the orthogonal Arnoldi vectors are recycled, therefore after solving a certain number of systems, the number of new Arnoldi vectors becomes less and less. This can be observed from the number of MV products used for solving each linear system (illustrated in Fig. 6.6a). It has a similar behavior as Algorithm 3 which recycles all the previously computed descent vectors. The method also has a similar problem as Algorithm 3, that is the dimension of the recycled Krylov subspace increase fast, see Fig. 6.6b. Since each new Arnoldi vector has to be orthogonalized to all the previous ones, the orthogonalization process becomes slower and slower, which affects the efficiency of the whole algorithm. Although Algorithms 4 and 5 are somewhat heuristic, they can remedy the the weakness of Algorithm 3, and also work quite well for

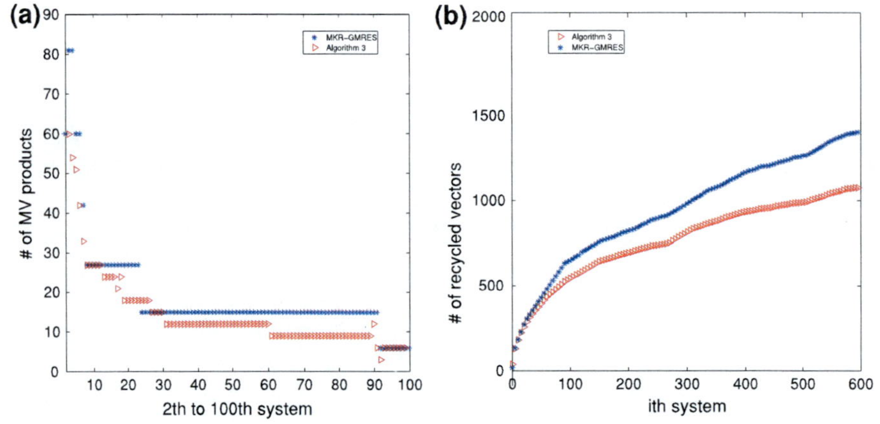

♯ of matrix-vector products ♯ of recycled vectors

Fig. 6.6 Results of MKR_GMRES. **a** ♯ of matrix–vector products **b** ♯ of recycled vectors certain systems. It is also possible to construct recycling method based on nested GMRES. Here, "nested" means the GMRES method is nested inside some other iteration solver and contributes to the inner iterations. We have studied this recycling technique in [9]. In this paper we have applied the algorithm to parametric model order reduction in system level simulation for Microelectromechanical systems (MEMS). The outer iteration of the recycling method is GCRO proposed in [4], and the inner iterations are done by GMRES. The recycled subspace is generated by Harmonic Ritz vectors of the coefficent matrix A and does not inflate with the number of linear systems as happens in the algorithms here. The original version of the recycling algorithm in [9] is proposed in [16], where the theoretical background of the method is analyzed in detail.

6.5 Conclusions

We have proposed variants of recycling descent vectors in GCR for solving sequences of linear systems with varying right-hand sides. The recycling methods are efficient for solving sequences of large-scale linear systems successively. The methods discussed here are limited to solving systems with identical coefficient matrix and different right-hand sides. The techniques proposed for controlling the number of recycled vectors are heuristic. Further research will focus on more robust recycling techniques (such as thick restarting) and on solving systems with different coefficient matrices.

Acknowledgements This research was supported by the Alexander von Humboldt-Foundation and by the Research Network *SyreNe—System Reduction for Nanoscale IC Design*, funded by the German Federal Ministry of Education and Science (BMBF), grant no. 03BEPAE1. Responsibility for the contents of this publication rests with the authors. We would like to thank Timo Reis (TU Berlin) for helpful discussions on the positive realness conditions for matrices arising from RLC circuits.

References

1. Ahuja, K.: Recycling Bi-Lanczos algorithms: BiCG, CGS, and BiCGSTAB. Master's thesis, Virginia Tech, Blacksburg, VA (2009)
2. Benner, P., Feng, L.H., Rudnyi, E.B.: Using the superposition property for model reduction of linear systems with a large number of inputs. In: Proceedings of the 18th International Symposium on Mathematical Theory of Networks and Systems, 12 (2008).
3. Davis, T.A.: Direct methods for sparse linear systems. Fundamentals of Algorithms, vol 2. Society for Industrial and Applied Mathematics (SIAM), Philadelphia (2006)
4. de Sturler, E.: Nested Krylov methods based on GCR. J. Comput. Appl. Math. **67**(1), 15–41 (1996)
5. Eisenstat, S.C., Elman, H.C., Schultz, M.H. (1983) Variational iterative methods for nonsymmetric systems of linear equations. SIAM J. Numer. Anal. **20**(2):345–357

6. Estévez-Schwarz, D., Tischendorf, C.: Structural analysis of electrical circuits and consequences for MNA. Int. J. Circuit Theory Appl. **28**(2), 131–162 (2000)
7. Feldmann, P., Freund, R.: Efficient linear circuit analysis by Padé approximation via the Lanczos process. IEEE Trans. Comput.-Aided Des. Integr. Circuits Syst. **14**, 639–649 (1995)
8. Feldmann, P., Liu, F.: Sparse and efficient reduced order modeling of linear subcircuits with large number of terminals. In: Proceedings of the International Conference on Computer-Aided Design, pp. 88–92 (2004)
9. Feng, L., Benner, P., Korvink, J.G.: Parametric model order reduction accelerated by subspace recycling. In: Proceedings of the 48th IEEE CDC and 28th Chinese Control Conf., Shanghai, P.R. China, December 16-18, 2009, pp. 4328–4333 (2009)
10. Freund, R.: SPRIM: structure-preserving reduced-order interconnect macromodeling. In: Tech. Dig. 2004 IEEE/ACM Intl. Conf. CAD, IEEE Computer Society Press, Los Alamitos, pp. 80–87 (2004)
11. Grimme, E.J.: Krylov projection methods for model reduction. Ph.D. thesis, University of Illinois, Urbana-Champaign (1997)
12. Kilmer, M., Miller, E., Rappaport, C.: A QMR-based projection technique for the solution of non-hermitian systems with multiple right hand sides. SIAM J. Sci. Comput. **23**(3), 761–780 (2001)
13. Li, P., Shi, W.: Model order reduction of linear networks with massive ports via frequency-dependent port packing. In: Proceedings of the Design Automation Conference, pp. 267–272 (2006)
14. Nakhla, N., Nakhla, M., Achar, R.: Sparse and passive reduction of massively coupled large multiport interconnects. In: Proceedings of the International Conference on Computer-Aided Design, pp. 622–626 (2007)
15. Odabasioglu, A., Celik, M., Pileggi, L.: PRIMA: passive reduced-order interconnect macromodeling algorithm. IEEE Trans. Comput.-Aided Des. Integr. Circuits Syst. **17**(8), 645–654 (1998)
16. Parks, M.L., de Sturler, E., Mackey, G., Johnson, D.D., Maiti, S.: Recycling Krylov subspaces for sequences of linear systems. SIAM J. Sci. Comput. **28**(5), 1651–1674 (2006)
17. Saad, Y.: Iterative Methods for Sparse Linear Systems, 2nd edn. SIAM Publications, Philadelphia (2003)
18. Simoncini, V., Gallopoulos, E.: An iterative method for nonsymmetric systems with multiple right-hand sides. SIAM J. Sci. Comput. **16**(4), 917–933 (1995)
19. Telichevesky, R., Kundert, K., White, J.: Efficient AC and noise analysis of two-tone RF circuits. In: Proceedings of the Design Automation Conference, pp. 292–297 (1996)
20. Ye, Z., Zhu, Z., Phillips, J.R.: Generalized Krylov recycling methods for solution of multiple related linear equation systems in electromagnetic analysis. In: Proceedings of the Design Automation Conference, pp. 682–687 (2008)

Chapter 7
Data-Driven Parameterized Model Order Reduction Using z-Domain Multivariate Orthonormal Vector Fitting Technique

Francesco Ferranti, Dirk Deschrijver, Luc Knockaert and Tom Dhaene

Abstract Efficient real-time design space exploration, design optimization and sensitivity analysis call for Parameterized Model Order Reduction (PMOR) techniques to take into account several design parameters, such as geometrical layout or substrate characteristics, in addition to time or frequency. This chapter presents a robust multivariate extension of the z-domain Orthonormal Vector Fitting technique. The new method provides accurate and compact rational parametric macromodels based on numerical electromagnetic simulations or measurements in either frequency-domain or time-domain. The technique can be seen as a *data-driven* PMOR method.

Keywords Parametric Macromodels · Vector Fitting · Approximation Algorithm · Rational Functions · Least-Squares

7.1 Introduction

Nowadays, full-wave electromagnetic methods [9, 11, 17] are widely used to simulate a variety of complex electromagnetic systems and are considered to be essential

F. Ferranti (✉) · D. Deschrijver · L. Knockaert · T. Dhaene
Department of Information Technology (INTEC), Ghent University-IBBT, 9000, Ghent, Belgium
e-mail: francesco.ferranti@intec.ugent.be

D. Deschrijver
e-mail: dirk.deschrijver@intec.ugent.be

L. Knockaert
e-mail: luc.knockaert@intec.ugent.be

T. Dhaene
e-mail: tom.dhaene@intec.ugent.be

P. Benner et al. (eds.), *Model Reduction for Circuit Simulation*,
Lecture Notes in Electrical Engineering, 74, DOI: 10.1007/978-94-007-0089-5_7,

for efficient design. The use of these methods usually results in the computation of a huge number of field (E,H) or circuit (i,v) unknowns, in the frequency-domain or time-domain, although users are usually only interested in a few of them at the input and output ports. These methods provide high accuracy, often at a significant cost in terms of memory storage and computing time. Therefore, Model Order Reduction (MOR) techniques are crucial to reduce the complexity of the model defined by the full-wave numerical method and the computational cost required by simulations, while retaining the important physical features of the original system [3, 7].

Efficient real-time design space exploration, design optimization and sensitivity analysis require the development of accurate parametric broadband macromodels that approximate the dynamic behavior of a system characterized by several design parameters, such as geometrical layout or substrate characteristics, in addition to time or frequency. These applications call for Parameterized Model Order Reduction (PMOR) techniques.

A frequency-domain technique called Multivariate Orthonormal Vector Fitting (MOVF) was presented in [4], to compute accurate rational parametric macromodels, based on parameterized frequency responses with a highly dynamic behavior. This technique can be seen as a *data-driven* PMOR method. Instead of reducing the size of the matrices of a parameterized state-space model directly (*model-based* PMOR), MOVF builds rational parametric macromodels with a reduced model complexity based on a set of input-output data samples. The goal of the macromodeling algorithm is to find a multivariate rational function which approximates a large set of $K + 1$ data samples $\left\{(s, \mathbf{g})_k,\ H(s, \mathbf{g})_k\right\}_{k=0}^{K}$ in a least-squares sense. These data samples depend on the complex frequency $s = j\omega$ and several additional parameters $\mathbf{g} = (g^{(n)})_{n=1}^{N}$ as design variables which describe e.g. the metallizations in an EM circuit (lengths, widths, ...) or the substrate features (thickness, dielectric constant, losses, ...). The proposed approach results in accurate and compact rational parametric macromodels of complex electromagnetic systems. A generalization of MOVF to include parameter derivatives in the modeling process was proposed in [6]. Parameter derivatives provide additional information about the underlying system and can often be simulated at a significantly lower computational cost than additional samples [5, 13, 15, 20]. The inclusion of derivatives can be useful to reduce the required amount of data samples, while preserving the accuracy of the results. In this chapter a new technique, the z-domain Multivariate Orthonormal Vector Fitting (ZD-MOVF) is described, representing the z-domain counterpart of [4]. It is a robust multivariate extension of the z-domain Orthonormal Vector Fitting technique (ZD-OVF) proposed in [14, 16, 21]. A microstrip example confirms the ability of the new algorithm to build parametric macromodels of dynamic systems with a good accuracy.

7.2 Background

In this section we explain the generation of z-domain data starting from frequency-domain or time-domain data and the choice of the λ parameter of the Tustin transform.

7.2.1 Generation of z-Domain Data

Microwave circuits and components can be characterized in frequency-domain or time-domain by numerical electromagnetic simulations or measurements. To obtain the corresponding parameterized z-domain response, $H_d(z, g)$, where z is the complex discrete frequency variable and g is a real design variable, a Tustin (bilinear) transform:

$$s \longrightarrow z = \frac{\lambda + s}{\lambda - s}, \quad \lambda \in \mathbb{R}^+ \tag{7.1}$$

can be used starting from frequency-domain data $H_c(s, g)$:

$$H_c(s, g) \longrightarrow H_d\left(z = \frac{\lambda + s}{\lambda - s}, g\right) \tag{7.2}$$

where c stands for continuous and d for discrete. If time-domain data is available, under the hypothesis of a negligible or absent aliasing in the sampling process, the frequency response of a continuous-time system can be computed by applying standard techniques, such as Fast Fourier Transform (FFT) algorithms on the data samples:

$$h_d([n], g) = h_c(nT_s, g) \tag{7.3}$$

where the real sequence $h_d([n], g)$ is equal to the signal in the time domain $h_c(t, g)$ at the equally spaced time samples nT_s and T_s is the sampling period. Once the parameterized frequency response is computed, the Tustin transform (7.1) is used as before. The obtained z-domain data can be normalized by discrete frequency z [14]. Once the parametric macromodel is computed in the z-domain, it can be converted back to the s-domain by using inverse Tustin transform.

7.2.2 Choice of λ of the Tustin Transform

The λ parameter of the Tustin transform can be freely chosen [19] under the constraint that it is not a real pole of the continuous-time system [1]. The numerical example in this letter shows that the algorithm is robust with respect to an arbitrary choice of λ, since its value does not affect the accuracy of the results over a wide range. To avoid harmful numerical conditions, extreme values of λ have to be discarded, such as very low (near zero) or very high (near infinity).

7.3 Parametric Macromodeling

To simplify the notation, the algorithm is only described for bivariate systems. The extension to the full multivariate formulation is straightforward. As in [4], the ZD-MOVF algorithm proposes to represent the parametric macromodel as the ratio of a bivariate numerator and denominator

$$R(z,g) = \frac{N(z,g)}{D(z,g)} = \frac{\sum_{p=0}^{P} \sum_{v=0}^{V} c_{pv} \chi_p(z) \psi_v(g)}{\sum_{p=0}^{P} \sum_{v=0}^{V} \tilde{c}_{pv} \chi_p(z) \psi_v(g)} \tag{7.4}$$

where P and V represent the maximum order of the corresponding basis functions $\chi_p(z)$ and $\psi_v(g)$ in the complex discrete frequency variable z and the real design variable g, respectively. To establish the coefficients c_{pv} and $\tilde{c}_{pv}$ of numerator and denominator in (7.4), the ZD-MOVF algorithm minimizes the Sanathanan–Koerner (SK) cost function [18] on a set of $K+1$ data samples $\{(z,g)_k, H_d(z,g)_k\}_{k=0}^{K}$. SK is an iterative procedure, in the first iteration step of the algorithm ($t=0$) Levi's cost function [12] is minimized to obtain an initial estimate of the coefficients c_{pv} and $\tilde{c}_{pv}$. In the following steps ($t = 1, \ldots, T$) of the SK iteration, the inverse of the previously estimated denominator $D^{(t-1)}(z,g)$ is used as an explicit least-squares weighting factor. A relaxed non-triviality constraint is added as an additional row in the system matrix [8], to avoid the trivial null solution and improve the convergence of the algorithm. Each equation is split in its real and imaginary parts, to ensure that the model coefficients $c_{pv}^{(t)}$, $\tilde{c}_{pv}^{(t)}$ are real. Scaling each column to unity length [7] is suitable to improve the numerical accuracy of the results.

7.4 Choice of Basis Functions

In this section we describe the choice of the basis functions for the discrete frequency and other parameters.

7.4.1 Discrete Frequency-Dependent Basis Functions

Based on a prescribed set of stable poles $\mathbf{a} = \{-a_p\}_{p=1}^{P}$, a set of partial fractions $\chi_p(z,\mathbf{a})$ is chosen, with $\chi_0(z) = 1$. To select the poles two steps are followed: first, they are chosen in the s-domain as complex conjugate pairs with small real parts and the imaginary parts linearly spaced over the frequency range of interest [7] and after that, the Tustin transform (7.1) is applied to map them from s- to z-domain. A linear combination of two fractions is chosen to ensure that the residues of $\chi_p(z,\mathbf{a})$ and $\chi_{p+1}(z,\mathbf{a})$ come in perfect conjugate pairs leading to real-valued time domain responses, i.e.:

$$\chi_p(z,\mathbf{a}) = z(z+a_p)^{-1} + z(z+a_{p+1})^{-1} \tag{7.5}$$

$$\chi_{p+1}(z, \mathbf{a}) = jz(z + a_p)^{-1} - jz(z + a_{p+1})^{-1} \tag{7.6}$$

To improve the numerical stability of the modeling algorithm, the Takenaka–Malmquist basis functions [10] can be used, as shown in [16]:

$$\chi_p(z, \mathbf{a}) = k_p \left(\prod_{i=1}^{p-1} \frac{1 + a_i^* z}{z + a_i} \right) \frac{(1 - z)|1 - a_p|}{(z + a_p)(z + a_{p+1})} \tag{7.7}$$

$$\chi_{p+1}(z, \mathbf{a}) = k_{p+1} \left(\prod_{i=1}^{p-1} \frac{1 + a_i^* z}{z + a_i} \right) \frac{(1 + z)|1 + a_p|}{(z + a_p)(z + a_{p+1})} \tag{7.8}$$

where

$$k_p = k_{p+1} = \frac{\sqrt{2}}{2} \sqrt{1 - |a_p|^2} \tag{7.9}$$

The orthonormal basis functions can improve the conditioning of the system equations and are less sensitive to the choice of the initial poles. Their use ensures a more numerically robust macromodeling procedure [3].

7.4.2 Parameter-Dependent Basis Functions

The parameter-dependent basis functions $\psi_v(g, \mathbf{b})$ are also chosen in partial fraction form as a function of jg, hence in rational form. The set of starting poles $\mathbf{b} = \{-b_v\}_{v=1}^{V}$ is composed by complex pairs with small real parts of opposite sign and imaginary parts linearly spaced over the parameter range of interest, provided that $\psi_0(g) = 1$. A linear combination of two fractions is used to ensure that $\psi_v(g, \mathbf{b})$ and $\psi_{v+1}(g, \mathbf{b})$ are real functions [4]:

$$\psi_v(g, \mathbf{b}) = (jg + b_v)^{-1} - (jg - (b_v)^*)^{-1} \tag{7.10}$$

$$\psi_{v+1}(g, \mathbf{b}) = j(jg + b_v)^{-1} + j(jg - (b_v)^*)^{-1} \tag{7.11}$$

7.5 Example: Double Folded Stub Microstrip Bandstop Filter

The double folded stub microstrip bandstop filter [2] under study is shown in Fig. 7.1. The substrate is 0.1270 mm thick with a relative dielectric constant ε_r equal to 9.9. The scattering parameters of the system are simulated by ADS-Momentum [1][1] in the s-domain and subjected to (7.2), to obtain the corresponding parametrized z-domain response.

[1] *Momentum EEsof EDA*, Agilent Technologies, Santa Rosa, CA.

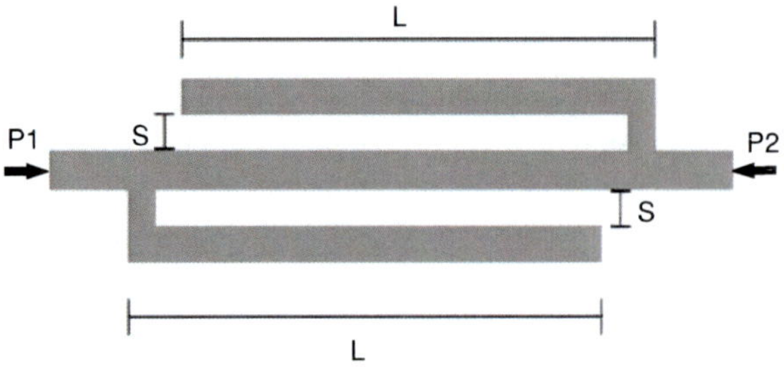

Fig. 7.1 Geometry of the double folded stub microstrip bandstop filter [2]

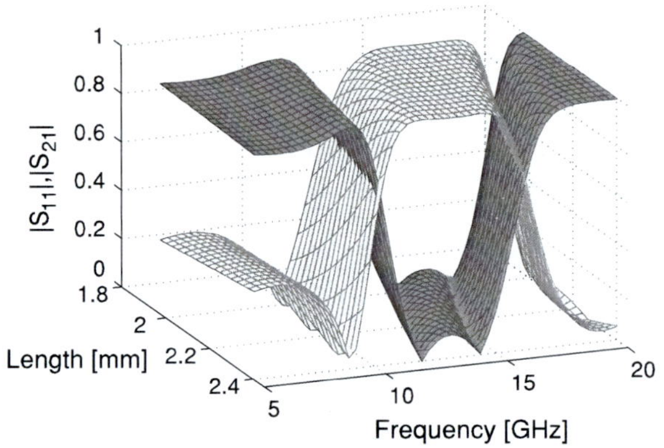

Fig. 7.2 Magnitude of the trivariate macromodels of S_{11} (light grey surface) and S_{21} (dark grey surface) for $S = 0.061$ mm

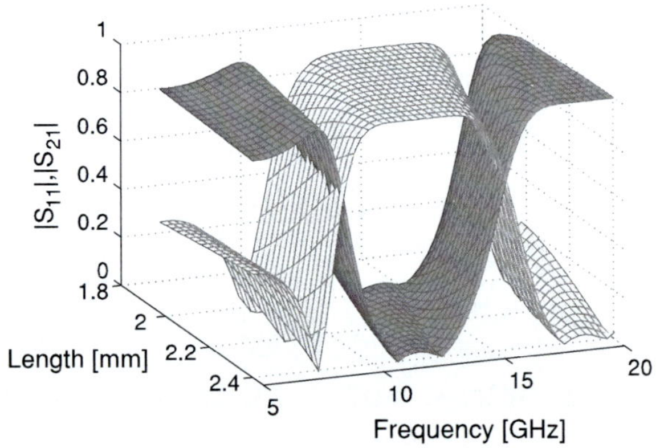

Fig. 7.3 Magnitude of the trivariate macromodels of S_{11} (light grey surface) and S_{21} (dark grey surface) for $S = 0.243$ mm

The parametric macromodels of scattering parameters S_{11} and S_{21} are built as functions of the varying length of each folded segment $L \in [1.98\text{–}2.40\,\text{mm}]$ and varying spacing between a folded stub and the main line $S \in [0.061\text{–}0.243$ mm] over the frequency range (5–20 GHz). The desired model accuracy is set to -60 dB, which corresponds to three significant digits. The initial data grid for S_{11} and S_{21} is of size $14 \times 10 \times 22$ samples (*L*, *S*, *freq*). The corresponding number of poles is chosen 6, 4 and 10 for both scattering parameters. Figures 7.2 and 7.3

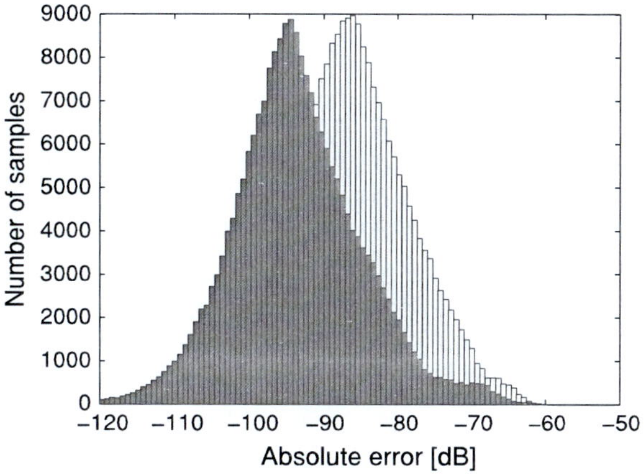

Fig. 7.4 Histogram: error distributions of the trivariate macromodels of S_{11} (light grey) and S_{21} (dark grey) over 226,500 validation samples

show the magnitude of the trivariate macromodels of S_{11} and S_{21} for the minimum and maximum values of the spacing variable S.

To compute the macromodels only 4 and 3 iterations of SK method discussed in Sect. 7.3 are needed and the maximum absolute error in the initial data grid corresponds to −62.84 and −67.54 dB, respectively. To confirm the quality of built macromodels a set of validation data samples is computed on a very dense grid of size $50 \times 30 \times 151$ samples. The histogram in Fig. 7.4 shows the number of validation samples that corresponds to a certain absolute error for both trivariate macromodels. Figure 7.4 shows that they have a good overall accuracy and the maximum absolute error over all the validation samples is bounded by −60.17 and −61.06 dB for S_{11} and S_{21} respectively. The choice of the λ parameter in the Tustin transform (7.1) does not influence the model accuracy over a broad range of values $[10^{-3}–10^{23}]$. It confirms that λ is free to choose and illustrates the robustness of the algorithm.

7.6 Conclusions

This chapter presents a robust multivariate extension of the z-domain Vector Fitting technique [14, 16, 21], for the calculation of accurate and compact parametric macromodels of high-speed components. By combining rational basis functions and the Sanathanan–Koerner least-squares estimator, the robustness of the method is ensured. An example illustrates the capability of the algorithm to model dynamic parameterized frequency responses with a good accuracy. Once the multivariate macromodeling process is completed, the resulting scalable behavior model can efficiently be employed in real-time design space exploration, fast optimization and sensitivity analysis.

Acknowledgements This work was supported by a grant of the Research Foundation-Flanders (FWO-Vlaanderen).

References

1. Al-Saggaf, U.M., Franklin, G.F.: Model reduction via balanced realizations: an extension and frequency weighting techniques. IEEE Trans. Autom. Control. **33**(7):687–692 (1988)
2. Bandler, J.W., Biernacki, R.M., Chen, S.H., Grobelny, P.A., Hemmers, R.H.: Space mapping technique for electromagnetic optimization. IEEE Trans. Microw. Theory Tech. **42**(12): 2536–2544 (1994)
3. Deschrijver, D., Haegeman, B., Dhaene, T.: Orthonormal vector fitting: a robust macromodeling tool for rational approximation of frequency domain responses. IEEE Trans. Adv. Packag. **30**(2):216–225 (2007)
4. Deschrijver, D., Dhaene, T., De Zutter, D.: Robust parametric macromodeling using multivariate orthonormal vector fitting. IEEE Trans. Microw. Theory Tech. **56**(7):1661–1667 (2008)
5. Dhaene, T., Deschrijver, D.: Generalised vector fitting algorithm for macromodelling of passive electronic components. IEE Electron. Lett. **41**(6):299–300 (2005)
6. Ferranti, F., Deschrijver, D., Knockaert, L., Dhaene, T.: Fast parametric macromodeling of frequency responses using parameter derivatives. IEEE Microw. Wirel. Compon. Lett. **18**(12):761–763 (2008)
7. Gustavsen, B., Semlyen, A.: Rational approximation of frequency domain responses by vector fitting. IEEE Trans. Power Deliv. **14**(3):1052–1061 (1999)
8. Gustavsen, B.: Improving the pole relocating properties of vector fitting. IEEE Trans. Power Deliv. **21**(3):1587–1592 (2006)
9. Harrington, R.F.: Field Computation by Moment Methods. Macmillan, NewYork (1968)
10. Heuberger, P.S.C., Van den Hof, P.M.J., Wahlberg, B.: Modelling and Identification with Rational Orthogonal Basis Functions. Springer-Verlag, London (2005)
11. Jin, J.M.: The Finite Element Method in Electromagnetics. 2nd (ed.) Wiley, New York (2002)
12. Levi, E.C.: Complex curve fitting. IRE Trans. Autom. Control. **AC-4**:37–44 (1959)
13. Liu, P., Li, Z-F., Han, G-B.: Application of asymptotic waveform evaluation to eigenmode expansion method for analysis of simultaneous switching noise in printed circuit boards (PCBs). IEEE Trans. Electromagn. Compat. **48**(3):485–492 (2006)
14. Mekonnen, Y.S., Schutt-Ainé, J. E.: Broadband macromodeling of sampled frequency data using z-domain vector fitting. IEEE Workshop Signal Propag. Interconnects. Camogli Genova, 45–48 (2007)
15. Nikolova, N.K., Li, Y., Li, Y., Bakr, M.H.: Sensitivity analysis of scattering parameters with electromagnetic time-domain simulators. IEEE Trans. Microw. Theory Tech. **54**(4):1598–1610 (2006)
16. Nouri, B., Achar, R., Nakhla, M., Saraswat, D.: *z*-Domain orthonormal vector fitting for macromodeling high-speed modules characterized by tabulated data. IEEE Workshop Signal Propag. Interconnects, SPI 2008, Avignon, pp 1–4 (2008)
17. Ruehli, A.E., Brennan, P.A.: Efficient capacitance calculations for three-dimensional multiconductor systems. IEEE Trans. Microw. Theory Tech. **21**(2):76–82 (1973)
18. Sanathanan, C., Koerner, J.: Transfer function synthesis as a ratio of two complex polynomials. IEEE Trans. Autom. Control **8**(1):56–58 (1963)
19. Sou, K.C., Megretski, A., Daniel, L.: Quasi-convex optimization approach to parameterized model order reduction. IEEE Trans. Comput-Aided Des. Integr. Circuits Syst. **27**(3):456–469 (2008)
20. Ureel, J., De Zutter, D.: A new method for obtaining the shape sensitivities of planar microstrip structures by a full-wave analysis. IEEE Trans. Microw. Theory Tech. **44**(2):249–260 (1996)
21. Wong, N., Lei, C.: IIR approximation of FIR filters via discrete-time vector fitting. IEEE Trans. Signal Process. **56**(3):1296–1302 (2008)

Chapter 8
Network Reduction by Inductance Elimination

Mark M. Gourary, Sergey G. Rusakov, Sergey L. Ulyanov and Michael M. Zharov

Abstract A new approach to model order reduction of passive linear circuits is proposed. It is based on the successive elimination of small inductances. Simultaneously with the inductance removal additional capacitances are connected to neighboring branches to provide first order accuracy of the reduced transfer functions. The passivity of the reduced circuit is easily verified. The proposed approach can be applied jointly with the widely used node elimination algorithm TICER.

8.1 Introduction

Model-order reduction (MOR) of linear interconnects is the effective tool to decrease computational efforts in the simulation of integrated circuits (IC) [1].

Fast MOR of RC interconnect can be achieved by the TICER method [2]. The TICER performs the reduction by the successive elimination of circuit nodes. An analysis of the TICER in comparison with other approaches can be found in [3]. TICER usually does not provide the circuit reduction ratio that is achieved in the moment-matching or balanced truncation approaches [4]. Note that in such cases the method can be implemented as a preprocessing step [5] that allows to obtain the essential speedup of more complicated methods.

The TICER elimination algorithm cannot be applied to nodes with connected inductances. Usually the inductances in IC interconnects are neglected and this is considered to be insignificant for the interconnect reduction. But the growth of

M. M. Gourary · S. G. Rusakov (✉) · S. L. Ulyanov · M. M. Zharov
IPPM RAS, 3 Sovetskaya, Moscow, Russia
e-mail: rusakov@ippm.ru

P. Benner et al. (eds.), *Model Reduction for Circuit Simulation*,
Lecture Notes in Electrical Engineering, 74, DOI: 10.1007/978-94-007-0089-5_8,

frequencies and/or power of IC increases the contributions of inductances into interconnect response which leads to a notable error caused by their neglect. Therefore it is desirable to extend the TICER approach to circuits with small inductances.

There are some proposals to take into account nodal inductances in the elimination algorithm. The special case of a node with two RL branches and a number of capacitances is considered in [6, 7]. In [8] high order admittances are obtained by the nodal elimination followed by optimization-based RLC synthesis procedure. Such an approach however seems to be too complicated and time-consuming. Paper [5] presents the rules for merging any two branches connected to the eliminating node. But the rules do not always save first order terms of Laplace admittances which results in an insufficient accuracy in the case of small inductances. One particular merging rule for LR-branch (at small L) and C-branch is equivalent to the inductance neglect.

The aim of this paper is the development of a first order elimination-based algorithm for circuits with small inductances. We propose an approach based on the simultaneous elimination of two circuit variables, i.e. nodal voltage and inductance current. The proposed approach generates a reduced circuit where the nodal inductance is deleted (shorted), and additional capacitances are connected to neighboring branches guaranteeing first order accuracy (Sect. 8.3). The circuit passivity can be verified by the criterion of the positive definiteness of the capacitance matrix. If the eliminated inductance is coupled with other circuit inductances (Sect. 8.4) then the reduced circuit contains couplings between inductances and capacitances (LC-couplings). It is shown that the couplings correspond to additional nonzero entries of the capacitance matrix of the Modified Nodal Analysis (MNA). This implies that the simulation of the reduced circuit with LC-couplings requires small corrections in the simulator code. The rules of inductance elimination in the presence of LC-couplings (Sect. 8.5) are also derived. The description of the algorithm (Sect. 8.6) and numerical experiments (Sect. 8.7) are presented.

8.2 Elimination of RC-Node by TICER

The elimination of an RC node (Fig. 8.1) is based on the expression for the star connection transform in the Laplace domain (we label branches by the same indexes as their outward nodes)

$$\tilde{y}_{ij} = \frac{y_i y_j}{Y_n} = \frac{(g_i + sc_i)(g_j + sc_j)}{G_n + sC_n}. \tag{8.1}$$

Here $y_i = g_i + sc_i$, $Y_n = \sum_{j \in S_n} y_j = G_n + sC_n$ are branch and nodal admittances respectively, S_n is the set of the indexes of outward nodes of RC-branches connected to the node n.

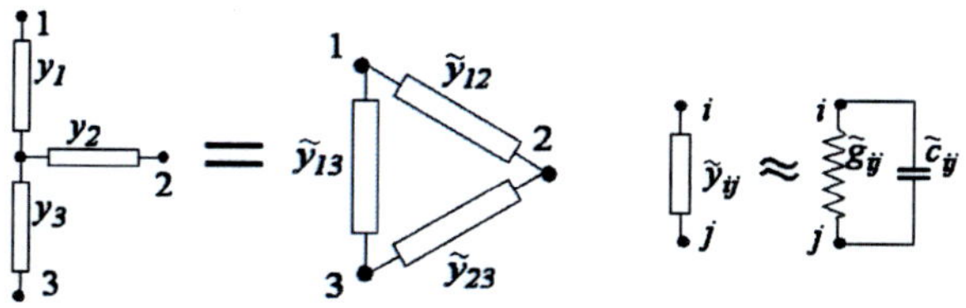

Fig. 8.1 Circuit transform under the elimination of an RC-node

If higher order terms of the Laplace variable s in the Taylor expansion of (8.1) are truncated then one obtains expressions for conductances and capacitances, which must be added to the circuit after node elimination:

$$\tilde{g}_{ij} = \frac{g_i g_j}{G_n}, \quad \tilde{c}_{ij} = \frac{g_i c_j + g_j c_i}{G_n} - \frac{g_i g_j}{G_n{}^2} C_n. \tag{8.2}$$

The method TICER [2] applies (8.2) without the last term in $\tilde{c}_{ij}$, and in this case all capacitances of the reduced circuit are positive. Thus the circuit passivity is provided. However, it has been proved in [9] that in spite of possible negative capacitances the application of (8.2) does not lead to the loss of circuit passivity. Hence, the exact first order expression (8.2) can be applied to the nodal elimination to achieve less error.

8.3 Inductance Elimination

Here we consider the case of a single inductance connected to an RC node (Fig. 8.2). To obtain the reduced circuit we perform simultaneous elimination of two circuit variables: nodal voltage and inductance current.

Let w_i be the current of the ith branch connected to node n

$$w_i = y_i(v_i - v_n), \quad i \in S_n. \tag{8.3}$$

The Kirchhoff Current Low (KCL) equation for node n allows us to obtain the inductance current as

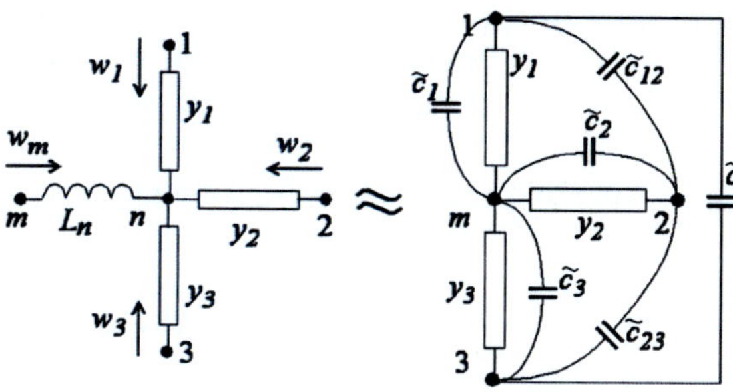

Fig. 8.2 Circuit transform under inductance elimination

$$w_m = -\sum_{j \in S_n} w_j = -\sum_{j \in S_n} y_j(v_j - v_n). \tag{8.4}$$

The inductance voltage is defined by the equation

$$v_n - v_m = sL_n w_m. \tag{8.5}$$

Substituting (8.4) into (8.5) we obtain

$$v_n(1 - sL_n Y_n) = v_m - sL_n \sum_{j \in S_n} y_j v_j. \tag{8.6}$$

The nodal variable derived from (8.6) by neglecting higher order terms is

$$v_n = v_m(1 + sL_n G_n) - sL_n \sum_{j \in S_n} g_j v_j. \tag{8.7}$$

After substitution of (8.7) into (8.3), neglecting higher order terms, and collecting terms in accordance to nodal voltages we obtain

$$w_i = y_i(v_i - v_m) + s\tilde{c}_i(v_i - v_m) + s \sum_{j \in S_n^i} \tilde{c}_{ij}(v_i - v_j), \tag{8.8}$$

where S_n^i is the set S_n without index i, and

$$\tilde{c}_i = -sL_n g_i G_n, \; \tilde{c}_{ij} = sL_n g_i g_j. \tag{8.9}$$

The expression for the ingoing current of the ith node (8.8) corresponds to the following reduced circuit (Fig. 8.2):

(a) node n is eliminated, and the inductance is shorted;
(b) old RC branches are saved in accordance with the first term of (8.8);
(c) new negative capacitances $\tilde{c}_i$ (8.9) are connected to each RC branch to define the current corresponding to the second term of (8.8);
(d) new capacitances $\tilde{c}_{ij}$ (8.9) are inserted between outward nodes of RC branches to define currents corresponding to the third term of (8.8).

Unlike the case of RC node elimination the passivity of the circuit after inductance elimination is not secured. So it is needed to test a passivity condition before the elimination is accepted. It is known that the circuit is passive if its conductance and capacitance matrices are nonnegative definite and the capacitance matrix is symmetric. In our case the conductance matrix is kept unchanged and the capacitance matrix is symmetric by the construction. Hence, the criterion for a symmetric matrix A to be nonnegative definite is applied to the capacitance matrix, i.e. $A_{ii} \geq 0$, $|A_{ij}| \leq \sqrt{A_{ii}A_{jj}}$. It is equivalent to the conditions on branch and nodal capacitances

$$C_i^{\mathrm{NEW}} \geq 0, \quad |c_{ij}^{\mathrm{NEW}}| \leq \sqrt{C_i^{\mathrm{NEW}} C_j^{\mathrm{NEW}}}, \tag{8.10}$$

where $c_{ij}^{\mathrm{NEW}} = c_{ij} + \tilde{c}_{ij}$.

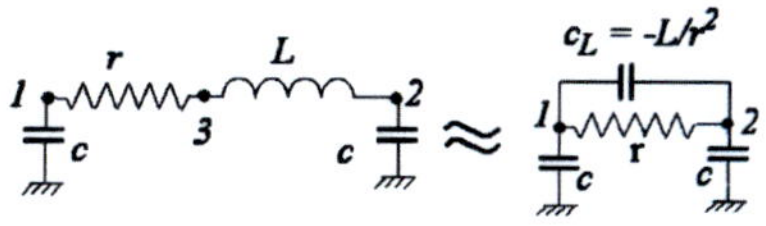

Fig. 8.3 RLC-section reducing

If (8.10) is not valid for the reduced circuit then the elimination must be rejected.

If the inductance L_n has a number of couplings with additional circuit iductances then each coupling must be processed in the same way.

A simple example of an inductance elimination is presented in Fig. 8.3. Conditions (8.10) for the reduced circuit yield $c \geq 2L/r^2$.

8.4 Elimination of Coupled Inductances

Here we consider the case that the inductance L_n is coupled with some other inductance L_k. This coupling is defined by the mutual inductance L_{nk} (Fig. 8.4).

In this case (8.3, 8.4) are valid, (8.5) is replaced by

$$v_n - v_m = sL_n w_m + sL_{nk} w_k \tag{8.11}$$

and the equation for the voltage of inductance L_k must be added:

$$u_k = sL_k w_k + sL_{nk} w_m. \tag{8.12}$$

Then the substitution of (8.4) into (8.11) leads to

$$v_n = v_m(1 + sL_n G_n) - sL_n \sum_{j \in S_n} g_j v_j + sL_{nk} w_k + O(s^2). \tag{8.13}$$

Substituting (8.13) into (8.3) we obtain a first order expression for the nodal current containing an additional term in comparison to (8.8)

$$w_i = \hat{w}_i + s\tilde{\theta}_{(i,m)k} w_k, \tag{8.14}$$

where $\hat{w}_i$ is the right hand side of (8.8),

$$\tilde{\theta}_{(i,m)k} = L_{nk} g_i. \tag{8.15}$$

The substitution of (8.4) into (8.12) leads to an expression for the inductance voltage containing a similar term:

$$u_k = sL_k w_k + s \sum_{j \in S_n} \tilde{\theta}_{(j,m)k} (v_j - v_m). \tag{8.16}$$

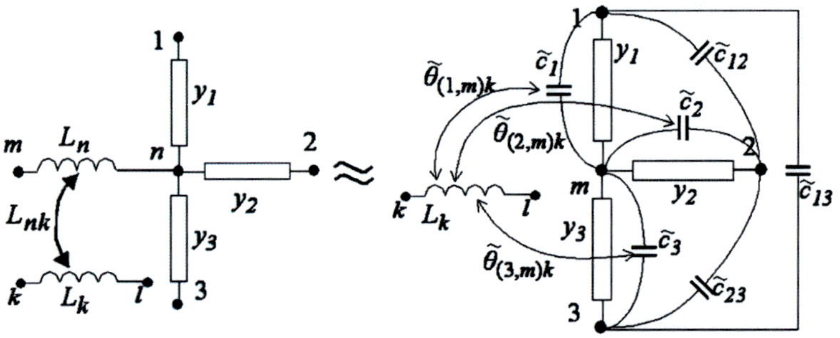

Fig. 8.4 Circuit transform by the elimination of the coupled inductance

The parameter $\tilde{\theta}_{(j,m)k}$ in (8.14, 8.16) corresponds to the coupling (Fig. 8.4) between RC- and L-branches (LC-coupling) that can be defined by the following time-domain expressions for the capacitance current (i_C) and the inductance voltage (u_L):

$$i_C = C\frac{du_C}{dt} + \theta\frac{di_L}{dt}, \quad u_L = L\frac{di_L}{dt} + \theta\frac{du_C}{dt}. \tag{8.17}$$

Our notation $\theta_{(i,m)k}$ refers to kth inductance coupled with the capacitance between nodes m, i. As far as we know there are no physical devices with such interaction. Formally it can be defined in the context of MNA approach as follows.

The MNA conductance and capacitance matrices of the RLCK circuit are

$$G^{\text{MNA}} = \begin{bmatrix} G & I \\ -I^T & 0 \end{bmatrix}, \; C^{\text{MNA}} = \begin{bmatrix} C & \Theta \\ \Theta^T & L \end{bmatrix}. \tag{8.18}$$

Here, submatrix I is the incidence matrix for the inductances. Submatrix Θ is the zero matrix for physically realizable RLCK circuits. With the Gaussian elimination the entries of Θ are replaced by nonzero values. Such nonzero values can be defined by LC-couplings

$$\Theta_{ik} = \sum_{j \in S_i} \theta_{(i,j)k}, \quad \theta_{(i,j)k} = -\theta_{(j,i)k}. \tag{8.19}$$

It can be seen from Fig. 8.4 and (8.14), (8.16) that the coupling of an eliminated inductance is replaced by LC-couplings with each RC-branch of the eliminated node. Due to the symmetry of C^{MNA} the passivity condition can be presented in the form

$$|\Theta_{ik}| \leq \sqrt{L_k C_i}, \quad |\Theta_{jk}| \leq \sqrt{L_k C_j}. \tag{8.20}$$

Standard SPICE-like simulators cannot process LC-couplings. Note that this capability can be easily implemented taking into account the corresponding time-domain expressions (8.17) and the MNA capacitance matrix (8.18).

8.5 Eliminations Under LC Couplings

The possible appearence of LC-couplings due to previous eliminations requires to analyze the case when the inductance to be eliminated has LC-coupling with any RC-branch (Fig. 8.5).

In this case, (8.5) must be replaced by

$$v_n - v_m = sL_n w_m + s\theta_{(k,l)n}(v_k - v_l). \tag{8.21}$$

After transformations similar to (8.6–8.8) we obtain

$$w_i = \hat{w}_i + sg_i\theta_{(k,l)n}(v_k - v_l) = \hat{w}_i + sg_i\theta_{(k,l)m}(v_k - v_i) + s(-g_i\theta_{(k,l)n})(v_l - v_i). \tag{8.22}$$

The current entering node m is defined as

$$w_m = -\sum_{j\in S_n} w_j = -\sum_{j\in S_n} \hat{w}_j + s(-G_n\theta_{(k,l)n})(v_k - v_m) + sG_n\theta_{(k,l)n}(v_l - v_m). \tag{8.23}$$

The last terms in (8.22, 8.23) represent the currents through the capacitances

$$\tilde{c}_{ki} = -\tilde{c}_{li} = g_i\theta_{(k,l)n}, \quad \tilde{c}_{lm} = -\tilde{c}_{km} = G_n\theta_{(k,l)n}. \tag{8.24}$$

Thus, after elimination of the inductance with LC-coupling the reduced circuit contains additional capacitances (8.24) as shown in Fig. 8.5. A special case of an elimination can be considered when the inductance is LC-coupled to one (kth) of its adjacent RC-branches. The resulting circuit contains capacitances

$$\tilde{c}_{ki} = -\tilde{c}_i = g_i\theta_{(k,n)n}\ (i \in S_n^k), \quad \tilde{c}_k = (g_k - G_n)\theta_{(k,n)n}. \tag{8.25}$$

These capacitances do not form additional entries of the capacitance matrix because shunt capacitances (8.9) are always created under inductance elimination.

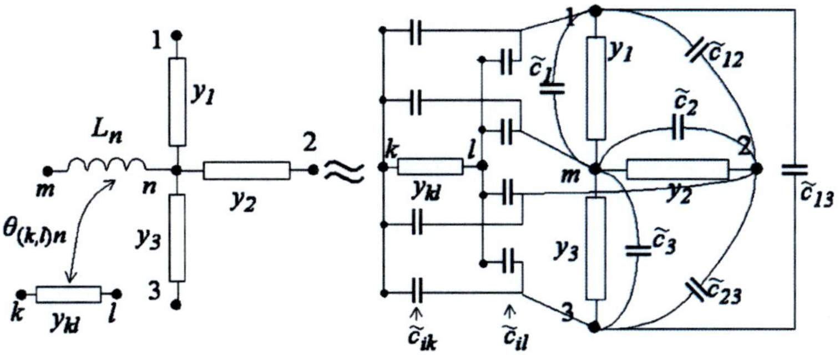

Fig. 8.5 Circuit transform by the elimination of the LC-coupled inductance

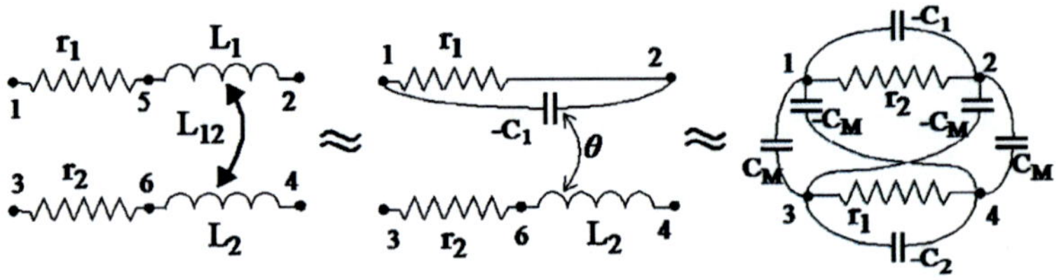

Fig. 8.6 Successive inductance eliminations transform inductively coupled branches into capacitively coupled branches. Here $\theta = L_{12}/r_1$, $c_1 = L_1/r_1^2$, $c_2 = L_2/r_2^2$, $c_M = L_{12}/r_1 r_2$

Note that if the circuit contains two coupled inductances then after their elimination the reduced circuit contains RC branches only. If all inductances of the circuit are eliminated then the reduced RLCK circuit is an RC circuit. An example of a reduction of two coupled LR branches by elimination of both inductances is presented in Fig. 8.6.

In the analysis presented above we ignored the nodes with multiple inductances. But this case does not require a special analysis because the presented algorithms are sufficient to provide the elimination of all small inductances. This can be explained in the following way. Since we do not consider DC undefined circuits containing inductance loops each connected graph of inductances is a tree that always has leaf nodes with only one inductance. Therefore if all inductances of that tree are sufficiently small the single-inductance algorithms will sequentially eliminate them.

LC-couplings must be taken into account in RC-node elimination (8.1, 8.2).

If the mth branch of node n which is to be eliminated is LC-coupled with the kth inductance (Fig. 8.7, $m = 1$), then the branch current is

$$w_m = y_m(v_m - v_n) + s\theta_{(m,n)k} w_k. \tag{8.26}$$

The currents of the other RC-branches are defined by (8.3), and the KCL for node n is

$$s\theta_{(m,n)k} w_k + \sum_{j \in S_n} y_j(v_j - v_n) = 0. \tag{8.27}$$

Substituting v_n derived from (8.27) into (8.2, 8.26), and neglecting higher order terms, we obtain

$$w_i = \sum_{j \in S_n^i} (\tilde{g}_{ij} + s\tilde{c}_{ij})(v_i - v_j) - s\theta_{(m,n)k} \frac{g_i}{G_n} w_k, \quad j \in S_n^m, \tag{8.28}$$

$$w_m = \sum_{j \in S_n^m} (\tilde{g}_{mj} + s\tilde{c}_{mj})(v_m - v_j) + s\theta_{(m,n)k} \frac{G_n - g_m}{G_n} w_k. \tag{8.29}$$

Here $\tilde{g}_{ij}$, $\tilde{c}_{ij}$ are defined by (8.2), and the second right hand terms (8.28, 8.29) correspond to LC-couplings in the reduced circuit (Fig. 8.7):

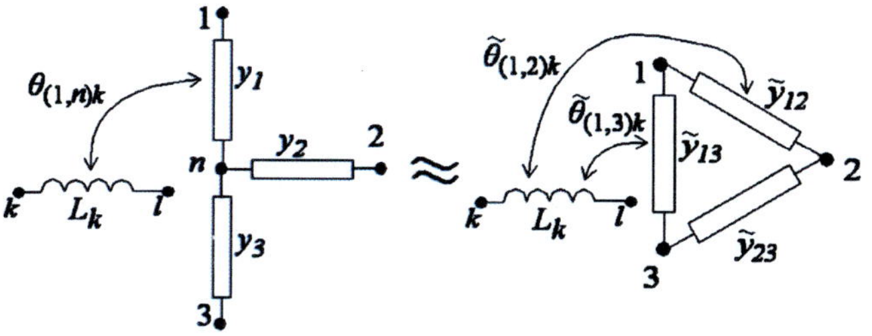

Fig. 8.7 Circuit transform by node elimination under LC-coupling

$$\tilde{\theta}_{(m,i)k} = \theta_{(m,n)k}\frac{g_i}{G_n}. \tag{8.30}$$

Thus after node elimination the branch coupling parameter is distributed among new branches adjacent to node m in proportion to their conductances.

8.6 Algorithmic Aspects

Presented above considerations allow us to extend the TICER reduction algorithm to include nodes with small inductances. The extention requires to adapt important details of the overall TICER algorithm for new types of eliminated nodes. First of all it is needed to provide an error control in the reduction algorithm.

The error due to the first order elimination is defined by the neglected higher order terms. For node elimination (8.2) one can evaluate the mth order term as $\delta_m^{(n)} = O(|s\tau_n|^m)$ [3], where $\tau_n = \tau_n^{RC} = C_n/G_n$ is the RC time-constant of the nth node. In TICER nodes to be eliminated are chosen by the condition $\tau_n < \tau_{\min}$ [3], where $\tau_{\min}$ is evaluated in accordance to the maximum operating frequency of interest $\tau_{\min} = a/f_{\max}$. Factor a is defined by the trade-off between the accuracy and the circuit reduction ratio.

To include the inductance elimination in the general algorithm it is necessary to define a nodal time-constant for the nodes with connected inductance. It can be shown that mth order terms neglected in the inductance elimination (8.7, 8.8) are evaluated as $\delta_m^{(n)} = O(|s\tau_n^{RC}|^k |s\tau_n^{RL}|^l)$, $(k + l = m)$, where τ_n^{RC} is defined earlier, and $\tau_n^{RL} = L_n G_n$ (RL time-constant). Thus, the nodal time-constant in this case can be evaluated as the worst-case value

$$\tau_n = \max(\tau_n^{RC}, \tau_n^{RL}). \tag{8.31}$$

This expression can be applied to each node if we assume $\tau_n^{RL} = 0$ for an RC node.

Another important aspect of TICER that can be extended for inductance eliminations is the restriction of the density of the reduced system. It is achieved by introducing user-supplied maximal nodal degree. The nodal degree is defined as the number of incident resistors (N_R) that provides no more than $N^R_{\text{new}} = N_R(N_R - 1)$ new entries in the conductance matrix. Pure capacitance branches are not considered. We think that it is desirable to take into account the entries in the superposition of conductance and capacitance matrices because simulations both in time and frequency domains require linear combination of the matrices. In this case the maximal number of new entries is evaluated by $N^{RC}_{\text{new}} = N_{RC}(N_{RC} - 1)$, where N_{RC} is the number of incident RC-branches.

The inductance elimination generates new entries (Fig. 8.2) due to capacitances $\tilde{c}_{ij}$ (8.9). The entry is not produced for branches with zero conductances, hence the maximal number of new entries is $N_R(N_R - 1)$. Besides $2N_R$ new entries are produced by each inductance coupled with the eliminated one (Fig. 8.4) in accordance with (8.15). The factor 2 appears because LC-coupling $\theta_{(i,j)k}$ corresponds with two entries $\Theta_{i,k}$, $\Theta_{j,k}$ (8.19). Each LC-coupling (Fig. 8.5) produces $2(N_R + 1)$ new entries due to (8.24). Thus the number of new entries after inductance elimination can be evaluated as

$$N^L_{\text{new}} = N_R(N_{LL} + 1) + 2N_{LC}(N_{RC} + 1). \tag{8.32}$$

Here N_{LL} is the number of inductance couplings of eliminated inductance, N_{LC} is the number of the LC couplings not counting the couplings with incident branches.

In the general case incident branches of an RC node can posess LC couplings with circuit inductances (Fig. 8.7). All couplings with one inductance produce N_R new entries in the matrix Θ. Thus the total number of new entries after RC-node elimination is

$$N^{RC}_{\text{new}} = N_{RC}(N_{RC} - 1) + N_R N_{LC}, \tag{8.33}$$

where N_{LC} is the number of inductances coupled with any RC-branch of eliminated node.

The TICER algorithm [3] extended to include inductance eliminations contains the following operations. The sorted queue of internal nodes with no more than one connected inductance is supported after each elimination step. The fields of the sorting key are the number of new entries (8.32, 8.33) and nodal time constant (8.31). Maximal values of the keys are user-defined data.

The elimination is performed for the first node in the queue by applying either (8.2, 8.30) for an RC-node or (8.9, 8.15, 8.24, 8.25) for the node with inductance. If passivity conditions (8.10, 8.20) are not satisfied then the elimination is ignored and the next node in the queue is processed. The algorithm stops if the queue is empty or all nodes in the queue do not satisfy passivity conditions.

Experiments show that computational efforts of inductance elimination are typically 2–4 times greater than efforts of RC-node (TICER) elimination. This is resulted from more complicated analysis of the circuit structure and the need of passivity verification.

8.7 Numerical Examples

Two circuit examples are presented below to illustrate the proposed inductance elimination algorithms.

The first example is an RLC line containing ten identical sections presented in Fig. 8.3 with parameters $r = 1\,\Omega, L = 0.4\,\text{pH}, C = 1\,\text{pF}$. Note that inductance value is slightly smaller than the maximum value guaranteeing passivity ($L \leq 0.5\,\text{pH}$).

Two reduced circuits were obtained:

- RC line using the elimination of all inductances by the proposed method;
- RC line using the neglect of all inductances.

The frequency- and time-domain response of the original RLC line as well as two reduced circuits are computed. The time-domain simulation was performed under a unit step excitation with t_{rise} = 3 psec.

The reduction error for each method was evaluated as the difference between the responses of the original and reduced circuits (Fig. 8.8). One can see that the elimination error is about tenfold less than the neglect error both in frequency (Fig. 8.8a) and time-domain (Fig. 8.8b).

The second example presents inductively coupled branches shown on the left of Fig. 8.6 with parameters $r_1 = r_2 = 5\,\Omega$, $L_1 = L_2 = 1\,\text{pH}$, $L_{12} = 0.5$ pH. Furthermore, load capacitances $1pF$ connected to nodes 2, 4 are added. Nodes 1 and 4 are assumed as input and output ports respectively. The reduced circuit is obtained in the form shown on the right of Fig. 8.6. Output frequency responses are shown in Fig. 8.9a. Time-domain crosstalk waveforms under an input pulse with unit magnitude and $t_{\text{pulse}} = 20\,\text{psec}, t_{\text{rise}} = 1\,\text{psec}, t_{\text{fall}} = 4\,\text{psec}$ are given in Fig. 8.9b. Note that in spite of large errors in the frequency response of the reduced circuit at $f >$ 100 GHz the error in the time domain crosstalk waveform is sufficiently small because time-constants of eliminated nodes $L_1/r_1 = L_2/r_2 = 0.2\,\text{psec}$ are essentially less then t_{rise} and t_{fall}.

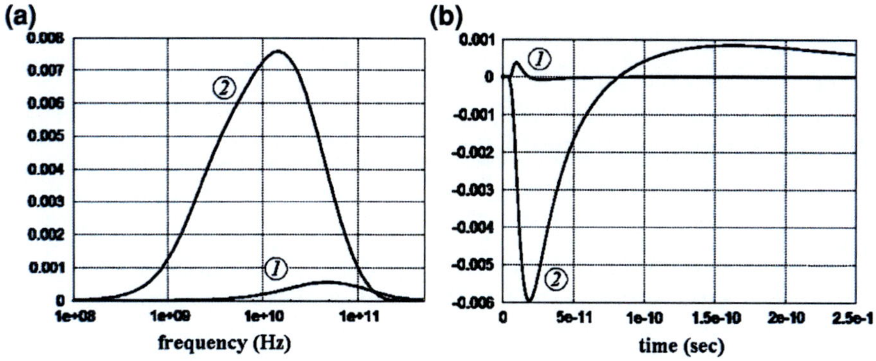

Fig. 8.8 RLC-line reduction errors due to the elimination (1) and the neglect (2)

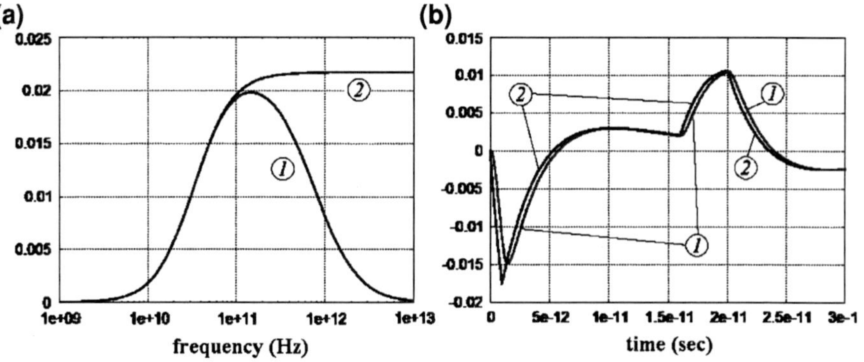

Fig. 8.9 Crosstalk frequency (a) and time-domain (b) responses in inductively coupled branches (1) and in the reduced circuit (2)

8.8 Conclusion

A new class of elimination algorithms for circuits with small inductances is presented in the paper. The proposed algorithms provide a reduction of an arbitrary RLCK circuit preserving first order accuracy of the circuit transfer functions. A passivity condition for the reduced circuit is given, which can be easily verified at each elimination step. Minor corrections in circuit simulators are required to perform the simulation of the reduced circuits. This work was supported by RFBR under grant 08-07-00171a.

References

1. Kamon, M., Marques, N., Massoud, Y., Silveira, L., White, J.: Interconnect analysis: from 3D structures to circuit models. In: IEEE/ACM design automation conference, pp. 910–914, New Orleans (1999)
2. Sheehan, B.N.: TICER: Realizable reduction of extracted RC circuits. In: Proceedings of International Conference on Computer-Aided Design, pp. 200-203, San Jose (1999)
3. Sheehan, B.N.: Realizable reduction of RC networks. IEEE Trans. Comput. Aided Des. Integr. Circuits Syst. **26**(8), 1393–1407 (2007)
4. Heres, P., Schilders, W.: Reduced order modelling—methods and constraints. Math. Ind. **5**, 205–211 (2004)
5. Amin, C.S., Chowdhury, M.H., Ismail, Y.I.: Realizable reduction of interconnect circuits including self and mutual inductances. IEEE Trans. Comput. Aided Des. Integr. Circuits Syst. **24**(2), 271 - 277 (2005)
6. Sheehan, B.N.: Branch merge reduction of RLCM networks. In: Proceedings of International Conference of Computer-Aided Design, pp. 658–664, San Jose (2003).
7. Rong, J., Chen, C.C.-P.: Realizable reduction for electromagnetically coupled RLMC interconnects. In: Proceedings of Design, Automation and Test in Europe Conference **12**, 1400–1401 (2004)

8. Qin, Z., Cheng, C-K.: Realizable parasitic reduction using generalized Y-δ transformation. In: Proceedings of IEEE/ACM Design Automation Conference, pp. 220–225, Anaheim (2003)
9. Zuochang, Y., Vasilyev, D., Zhenhai, Z., Phillips, J. R.: Sparse Implicit Projection (SIP) for reduction of general many-terminal networks. In: Proceedings of International Conference on Computer-Aided Design, pp. 736–743, San Jose (2008)

Chapter 9
Simulation of Coupled Oscillators Using Nonlinear Phase Macromodels and Model Order Reduction

Davit Harutyunyan and Joost Rommes

Abstract Oscillators are used in many integrated RF circuits. Since their behavior is highly nonlinear, full system simulation can be expensive. Furthermore, the behavior of an oscillator can be (un)intendedly perturbed by that of other components and oscillators. We apply a method to build nonlinear phase macromodels of voltage controlled oscillators and show how these can be used to predict the behavior of oscillators under perturbation. Model order reduction techniques are used to decrease simulation times. Numerical results for realistic design illustrate the proposed approach.

Keywords Behavioral modeling · Circuit simulation · Injection locking · Phase noise · Pulling · Voltage-Controlled Oscillators (VCOs)

9.1 Introduction

The request for more functionality on a smaller physical area makes the design of modern RF (radio frequency) integrated circuits increasingly more complicated. Modern RF chips for mobile devices, for instance, typically have an FM

D. Harutyunyan
Department of Mathematics and Computer Science, CASA Group, University of Eindhoven, Eindhoven, The Netherlands
e-mail: d.harutyunyan@tue.nl

J. Rommes (✉)
NXP Semiconductors, Corporate I&T/DTF, HTC 37 WY4-01,
5656 AE Eindhoven, The Netherlands
e-mail: joost.rommes@nxp.com

P. Benner et al. (eds.), *Model Reduction for Circuit Simulation*,
Lecture Notes in Electrical Engineering, 74, DOI: 10.1007/978-94-007-0089-5_9,

radio, Blue-tooth, and GPS on one chip. Each of these functionalities are implemented with Voltage Controlled Oscillators (VCOs), that are designed to oscillate at certain different frequencies. Such oscillators are influenced by unintended (parasitic) signals coming from other blocks (such as Power Amplifiers) or from other oscillators, via for instance (unintended) inductive coupling through the substrate. A possibly undesired consequence of the perturbation is that the oscillators lock to a frequency different than designed for, or show pulling, in which case the oscillators are perturbed from their free running orbit without locking. This makes floor planning, i.e., determining the locations for the functional blocks, one of the most challenging tasks in RF chip design.

Our motivation comes from the design of RF systems, where oscillators play an important role [4, 8, 12, 21] in, for instance, high-frequency phase locked loops (PLLs). The nonlinear dynamics of interest include changes in the frequency spectrum of the oscillator due to small noise signals (an effect known as jitter [8]), which may lead to pulling or locking of the oscillator to a different frequency and may cause the oscillator to malfunction. Since both phase and amplitude dynamics are strongly nonlinear and spread over separated time scales [17], simulation is difficult. Accurate simulation requires very small time steps during time integration, resulting in unacceptable simulation times that block the design flow. Furthermore, transient simulation only gives limited understanding of the causes and mechanisms of the pulling and locking effects. Oscillators appear in many other physical systems and applications, see e.g. [2, 18].

Here we use the nonlinear phase macromodel introduced and developed in [8, 10, 11, 16, 17, 25]. Contrary to linear macromodels [1, 16, 21], the nonlinear phase macromodel is able to capture nonlinear effects such as injection locking. Because the macromodel replaces the original oscillator system by a single scalar equation, simulation times are decreased while the nonlinear oscillator effects can still be studied without loss of accuracy. We use the macromodels to predict the behavior of inductively coupled oscillators.

In some applications one exploits the coupling of oscillators. To reduce clockskew (clocksignals becoming out of phase), for instance, oscillators can be coupled via transmission lines [9]. Since accurate models for transmission lines can be large, this may lead to increased simulation times. We show how model order reduction techniques [3, 5, 22] can be used to decrease simulation times without unacceptable loss of accuracy.

The paper is organized as follows. Section 9.2 gives a summary of the phase noise theory. In Sect. 9.3 we show how the phase noise theory can be used to analyze oscillator–balun coupling. In Sect. 9.4, we explain the coupling of oscillators via transmission lines. Application of model order reduction techniques in simulation of coupled oscillators is described in Sect. 9.5 Numerical results are presented in Sect. 9.6 and Sect. 9.7 concludes.

9.2 Phase Noise Analysis of Oscillators

A general free-running oscillator is described as an autonomous system of differential equations:

$$\frac{\mathrm{d}q(x)}{\mathrm{d}t} + j(x) = 0, \tag{9.1a}$$

$$x(0) = x(T), \tag{9.1b}$$

where $x(t) \in \mathbb{R}^n$ are the state variables, T is the period of the free running oscillator, which is in general unknown, $q, j : \mathbb{R}^n \rightarrow \mathbb{R}^n$ are (nonlinear) functions and n is the system size. The solution of (9.1a, b) is called periodic steady state (PSS) and is denoted by x_{pss}. Although finding a PSS solution can be a challenging task in itself, we will not discuss this in the present paper and refer to, for example, [6, 10, 13–15, 23].

A general oscillator under perturbation can be expressed as a system of differential equations

$$\frac{\mathrm{d}q(x)}{\mathrm{d}t} + j(x) = b(t), \tag{9.2}$$

where $b(t) \in \mathbb{R}^n$ are perturbations to the free running oscillator. For small perturbations $b(t)$ it can be shown [8] that the solution of (9.2) can be approximated by

$$x_p(t) = x_{pss}(t + \alpha(t)), \tag{9.3}$$

where $\alpha(t) \in \mathbb{R}$ is called the phase shift, which satisfies the following scalar nonlinear differential equation:

$$\dot{\alpha}(t) = V^T(t + \alpha(t)) \cdot b(t), \tag{9.4a}$$

$$\alpha(0) = 0, \tag{9.4b}$$

with $V(t) \in \mathbb{R}^n$ being the perturbation projection vector (PPV) [8] of (9.2). The PPV is a periodic function with the same period as the oscillator and can efficiently be computed directly from the PPS solution, see e.g. [7].

9.3 Oscillator Coupled to a Balun

In this section we consider the mathematical model of an oscillator inductively coupled to a balun. A balun is an electrical transformer that can transform balanced signals to unbalanced signals and vice versa. A schematic view of an LC oscillator coupled to a balun with mutual inductors is given in Fig. 9.1. The corresponding mathematical model is given by the following set of equations:

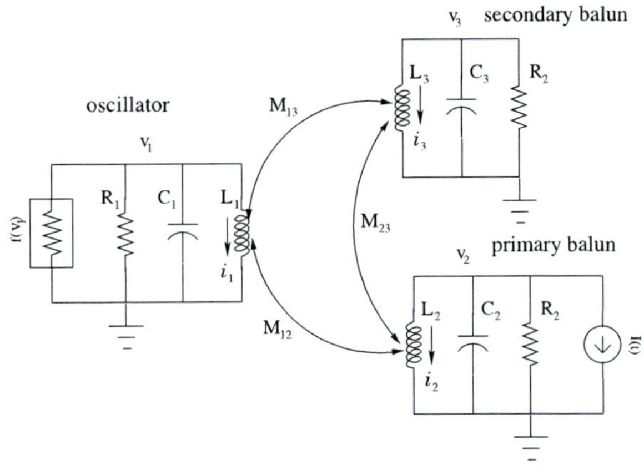

Fig. 9.1 Oscillator coupled with a balun

$$C_1\frac{\mathrm{d}v_1(t)}{\mathrm{d}t}+\frac{v_1(t)}{R_1}+i_1(t)+S\tanh\left(\frac{G_n}{S}v_1(t)\right)=0, \tag{9.5a}$$

$$L_1\frac{\mathrm{d}i_1(t)}{\mathrm{d}t}+M_{12}\frac{\mathrm{d}i_2(t)}{\mathrm{d}t}+M_{13}\frac{\mathrm{d}i_3(t)}{\mathrm{d}t}-v_1(t)=0, \tag{9.5b}$$

$$C_2\frac{\mathrm{d}v_2(t)}{\mathrm{d}t}+\frac{v_2(t)}{R_2}+i_2(t)+I(t)=0, \tag{9.5c}$$

$$L_2\frac{\mathrm{d}i_2(t)}{\mathrm{d}t}+M_{12}\frac{\mathrm{d}i_1(t)}{\mathrm{d}t}+M_{23}\frac{\mathrm{d}i_3(t)}{\mathrm{d}t}-v_2(t)=0, \tag{9.5d}$$

$$C_3\frac{\mathrm{d}v_3(t)}{\mathrm{d}t}+\frac{v_3(t)}{R_3}+i_3(t)=0, \tag{9.5e}$$

$$L_3\frac{\mathrm{d}i_3(t)}{\mathrm{d}t}+M_{13}\frac{\mathrm{d}i_1(t)}{\mathrm{d}t}+M_{23}\frac{\mathrm{d}i_2(t)}{\mathrm{d}t}-v_3(t)=0, \tag{9.5f}$$

where $M_{ij}=k_{ij}\sqrt{L_iL_j},\ i,j=1,2,3,\ i<j$ is the mutual inductance and k_{ij} is the coupling factor. The parameters of the nonlinear resistor are $S=1/R_1$ and $G_n=-1.1/R_1$ and the current injection in the primary balun is denoted by $I(t)$.

For small coupling factors we can consider $M_{12}\frac{\mathrm{d}i_2(t)}{\mathrm{d}t}+M_{13}\frac{\mathrm{d}i_3(t)}{\mathrm{d}t}$ in (9.5b) as a small perturbation to the oscillator and apply the phase shift macromodel to (9.5a)–(9.5b). Then the reduced model corresponding to (9.5a)–(9.5b) is

$$\frac{\mathrm{d}\alpha(\mathrm{t})}{\mathrm{d}t}=V^T(t+\alpha(t))\cdot\begin{pmatrix}0\\ -M_{12}\frac{\mathrm{d}i_2(t)}{\mathrm{d}t}-M_{13}\frac{\mathrm{d}i_3(t)}{\mathrm{d}t}\end{pmatrix}, \tag{9.6}$$

where V is the PPV of the oscillator. The balun is described by a linear circuit (9.5c)–(9.5f) which can be written in a more compact form:

$$E\frac{\mathrm{d}x(t)}{\mathrm{d}t} = Ax(t) + Bu(t), \tag{9.7}$$

where

$$E = \begin{pmatrix} C_2 & 0 & 0 & 0 \\ 0 & L_2 & 0 & M_{23} \\ 0 & 0 & C_3 & 0 \\ 0 & M_{23} & 0 & L_3 \end{pmatrix}, \quad A = \begin{pmatrix} -\frac{1}{R_2} & -1 & 0 & 0 \\ 1 & 0 & 0 & 0 \\ 0 & 0 & -\frac{1}{R_3} & 0 \\ 0 & 0 & 1 & 0 \end{pmatrix}, \tag{9.8a}$$

$$B = \begin{pmatrix} -1 & 0 \\ 0 & -M_{12} \\ 0 & 0 \\ 0 & -M_{13} \end{pmatrix}, \quad x(t) = \begin{pmatrix} v_2(t) \\ i_2(t) \\ v_3(t) \\ i_3(t) \end{pmatrix}, \quad u(t) = \begin{pmatrix} I(t) \\ \frac{\mathrm{d}i_1(t)}{\mathrm{d}t} \end{pmatrix}. \tag{9.8b}$$

With these notations (9.6) and (9.7) can be written in the following form

$$\frac{\mathrm{d}\alpha(t)}{\mathrm{d}t} = V^T(t+\alpha(t)) \cdot \begin{pmatrix} 0 \\ \frac{\mathrm{d}y(t)}{\mathrm{d}t} \end{pmatrix} \tag{9.9a}$$

$$E\frac{\mathrm{d}x(t)}{\mathrm{d}t} = Ax(t) + Bu(t), \tag{9.9b}$$

$$y(t) = \mathcal{C}^T x, \tag{9.9c}$$

where $\mathcal{C}^T = (0, -M_{12}, 0, -M_{13})$ and $i_1(t)$ is computed by using (9.3).

9.4 Oscillator Coupling to a Transmission Line

In some applications oscillators are coupled via transmission lines. By coupling oscillators via transmission lines, for instance, one can reduce the clockskew in clock distribution networks [9]. Accurate models for transmission lines may contain up to thousands or millions of RLC components [26]. Furthermore, the oscillators or the components that perturb (couple to) the oscillators can consist of many RLC components, for instance when ones takes into account parasitic effects. Since simulation times usually increase with the number of elements, one would like to limit the number of (parasitic) components as much as possible, without losing accuracy.

For the sake of simplicity in this paper we consider an RC transmission line coupled to an oscillator, see Fig. 9.2. Using phase macromodel for oscillator and by applying Kirchhoff's current law to the transmission line circuit, we obtain the following set of differential equations:

$$\frac{\mathrm{d}\alpha(t)}{\mathrm{d}t} = V^T(t+\alpha(t)) \cdot \begin{pmatrix} \frac{y(t)-v(t)}{R_1} \\ 0 \end{pmatrix} \tag{9.10a}$$

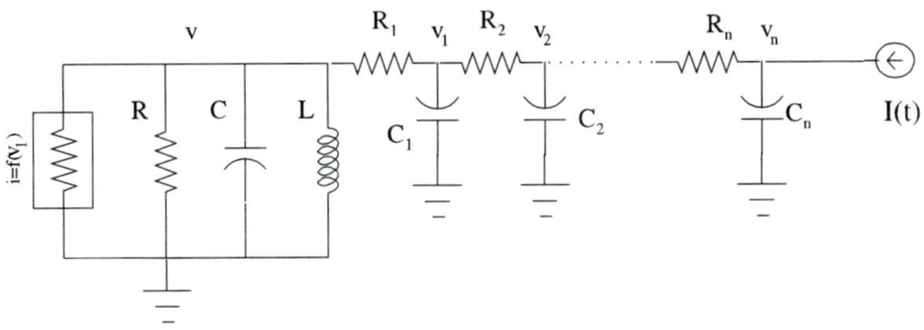

Fig. 9.2 Oscillator coupled to a transmission line

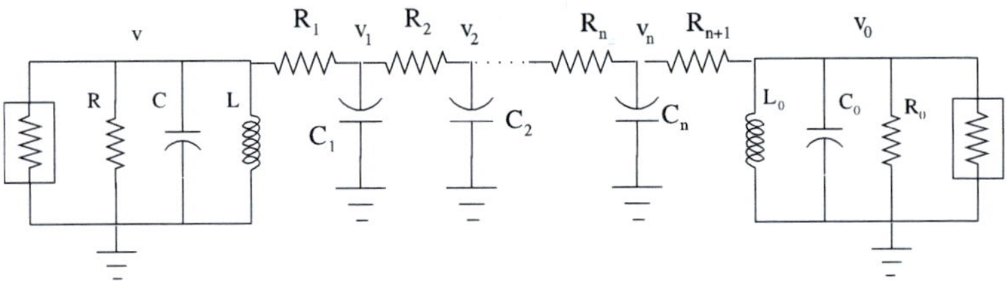

Fig. 9.3 Two oscillators coupled via a transmission line

$$E\frac{dx(t)}{dt} = Ax(t) + Bu(t), \tag{9.10b}$$

$$y(t) = \mathcal{C}^T x, \tag{9.10c}$$

where

$$E = \text{diag}(C_1, C_2, \ldots, C_n), \quad A = \text{tridiag}\left(\frac{1}{R_i}, -\frac{1}{R_i} - \frac{1}{R_{i+1}}, \frac{1}{R_{i+1}}\right), \tag{9.11a}$$

$$B = \begin{pmatrix} \frac{1}{R_1} & 0 \\ 0 & 0 \\ \vdots & \vdots \\ 0 & 1 \end{pmatrix}, \quad x(t) = \begin{pmatrix} v_1(t) \\ v_2(t) \\ \vdots \\ v_n(t) \end{pmatrix}, \quad u(t) = \begin{pmatrix} v(t) \\ I(t) \end{pmatrix}, \quad \mathcal{C} = \begin{pmatrix} 1 \\ 0 \\ \vdots \\ 0 \end{pmatrix}. \tag{9.11b}$$

In a similar way the phase macromodel of two oscillators coupled via a transmission line, see Fig. 9.3, is given by the following equations:

$$\frac{d\alpha_1(t)}{dt} = V_1^T(t + \alpha_1(t)) \cdot \begin{pmatrix} \frac{v_1(t) - v(t)}{R_1} \\ 0 \end{pmatrix} \tag{9.12a}$$

$$E\frac{dx(t)}{dt} = Ax(t) + Bu(t), \tag{9.12b}$$

$$\frac{\mathrm{d}\alpha_2(t)}{\mathrm{d}t} = V_2^T(t+\alpha_2(t)) \cdot \begin{pmatrix} \frac{v_n(t)-v_0(t)}{R_{n+1}} \\ 0 \end{pmatrix}, \tag{9.12c}$$

where $\alpha_1(t)$ and $\alpha_2(t)$ (V_1 and V_2) are phase shifts (PPV's) of the corresponding oscillator. The matrices E, A and x are given by (9.11a, b) and

$$B = \begin{pmatrix} \frac{1}{R_1} & 0 \\ 0 & 0 \\ \vdots & \vdots \\ 0 & \frac{1}{R_{n+1}} \end{pmatrix}, \quad u(t) = \begin{pmatrix} v(t) \\ v_0(t) \end{pmatrix}. \tag{9.13}$$

9.5 Model Order Reduction

Model order reduction (MOR) techniques [3, 5, 22] can be used to reduce the number of elements significantly. Here we show how model order reduction can be used for the analysis of oscillator perturbation effects as well. Since the main focus is to show how MOR techniques can be used (and not which technique is the most suitable), we limit the discussion here to balanced truncation [20]. For other methods, see, e.g., [3, 5, 22].

Given any dynamical system (A, B, C)

$$E\frac{\mathrm{d}x(t)}{\mathrm{d}t} = Ax(t) + Bu(t), \quad y(t) = C^T x,$$

and assuming $E = I$, balanced truncation [20] consists of first computing a balancing transformation $V \in \mathbb{R}^{n\times n}$, where B and C are input-to-state and state-to-output vectors, respectively. The balanced system (V^TAV, V^TB, V^TC) has the nice property that the Hankel singular values[1] are easily available. A reduced order model can be constructed by selecting the columns of V that correspond to the $k < n$ largest Hankel singular values. With $V_k \in \mathbb{R}^{n\times k}$ having as columns these k columns, the reduced order model (of order k) becomes $(V_k^TAV_k, V_k^TB, V_k^TC)$. If $E \neq I$ is nonsingular, balanced truncation can be applied to $(E^{-1}A, E^{-1}B, C)$. For more details on balanced truncation, see [5, 20, 22].

In this paper we apply model order reduction to linear circuits that are coupled to oscillators, and the relevant equations for each problem describing linear circuits have the form of (9.9b)–(9.9c). For each problem the corresponding matrices E, A, B, and C can be identified readily, see (9.8a, b), (9.11a, b), (9.13) and note $C \equiv \mathcal{C}$. We use Matlab [19] implementation for balanced truncation to obtain reduced order models:

[1] Similar to singular values of matrices, the Hankel singular values and corresponding vectors can be used to identify the dominant subspaces of the system's statespace: the larger the Hankel singular value, the more dominant.

```
sys   = ss( -E\A, -E\B, C', 0 );
[hsv, baldata] = hsvd(sys); % Hankel singular values
mor_dim = nnz((hsv>1e-10)); % choose largest singular
                            % values where mor_dim is
                            % the dimension of the reduced system
rsys = balred(sys,mor_dim,'Elimination','Truncate',...
                   'Balancing',baldata) ; %truncate
```

Note that we can apply balanced truncation because E is nonsingular. It is well known that in many cases in circuit simulation the system is a descriptor system and hence E is singular. Although generalizations of balanced truncation to descriptor systems exist [22, 24], other MOR techniques such as Krylov subspace methods and modal approximation might be more appropriate. We refer the reader to [3, 5, 22] for a good introduction to such techniques and MOR in general.

9.6 Numerical Experiments

In all the numerical experiments the simulations are run until $T_{\text{final}} = 6 \times 10^{-7}$ s with fixed time step $\tau = 10^{-11}$. In this section all the numerical results done with the phase macromodel combined with MOR technique are called macromodel-MOR simulation.

We compare our results with simulation results of the full circuit (no macro-modeling, no model order reduction), hereafter full-simulation. Because the full circuit represents a stiff ODE, we use the Matlab built-in ODE solver *ode15s* with relative tolerance set to 10^{-7} (larger values of relative tolerance result in incorrect solutions) to achieve a comparable accuracy with the macromodel-MOR simulation results. In all experiments we observed that the simulation time of the macromodel-MOR technique is typically five times faster than full-simulation times.

9.6.1 Oscillator Coupled to a Balun

Consider an oscillator coupled to a balun as shown in Fig. 9.1 with the following parameters values:

Oscillator	Primary balun	Secondary balun
$L_1 = 0.64 \times 10^{-9}$ H	$L_2 = 1.10 \times 10^{-9}$ H	$L_3 = 3.60 \times 10^{-9}$ H
$C_1 = 1.71 \times 10^{-12}$ F	$C_2 = 4.00 \times 10^{-12}$ F	$C_3 = 1.22 \times 10^{-12}$ F
$R_1 = 50\,\Omega$	$R_2 = 40\,\Omega$	$R_2 = 60\,\Omega$

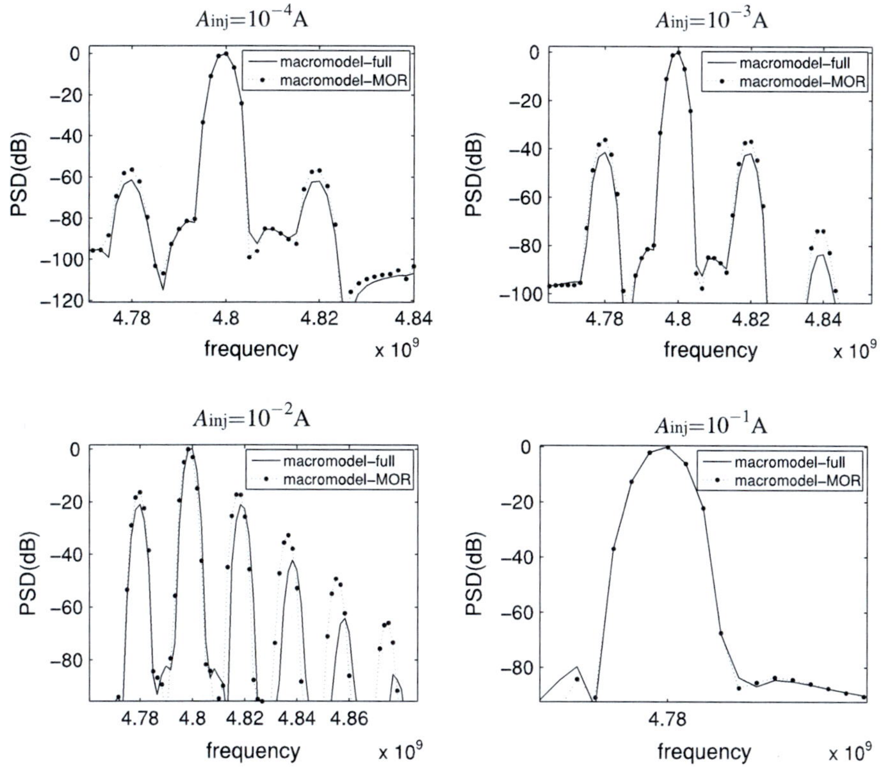

Fig. 9.4 Comparison of the output spectrum of the oscillator coupled to a balun obtained by the macromodel-full and the macromodel-MOR simulations for an increasing injected current amplitude A_{inj} and an offset frequency $f_{\text{off}} = 20\,\text{MHz}$

The coefficients of the mutual inductive couplings are $k_{12} = 10^{-3}, k_{13} = 5.96 \times 10^{-3}, k_{23} = 9.33 \times 10^{-3}$. The injected current in the primary balun is of the form

$$I(t) = A_{\text{inj}} \sin(2\pi(f_0 - f_{\text{off}})t), \tag{9.14}$$

where $f_0 = 4.8\,\text{GHz}$ is the oscillator's free running frequency, f_{off} is the offset frequency and A_{inj} is the current amplitude.

Simulation results of (9.9a, b, c) are shown in Fig. 9.4, where the frequency is plotted versus power spectral density (PSD[2]). The obtained results by the macromodel-MOR technique with mor_dim = 2 provide a good approximation to the full-simulation results. We note that for the injected current with $A_{\text{inj}} = 10^{-1}$ A the oscillator is locked to the injected signal. Similar results are also obtained for the balun.

[2] Matlab code for plotting the PSD is given in [12].

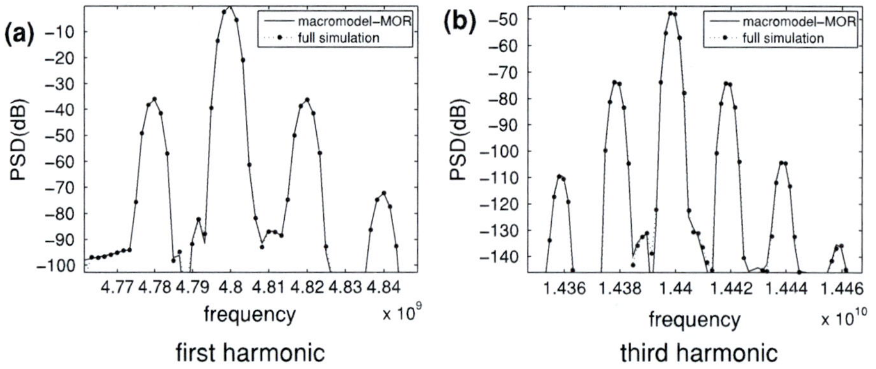

Fig. 9.5 Comparison of the output spectrum around the first and third harmonics of the oscillator coupled to a transmission line, cf. Fig. 9.2. **a** First harmonic. **b** Third harmonic

9.6.2 Oscillators Coupled with Transmission Lines

In this section we consider two academic examples, where transmission lines are modeled with RC components.

Single Oscillator Coupled to a Transmission Line

Let us consider the same oscillator as given in the previous section, now coupled to a transmission line, see Fig. 9.2. The size of the transmission line is $n = 100$ with the following parameters: $C_1 = \cdots = C_n = 10^{-2}\,\text{pF}, R_1 = 40\,\text{k}\Omega, R_2 = \cdots = R_n = 1\,\Omega$. The injected current has the form (9.14) with $A_{\text{inj}} = 10^{-2}\,\text{A}$ and $f_{\text{off}} = 20\,\text{MHz}$. Dimension of the reduced system is mor_dim = 18. Simulation results of (9.10a, b, c) around the first and third harmonics (this oscillator does not have a second harmonic) are shown in Fig. 9.5. The macromodel-MOR method, using techniques described in Sect. 9.5, gives a good approximation to the full simulation results.

Two LC Oscillators Coupled via a Transmission Line

For this experiment we consider two LC oscillators coupled via a transmission line with the mathematical model given by (9.12a, b, c). The first oscillator has a free running frequency $f_1 = 4.8\,\text{GHz}$ and is described in Sect. 9.6.1. The second LC oscillator has the following parameter values: $R_0 = 50\,\Omega, L_0 = 0.64\,\text{nH}$, $C_0 = 1.87\,\text{pF}$ and a free running frequency $f_2 = 4.6\,\text{GHz}$. The size of the transmission line is $n = 100$ with the following parameters: $C_1 = \cdots = C_n = 10^{-2}\,\text{pF}, R_1 = R_{n+1} = 4\,\text{k}\Omega, R_2 = \cdots = R_n = 0.001\,\Omega$. Dimension of the reduced system is mor_dim = 16. Numerical simulation results are given in Fig. 9.6. As can be seen from the spectra of Fig. 9.6, the macromodel-MOR and full simulation

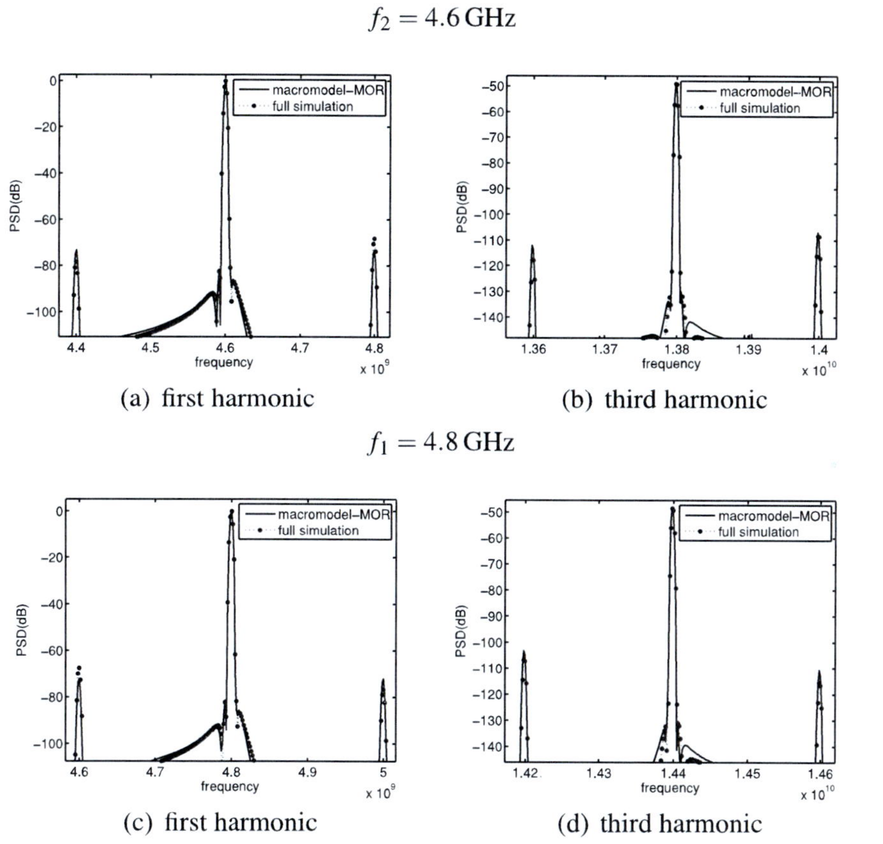

Fig. 9.6 Comparison of the output spectrum around the first and third harmonics of two oscillators coupled via a transmission line. $f_2 = 4.6\,\text{GHz}$: **a** first harmonic, **b** third harmonic; $f_1 = 4.8\,\text{GHz}$: **c** first harmonic, **d** third harmonic

results match well: with both simulation methods the carrier and beat frequencies are the same with a comparable PSD.

9.7 Conclusion

In this paper we have used nonlinear phase macromodels to accurately predict the behavior of individual and mutually coupled voltage controlled oscillators under perturbation. Several types of coupling have been described, including oscillator–balun coupling. For small perturbations, the nonlinear phase macromodels can also be used to produce results with accuracy comparable to full circuit simulations [11]. In addition, model order reduction techniques have been applied to transmission lines that couple oscillators. With these techniques, reduced-order models could be obtained that decreased simulation times while preserving the required accuracy.

Acknowledgment This work was supported by EU Marie-Curie project O-MOORE-NICE! FP6 MTKI-CT-2006-042477.

References

1. Adler, R.: A study of locking phenomena in oscillators. In: Proceedings of the I.R.E. and Waves and Electrons, vol. 34, pp. 351–357 (1946)
2. Agarwal, S., Roychowdhury, J.: Efficient multiscale simulation of circadian rhythms using automated phase macromodelling techniques. In: Proceedings of the Pacific Symposium on Biocomputing, vol. 13, pp. 402–413 (2008)
3. Antoulas, A.C.: Approximation of Large-Scale Dynamical Systems. SIAM, Philadelphia, PA (2005)
4. Banai, A., Farzaneh, F.: Locked and unlocked behaviour of mutually coupled microwaveoscillators. In: IEE Proceedings of the Antennas and Propagation, vol. 147, pp. 13–18 (2000)
5. Benner, P., Mehrmann, V., Sorensen, D. (eds.): Dimension Reduction of Large-Scale Systems, Lecture Notes in Computational Science and Engineering, vol. 45. Springer, Berlin (2005)
6. Ciarlet, P.G., Schilders, W.H.A., ter Maten, E.J.W. (eds.): Numerical Methods in Electromagnetics, Handbook of Numerical Analysis, vol. 13. Elsevier, Amsterdam (2005)
7. Demir, A., Long, D., Roychowdhury, J.: Computing phase noise eigenfunctions directly from steady-state Jacobian matrices. In: IEEE/ACM International Conference on Computer Aided Design, 2000. ICCAD-2000, pp. 283–288 (2000)
8. Demir, A., Mehrotra, A., Roychowdhury, J.: Phase noise in oscillators: a unifying theory and numerical methods for characterization. IEEE Trans. Circ. Syst. I **47**(5), 655–674 (2000)
9. Galton, I., Towne, D.A., Rosenberg, J.J., Jensen, H.T.: Clock distribution using coupled oscillators. In: IEEE International Symposium on Circuits and Systems, vol. 3, pp. 217–220 (1996)
10. Günther, M., Feldmann, U., ter Maten, J.: Modelling and discretization of circuit problems. In: Handbook of Numerical Analysis, vol. XIII, pp. 523–659. North-Holland, Amsterdam (2005)
11. Harutyunyan, D., Rommes, J., Termaten, E.J.W., Schilders, W.H.A.: Simulation of mutually coupled oscillators using nonlinear phase macromodels. IEEE Transactions on Computer-Aided Design of Integrated Circuits and Systems,Vol. 28(**10**) pp. 1456-1466 (2009)
12. Heidari, M.E., Abidi, A.A.: Behavioral models of frequency pulling in oscillators. In: IEEE International Behavioral Modeling and Simulation Workshop, pp. 100–104 (2007)
13. Houben, S.H.J.M.: Circuits in motion: the numerical simulation of electrical oscillators. Ph.D. thesis, Technische Universiteit Eindhoven (2003)
14. Kevenaar, T.A.M.: Periodic steady state analysis using shooting and wave-form-Newton. Int. J. Circ. Theory Appl. **22**(1), 51–60 (1994)
15. Kundert, K., White, J., Sangiovanni-Vincentelli, A.: An envelope-following method for the efficient transient simulation of switching power and filter circuits. In: IEEE International Conference on Computer-Aided Design, 1988. ICCAD-88. Digest of Technical Papers, pp. 446–449 (1988)
16. Lai, X., Roychowdhury, J.: Capturing oscillator injection locking via nonlinear phase-domain macromodels. IEEE Trans. Micro. Theory Tech. **52**(9), 2251–2261 (2004)
17. Lai, X., Roychowdhury, J.: Fast and accurate simulation of coupled oscillators using nonlinear phase macromodels. In: Microwave Symposium Digest, 2005 IEEE MTT-S International, pp. 871–874 (2005)

18. Lai, X., Roychowdhury, J.: Fast simulation of large networks of nanotechnological and biochemical oscillators for investigating self-organization phenomena. In: Proceedings of the IEEE Asia South-Pacific Design Automation Conference, pp. 273–278 (2006)
19. Mathworks: Matlab 7 (2009). URL http://www.mathworks.com/
20. Moore, B.C.: Principal component analysis in linear systems: controllability, observability and model reduction. IEEE Trans. Autom. Control **26**(1), 17–32 (1981)
21. Razavi, B.: A study of injection locking and pulling in oscillators. IEEE J. Solid-State Circ. **39**(9), 1415–1424 (2004)
22. Schilders, W.H.A., van der Vorst, H.A., Rommes, J. (eds.): Model Order Reduction: Theory, Research Aspects and Applications, Mathematics in Industry, vol. 13. Springer, Berlin (2008)
23. Semlyen, A., Medina, A.: Computation of the periodic steady state in systems with nonlinear components using a hybrid time and frequency domain method. IEEE Trans. Power Syst. **10**(3), 1498–1504 (1995)
24. Stykel, T.: Gramian based model reduction for descriptor systems. Math. Control Signals Syst. **16**, 297–319 (2004)
25. Wan, Y., Lai, X., Roychowdhury, J.: Understanding injection locking in negative resistance LC oscillators intuitively using nonlinear feedback analysis. In: Proceedings of the IEEE Custom Integrated Circuits Conference, pp. 729–732 (2005)
26. Yanzhu, Z., Dingyu, X.: Modeling and simulating transmission lines using fractional calculus. In: International Conference on Wireless Communications, Networking and Mobile Computing, 2007. WiCom 2007, pp. 3115–3118 (2007). doi: 10.1109/WICOM.2007.773

Chapter 10
POD Model Order Reduction of Drift-Diffusion Equations in Electrical Networks

Michael Hinze, Martin Kunkel and Morten Vierling

Abstract We consider integrated circuits with semiconductors modelled by modified nodal analysis and 1D drift-diffusion equations. The drift-diffusion equations are discretized in space using finite element methods. The space discretization yields a high dimensional differential-algebraic equation. We show how POD methods can be used to reduce the dimension of the model. We compare reduced and fine models and give numerical results for a basic network with one diode. Furthermore we discuss an adaptive approach to construct POD models which are valid over certain parameter ranges. Finally, numerical investigations for the reduction of a 4-diode rectifier network are presented, which clearly indicate that POD model reduction delivers surrogate models for the diodes involved, which depend on the position of the semiconductor in the network.

Keywords Model Order Reduction · Reduced Basis Methods · Mixed FiniteElement Methods · Drift-Diffusion Equations · Integrated Circuits

M. Hinze · M. Vierling
Department of Mathematics, University of Hamburg, Bundesstr. 55, 20146, Hamburg, Germany
e-mail: michael.hinze@uni-hamburg.de

M. Vierling
e-mail: morten.vierling@uni-hamburg.de

M. Kunkel (✉)
Universität der Bundeswehr München, LRT1, Werner-Heisenberg-Weg 39
85577, Neubiberg, Germany
e-mail: martin.kunkel@unibw.de

P. Benner et al. (eds.), *Model Reduction for Circuit Simulation*,
Lecture Notes in Electrical Engineering, 74, DOI: 10.1007/978-94-007-0089-5_10,

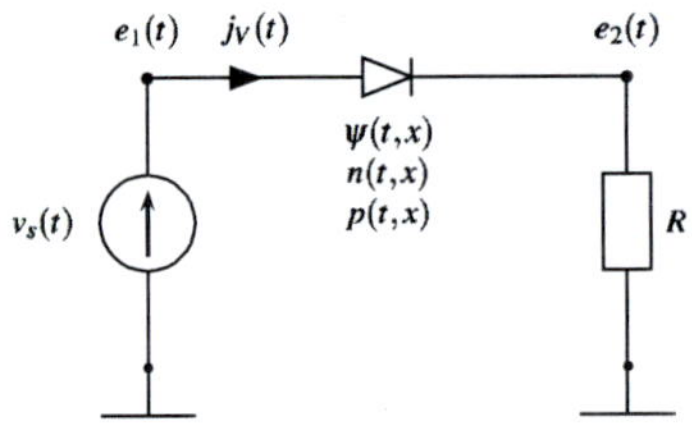

Fig. 10.1 Basic test circuit with one diode. The network is described by
$A_V = (1, \; 0)^\top,$
$A_S = (-1, \; 1)^\top,$
$A_R = (0, \; 1)^\top,$
$g(A_R^\top e, t) = \frac{1}{R} e_2(t)$

10.1 Introduction

In this article we investigate a POD-based model order reduction for semiconductors in electrical networks. Electrical networks can be efficiently modelled by a differential-algebraic equation (DAE) which is obtained from modified nodal analysis. Denoting by e the node potentials and by j_L and j_V the currents of inductive and voltage source branches, the DAE reads (see [9, 19])

$$A_C \frac{d}{dt} q_C(A_C^\top e, t) + A_R g(A_R^\top e, t) + A_L j_L + A_V j_V = -A_I i_s(t), \tag{10.1}$$

$$\frac{d}{dt} \phi_L(j_L, t) - A_L^\top e = 0, \tag{10.2}$$

$$A_V^\top e = v_s(t). \tag{10.3}$$

Here, the incidence matrix $A = [A_R, A_C, A_L, A_V, A_I]$ represents the network topology and q_C, g and ϕ_L are continuously differentiable functions defining the voltage-current relations of the network components. The continuous functions v_s and i_s are the voltage and current sources. For example consider the network in Fig. 10.1. Under the assumption that the Jacobians

$$D_C(e,t) := \frac{\partial q_C}{\partial e}(e,t), \quad D_G(e,t) := \frac{\partial g}{\partial e}(e,t), \quad D_L(j,t) := \frac{\partial \phi_L}{\partial j}(j,t)$$

are positive definite, analytical properties (e.g. the index) of DAE (10.1)–(10.3) are investigated in [5] and [6]. In linear networks, the matrices D_C, D_G and D_L are positive definite diagonal matrices with capacitances, conductivities and inductances on the diagonal.

Often semiconductors themselves are modelled by electrical networks. These models are stored in a library and are stamped into the surrounding network in order to create a complete model of the integrated circuit. Here we use a different approach which uses the transient drift-diffusion equations as a continuous model for semiconductors. Advantages are the higher accuracy of the model and fewer model parameters. On the other hand, numerical simulations are more expensive. For a comprehensive overview of the drift-diffusion equations we refer to [1, 2, 4, 11, 15]. Using the notation introduced there, we have the following system of partial differential equations for the electrostatic potential $\psi(t,x)$, the electron and hole concentrations $n(t,x)$ and $p(t,x)$ and the current densities $J_n(t,x)$ and $J_p(t,x)$:

$$
\begin{aligned}
\operatorname{div}(\varepsilon \operatorname{grad}\psi) &= q(n-p-C),\\
-q\partial_t n + \operatorname{div} J_n &= qR(n,p,J_n,J_p),\\
q\partial_t p + \operatorname{div} J_p &= -qR(n,p,J_n,J_p),\\
J_n &= \mu_n q(U_T \operatorname{grad} n - n \operatorname{grad}\psi),\\
J_p &= \mu_p q(-U_T \operatorname{grad} p - p \operatorname{grad}\psi),
\end{aligned}
$$

with $(t,x) \in [0,T] \times \Omega$ and $\Omega \subset \mathbb{R}^d$. The nonlinear function R describes the rate of electron/hole recombination, q is the elementary charge, ε the dielectricity, μ_n and μ_p are the mobilities of electrons and holes. The temperature is assumed to be constant which leads to a constant thermal voltage U_T. The function C is the time independent doping profile. Note that we do not formulate into quasi-Fermi potentials since the additional non-linearities would imply higher online simulation time for the reduced model.

We specify boundary conditions for Ohmic contacts $\Gamma_O \subset \Gamma := [0,T] \times \partial\Omega$ with normal vector ν:

$$
\psi(t,x) = \psi_{bi}(x) + e(x,t) = U_T \log\left(\frac{\sqrt{C(x)^2 + 4n_i^2} + C(x)}{2n_i}\right) + e(x,t), \tag{10.4}
$$

$$
n(t,x) = \frac{1}{2}\left(\sqrt{C(x)^2 + 4n_i^2} + C(x)\right), \tag{10.5}
$$

$$
p(t,x) = \frac{1}{2}\left(\sqrt{C(x)^2 + 4n_i^2} - C(x)\right), \tag{10.6}
$$

for $(t,x) \in \Gamma_O$. Here, e denotes the applied potential at the interface boundaries, $\psi_{bi}(x)$ is the build-in potential and n_i the constant intrinsic concentration. At isolated boundaries $\Gamma_I \subset \Gamma$ it holds grad $\psi \cdot \nu = 0$, $J_n \cdot \nu = 0$ and $J_p \cdot \nu = 0$.

The coupling between the drift-diffusion equations and the electrical network is established by the applied potential $e(x,t)$ at the interfaces and the current through the semiconductor which can be calculated as

$$
j_S = \int_{\Gamma_O} (J_n + J_p - \varepsilon\partial_t \operatorname{grad}\psi) \cdot \nu \, d\sigma, \tag{10.7}
$$

e.g. the current is the integral over the current density $J_n + J_p$ plus the displacement current. The complete model forms a partial differential-algebraic equation

(PDAE). The analytical and numerical analysis of such systems is subject to current research, see [3, 8, 17, 19].

This paper is organized as follows. In Sect. 10.2 we present the model for the complete coupled system, meaning the network including semiconductors. The coupled model then is simulated using finite-element methods in Sect. 10.3. This gives us so-called snapshots $y_i := y(t_i)$, $i = 1, \ldots, k$, which represent the state of the circuit and the semiconductors at time t_i. Based on these snapshots and POD we construct a reduced model in Sect. 10.4. A numerical investigation of the model is presented in Sect. 10.5 where also advantages and shortcomings of our approach are discussed.

10.2 Complete Coupled System

In the present section we develop the complete system in 1D. The n_s semiconductors are diodes with length L_k, $k = 1, \ldots, n_s$. For the sake of simplicity we assume that the contacts are Ohmic, and that the dielectricities ε_k are constant over the whole domain $\Omega := [0, L_k]$. Furthermore we focus on the Shockley-Read-Hall recombination

$$R(n,p) = \frac{np - n_i^2}{\tau_p(n+n_i) + \tau_n(p+n_i)}$$

which does not depend on the current densities. τ_n and τ_p are the average lifetimes of electrons and holes.

The simulation of the complete coupled system is expensive and numerically difficult due to bad scaling of the drift-diffusion equations. The numerical issues can be significantly reduced by the unit scaling procedure discussed in [15], e.g. by substituting

$$x = L\tilde{x}, \quad \psi = U_T\tilde{\psi}, \quad n = \|C\|_\infty \tilde{n}, \quad p = \|C\|_\infty \tilde{p}, \quad C = \|C\|_\infty \tilde{C},$$
$$J_n = \frac{qU_T\|C\|_\infty}{L}\mu_n\tilde{J}_n, \quad J_p = \frac{qU_T\|C\|_\infty}{L}\mu_p\tilde{J}_p, \quad n_i = \tilde{n}_i\|C\|_\infty.$$

The scaled complete coupled system is constructed as follows. (We neglect the tilde-sign over the scaled variables.) The current through the diodes must be considered in Kirchhoff's current law. Consequently, the term $A_S j_S$ is added to Eq. 10.1, e.g.

$$A_C \frac{d}{dt} q_C(A_C^\top e, t) + A_R g(A_R^\top e, t) + A_L j_L + A_V j_V + A_S j_S = -A_I i_s(t), \tag{10.8}$$

$$\frac{d}{dt}\phi_L(j_L, t) - A_L^\top e = 0, \tag{10.9}$$

$$A_V^\top e = v_s(t). \tag{10.10}$$

The voltage–current relation for the semiconductor k is established by the couplings

$$j_{S,k}(t) = a\frac{qU_T\|C\|_\infty}{L}\left(\mu_n J_n(t,1) + \mu_p J_p(t,1)\right) - a\frac{\varepsilon U_T}{L}\partial_{tx}\psi(t,1), \tag{10.11}$$

$$\psi(t,0) = \frac{\psi_{bi}(0)}{U_T}, \tag{10.12}$$

$$\psi(t,1) = \frac{\psi_{bi}(L) + A_{S,k}^\top e(t)}{U_T}, \tag{10.13}$$

which is basically the evaluation of Eqs. 10.4 and 10.7. The left boundary is chosen as reference terminal. a denotes the area size of the contact of the diode. In order to simplify the presentation, we neglect the index k for the semiconductors wherever possible. The scaled drift-diffusion equations for each semiconductor read

$$\lambda\partial_{xx}\psi = n - p - C, \tag{10.14}$$

$$-\partial_t n + \nu_n\partial_x J_n = R(n,p), \tag{10.15}$$

$$\partial_t p + \nu_p\partial_x J_p = -R(n,p), \tag{10.16}$$

$$J_n = \partial_x n - n\partial_x\psi, \tag{10.17}$$

$$J_p = -\partial_x p - p\partial_x\psi \tag{10.18}$$

with $\lambda := \frac{\varepsilon U_T}{L^2 q\|C\|_\infty}$, $\nu_n := \frac{U_T\mu_n}{L^2}$ and $\nu_p := \frac{U_T\mu_p}{L^2}$. Finally we evaluate the boundary conditions (10.5) and (10.6) at the Ohmic contacts, e.g.

$$n(t,0) = \frac{1}{2}\left(\sqrt{C(0)^2 + 4n_i^2} + C(0)\right), \quad n(t,1) = \frac{1}{2}\left(\sqrt{C(1)^2 + 4n_i^2} + C(1)\right),$$
$$p(t,0) = \frac{1}{2}\left(\sqrt{C(0)^2 + 4n_i^2} - C(0)\right), \quad p(t,1) = \frac{1}{2}\left(\sqrt{C(1)^2 + 4n_i^2} - C(1)\right)$$

for all $t \in [0, T]$.

Note, that in 1D simulations, there are no boundary conditions for J_n and J_p, since all boundaries are connected to the network, e.g. $\Gamma_O = \{0, 1\}$ and $\Gamma_I = \emptyset$.

10.3 Simulation of the Full System

Classical approaches for simulation of drift-diffusion equations (e.g. Gummel iterations [7], a nonlinear Gauss-Seidel approach) approximate J_n and J_p by piecewise constant functions and then solve equations (10.15) and (10.16) with respect to n and p explicitly. This helps reducing the computational effort and increases the numerical stability. For the proposed model order reduction this method has the disadvantage of introducing additional non-linearities, since the discrete solution of n and p is build up in terms of $\exp\psi$, see [17].

Here we consider two complementary finite element approaches to the drift-diffusion equations, namely standard Galerkin and mixed finite element methods. We start with a standard Galerkin approach. For the sake of simplicity we use an equally distributed finite element mesh with N elements and mesh width $h := 1/N$. The functions ψ, n and p are approximated by piecewise linear and globally continuous functions, J_n and J_p are approximated by piecewise constant functions, e.g.

$$\psi(t,x) := \sum_{i=0}^{N} \psi_i(t)\phi_i(x), \quad n(t,x) := \sum_{i=0}^{N} n_i(t)\phi_i(x), \quad p(t,x) := \sum_{i=0}^{N} p_i(t)\phi_i(x),$$
$$J_n(t,x) := \sum_{i=1}^{N} J_{n,i}(t)\varphi_i(x), \quad J_p(t,x) := \sum_{i=1}^{N} J_{p,i}(t)\varphi_i(x),$$

where the functions $\{\phi_i\}$ and $\{\varphi_i\}$ are the corresponding ansatz or trial functions. For ψ, n and p only the interior coefficients, e.g. $\psi(t) = (\psi_1, \ldots, \psi_{N-1})^\top$, are variable, the coefficients corresponding to the boundary elements are given by the Dirichlet boundary conditions. Note that the time is not discretized at this point which refers to the so-called method of lines. The finite element method leads to the following DAE for the unknown vector-valued functions of time ψ, n, p, J_n, J_p for each semiconductor:

$$\begin{aligned}
0 &= \lambda S\psi(t) + Mn(t) - Mp(t) - C_h + b_\psi(e(t)), \\
-M\dot{n}(t) &= -\nu_n D^\top J_n(t) + hR(n(t),p(t)), \\
M\dot{p}(t) &= -\nu_p D^\top J_p(t) - hR(n(t),p(t)), \\
0 &= hJ_n(t) + Dn(t) - \operatorname{diag}\big(Bn(t) + \tilde{b}_n\big)\, D\psi(t) + b_n, \\
0 &= hJ_p(t) - Dp(t) - \operatorname{diag}\big(Bp(t) + \tilde{b}_p\big)\, D\psi(t) + b_p,
\end{aligned} \tag{10.19}$$

where $S, M \in \mathbb{R}^{(N-1)\times(N-1)}$ and $D, B \in \mathbb{R}^{N\times(N-1)}$ are assembled finite element matrices. $\operatorname{diag}(v)$ denotes the operator which takes a vector v and creates a diagonal matrix with v on the diagonal. The vectors $b_\psi(e(t))$, b_n, $\tilde{b}_n$, b_p and $\tilde{b}_p$ implement the boundary conditions imposed on ψ, n and p, which are given by

$$\begin{aligned}
\psi_0(t) &= \frac{\psi_{bi}(0)}{U_T}, \quad \psi_N(t) = \frac{\psi_{bi}(L) + A_{S,k}^\top e(t)}{U_T}, \\
n_0(t) &= \frac{1}{2}\left(\sqrt{C(0)^2 + 4n_i^2} + C(0)\right), \quad n_N(t) = \frac{1}{2}\left(\sqrt{C(1)^2 + 4n_i^2} + C(1)\right), \\
p_0(t) &= \frac{1}{2}\left(\sqrt{C(0)^2 + 4n_i^2} - C(0)\right), \quad p_N(t) = \frac{1}{2}\left(\sqrt{C(1)^2 + 4n_i^2} - C(1)\right).
\end{aligned}$$

Discretization of the coupling condition for the current completes the discretized system:

$$j_{S,k}(t) = \frac{aqU_T\|C\|_\infty}{L}\left(\mu_n J_{n,N}(t) + \mu_p J_{p,N}(t)\right) - \frac{a\varepsilon U_T}{Lh}\left(\dot{\psi}_N(t) - \dot{\psi}_{N-1}(t)\right),$$

Supposing that the fluxes J_n and J_p as well as the gradient of ψ play a dominant role in (10.14)–(10.18) one might argue, that they should be resolved better than as piecewise constant functions. This directly leads to the Raviart-Thomas finite element approach, where the concentrations n and p and the potential ψ are approximated by piecewise constant functions, but the fluxes $\nabla\psi$, J_n and J_p are elements of the Raviart-Thomas space of order zero. In one space dimension, the Raviart-Thomas approach leads to piecewise linear, continuous approximations to these functions, so that the ansatz space is just $\operatorname{span}\{\phi_0, \ldots, \phi_N\}$. The ansatz space for ψ, n and p is given by $\operatorname{span}\{\varphi_1, \ldots, \varphi_N\}$. The idea is now not to discretize (10.14)–(10.18) separately, but instead to discretize variable-flux pairs together. For example the Eqs. 10.15, 10.17

$$\begin{aligned} -\partial_t n + \nu_n \partial_x J_n &= R(n,p), \\ J_n &= \partial_x n - n\partial_x \psi, \end{aligned}$$

define the variable n and its flux J_n. The first equation is tested with φ, the second is tested with ϕ and integrated by parts to obtain

$$\begin{aligned} -\int_\Omega (\partial_t n)\varphi + \nu_n \int_\Omega \varphi(\partial_x J_n) &= \int_\Omega R(n,p)\varphi, \\ \int_\Omega J_n\phi &= -\int_\Omega n(\partial_x\phi) - \int_\Omega n(\partial_x\psi)\phi + [n\phi]_0^1. \end{aligned}$$

This formulation avoids space-derivatives of n. The pairs of Eqs. 10.16, 10.18 and 10.14, 10.20 are treated in a similar way. By using $\{\varphi_i\}$ and $\{\phi_i\}$ as test and trial functions and by introducing a new variable g_ψ and the equation

$$g_\psi = \partial_x \psi, \tag{10.20}$$

we arrive at the following finite dimensional system of equations:

$$\begin{aligned} 0 &= \lambda D g_\psi(t) + hn(t) - hp(t) + \hat{C}_h \\ -h\dot{n}(t) &= \nu_n D J_n(t) + hR(n(t), p(t)) \\ h\dot{p}(t) &= \nu_p D J_p(t) - hR(n(t), p(t)) \\ 0 &= -D^\top \psi(t) + M g_\psi(t) - \hat{b}_\psi(e(t)) \\ 0 &= -D^\top n(t) + M J_n(t) + \hat{\mathbf{B}}(g_\psi(t), n(t)) - \hat{b}_n \\ 0 &= -D^\top p(t) + M J_p(t) + \hat{\mathbf{B}}(g_\psi(t), p(t)) - \hat{b}_p \end{aligned}$$

Table 10.1 Diode model parameters

Parameter	Value
L	10^{-4} [cm]
U_T	0.0259 [V]
μ_n	1, 350 [cm^2/Vsec]
μ_p	480 [cm^2/Vsec]
a	10^{-5} [cm^2]
ε	1.03545×10^{-12} [F/cm]
n_i	1.4×10^{10} [1/cm^3]
τ_n	330×10^{-9} [sec]
τ_p	33×10^{-9} [sec]
$C(x),\ x<L/2$	-9.94×10^{15} [1/cm^3]
$C(x),\ x\geq L/2$	4.06×10^{18} [1/cm^3]

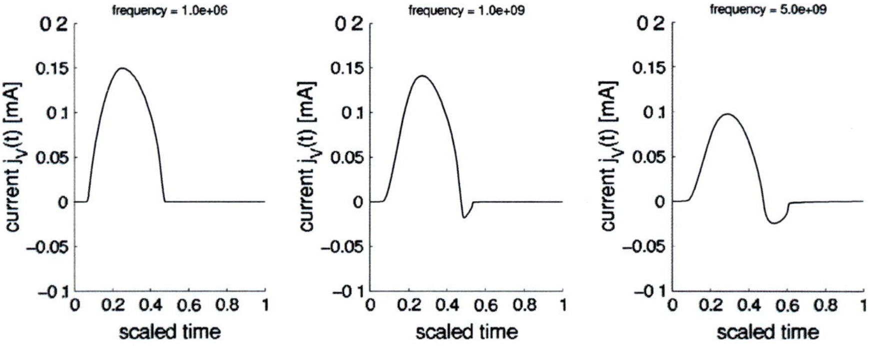

Fig. 10.2 Current j_V through the basic network for input frequencies 1 MHz, 1 GHz and 5 GHz. The capacitive effect is clearly demonstrated

with sparse matrices $D \in \mathrm{R}^{N\times(N+1)}$, $M \in \mathbb{R}^{(N+1)\times(N+1)}$, a bilinear map $\hat{\mathbf{B}} : \mathbb{R}^{(N+1)} \times \mathbb{R}^N \to \mathbb{R}^{N+1}$, and vectors $\hat{C}_h \in \mathbb{R}^N$ and $\hat{b}_\psi(e(t)), \hat{b}_n, \hat{b}_p \in \mathbb{R}^{N+1}$.

The discretized equations are implemented in MATLAB, and the DASPK software package [13] is used to integrate the high dimensional DAE. Initial values are stationary states obtained by setting all time derivatives to 0. In order to solve the Newton systems which arise from the BDF method efficiently, one may reorder the variables of the sparse system with respect to minimal bandwidth. Then, one can use the internal DASPK routines for the solution of the linear systems. Alternatively one can implement the preconditioning subroutine of DASPK using a direct sparse solver. Note that for both strategies we only need to calculate the reordering matrices once, since the sparsity structure remains constant.

A basic test circuit with one diode is depicted in Fig. 10.1, where the model parameters are presented in Table 10.1. The input $v_s(t)$ is chosen to be sinusoidal with amplitude 5 V. The numerical results in Fig. 10.2 show the capacitive effect of the diode for high input frequencies. Similar results are obtained in [16] using the simulator MECS.

10.4 Model Reduction

Now we want to reduce the computational effort of repeated dynamical simulations by applying proper orthogonal decomposition (POD) to the drift-diffusion equations. The POD reduction procedure is formulated in the Appendix, and from here onwards is used with $X := L^2(\Omega)$. As an example we discuss the reduction based on the standard finite element approximation (10.19), Raviart-Thomas elements are treated analogously.

The test functions for a standard Dirichlet problem are expected to vanish at the boundary. Hence, in the standard finite element case, before performing the POD of the state space, we relate the solution to a reference state (not necessarily a solution), e.g. a stationary solution or a mean value, that fulfills the boundary conditions. The functions

$$\tilde{\psi} = \psi - \psi_r, \quad \tilde{n} = n - n_r, \quad \tilde{p} = p - p_r$$

then satisfy homogeneous boundary conditions. The reference ψ_r is an exception in so far, that in general it cannot be a stationary state, since ψ underlies varying boundary conditions. Here, we use the stationary state, scaled in such a way that the boundary conditions are satisfied, e.g.

$$\psi_r(t,x) = \frac{\psi(t,1)}{\psi(0,1)}\psi_r(0,x) = \frac{A_S^\top e(t) + \psi_{bi}}{A_S^\top e(0) + \psi_{bi}}\psi_r(0,x).$$

We further set

$$\tilde{J}_n = J_n - J_n^r, \quad \tilde{J}_p = J_p - J_p^r.$$

In the case of the Raviart-Thomas approach the relation to a reference state is not necessary, since the boundary values are included more naturally through the variational formulation.

The time-snapshot POD procedure described in the Appendix delivers Galerkin ansatz spaces for $\tilde{\psi}$, $\tilde{n}$, $\tilde{p}$, $\tilde{J}_n$ and $\tilde{J}_p$. This leads to the ansatz

$$\psi^{\mathrm{POD}}(t) = \psi_r(t) + U_\psi H_\psi(t),$$

$$n^{\mathrm{POD}}(t) = n_r(t) + U_n H_n(t), \quad p^{\mathrm{POD}}(t) = p_r(t) + U_p H_p(t),$$

$$J_n^{\mathrm{POD}}](t) = J_n^r(t) + U_{J_n} H_{J_n}(t), \quad J_p^{\mathrm{POD}}(t) = J_p^r(t) + U_{J_p} H_{J_p}(t).$$

The injection matrices

$$U_\psi \in \mathbb{R}^{(N-1)\times k_\psi}, \quad U_n \in \mathbb{R}^{(N-1)\times k_n}, \quad U_p \in \mathbb{R}^{(N-1)\times k_p},$$

$$U_{J_n} \in \mathbb{R}^{N\times k_{J_n}}, \quad U_{J_p} \in \mathbb{R}^{N\times k_{J_p}},$$

contain the (time independent) POD-functions as columns as in Eq. 10.21, the vectors $H_{(\cdot)}$ the corresponding time-variant coefficients. The number $k_{(\cdot)}$ is the respective number of POD basis functions included. Assembling the POD system then leads to a DAE system similar to (10.19),

$$\begin{aligned} 0 &= \lambda \tilde{S} H_\psi + U_\psi^\top (M(U_n H_n - U_p H_p) - C_h) + c_1 + \tilde{c}_1(e), \\ -\tilde{M}_{nn} \dot{H}_n &= -v_n \tilde{D}_n^\top H_{J_n} + h U_n^\top R\big(n_r + U_n H_n, p_r + U_p H_p\big) + c_2, \\ \tilde{M}_{pp} \dot{H}_p &= -v_p \tilde{D}_p^\top H_{J_p} - h U_p^\top R\big(n_r + U_n H_n, p_r + U_p H_p\big) + c_3, \\ 0 &= h H_{J_n} + \tilde{D}_n H_n + \mathbf{B}(H_n, H_\psi) + L_n H_\psi + c_4, \\ 0 &= h H_{J_p} - \tilde{D}_p H_p + \mathbf{B}(H_p, H_\psi) + L_p H_\psi + c_5. \end{aligned}$$

Here,

$$\begin{aligned} \tilde{S} = U_\psi^\top S U_\psi, \quad \tilde{M}_{nn} = U_n^\top M U_n, \quad \tilde{M}_{pp} = U_p^\top M U_p, \\ \tilde{D}_n = U_{J_n}^\top D U_n, \quad \tilde{D}_p = U_{J_p}^\top D U_p, \end{aligned}$$

where the matrices M, S, and D and the vector C_h are the same as in Eq. 10.19. The constant vectors c_i and matrices L_n, L_p arise from the reference states, and can be computed offline together with the other matrices and the bi-linear map $\mathbf{B}$; c_2 and c_3 vanish in case of stationary reference states n_r and p_r. The vector $\tilde{c}_1(e(t))$ includes the boundary condition on ψ. Note, that all matrix-matrix multiplications, such as $U_\psi^\top M U_n$ can be calculated in the offline-phase. The nonlinear recombination function R must be evaluated online and cannot be reduced. Also the bi-linear $\mathbf{B}$ cannot be reduced completely. Besides the linear terms one can also expect a performance gain in the solution of the linear systems in BDF method, which are now relatively small.

10.5 Numerical Investigation

Figure 10.3 shows the development of the error between the reduced and the unreduced numerical solutions, plotted over the neglected information Δ (see (10.22)), which is measured by the relative error between the non-reduced states ψ, n, p, J_n, J_p and their projections onto the respective reduced state space. The number of POD basis functions for each variable is chosen so that the indicated approximation quality is reached, i.e. $\Delta := \Delta_\psi \simeq \Delta_n \simeq \Delta_p \simeq \Delta_{J_n} \simeq \Delta_{J_p}$. Since we compute all POD basis functions anyway, this procedure does not involve any additional costs.

In Fig. 10.4 the simulation times are plotted versus the neglected information Δ. As one also can see, the simulation based on standard finite elements takes twice as long as that based on RT elements. However, this difference is not observed for the simulation of the corresponding reduced models.

Fig. 10.3 L_2 error of j_V between reduced and unreduced problem, both for standard and Raviart-Thomas FEM

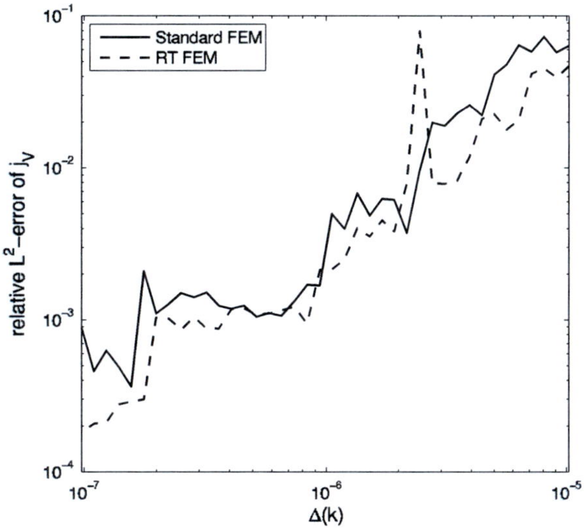

Fig. 10.4 Time consumption for simulation runs for Fig. 10.3. The fine lines indicate the time consumption for the simulation of the original full system

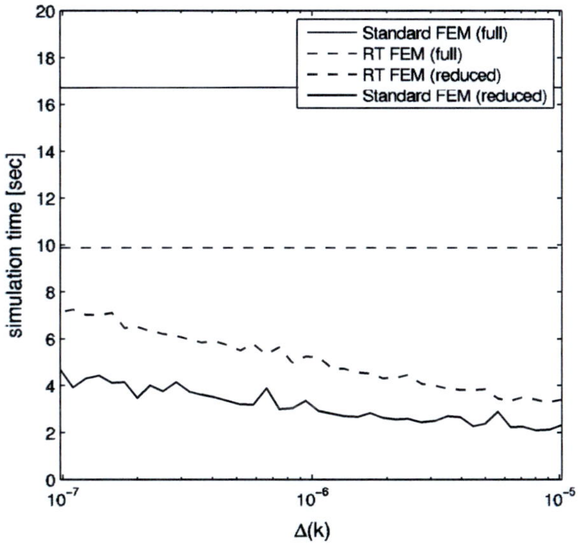

The whole POD approach in the present situation makes sense only, if the singular values would decay rapidly. Else one would have to use too many POD basis functions and would end up with a rather large and dense POD Galerkin system. To illustrate, that the singular values are indeed decaying exponentially, Fig. 10.5 shows the total number of singular vectors $k = k_\psi + k_n + k_p + k_{J_n} + k_{J_p}$ required to undercut a given state space cut-off error Δ. While the number of singular vectors included increases only linearly, the cut-off error tends to zero exponentially.

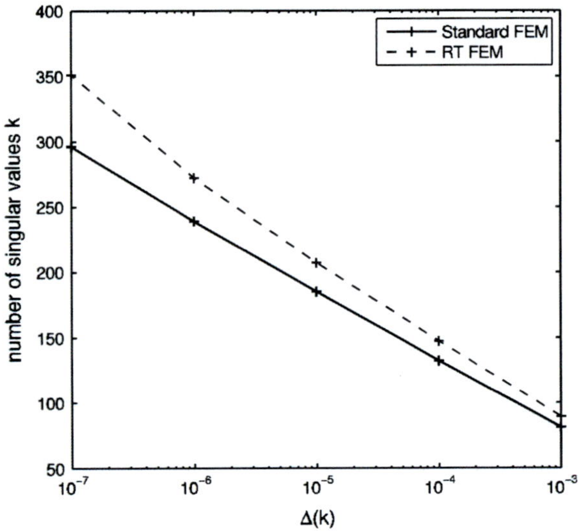

Fig. 10.5 The number of required singular values grows only logarithmically with the requested accuracy

Although POD model order reduction often works well, it is clearly a drawback of the method that the reduced system depends on the inputs and on the parameters of the system under consideration. A possible remedy consists in performing simulations over a certain input and/or parameter range and then to collect all simulations in a global snapshot matrix $Y := [Y^1, Y^2, \ldots]$. Here, each Y^i represents the snapshots taken for a certain input resp. parameter. In this context now the question pops up which inputs/parameters to choose in order to obtain a reduced model, which is valid over the whole input/parameter range. Let us elaborate on this complex of questions with the following example.

For the basic circuit we choose the frequency of the input voltage v_s as parameter. Let the parameter space be the interval $[10^8\,\text{Hz}, 10^{12}\,\text{Hz}]$. We start the investigation with a reduced model where the snapshot matrix is created from the simulation of the full model at a frequency of 10^{10} Hz. The difference between simulations of the full model and the reduced model is the reduction error plotted in Fig. 10.6 (dashed line). A second reduced model is constructed by adding snapshots from the full simulation at a frequency of 10^8 Hz, which is the frequency for which the error is maximal. We do not alter the number of considered singular values. One can see that the error is significantly reduced in the second model (dotted line). A third model is constructed analogously (solid line).

Of course this adaptive reduction method only works in theory, since it is based on full model integrations over the whole parameter space. For practical purposes we will need to develop a-posterior error estimators in the parameter space. It may be possible to apply the methods of [12] to the problem under consideration. In the present situation also a parameter POD along the lines of [10] could be applied. The parameter snapshots are chosen as simulations of the quasi-stationary drift-diffusion model related to the respective parameter value (here: frequency of sinusoidal input voltage source).

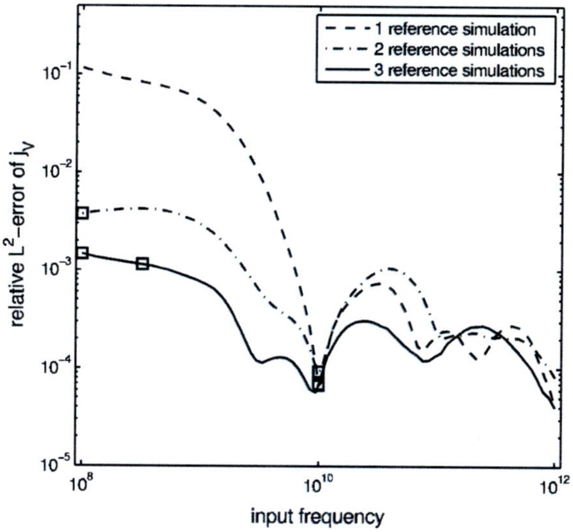

Fig. 10.6 Reduction error plotted over the frequency parameter space. The reduced model was created based on 1 to 3 full simulations at the reference frequencies. The reference frequencies are marked by squares. The number of singular values used for the reduction is the same for all three models

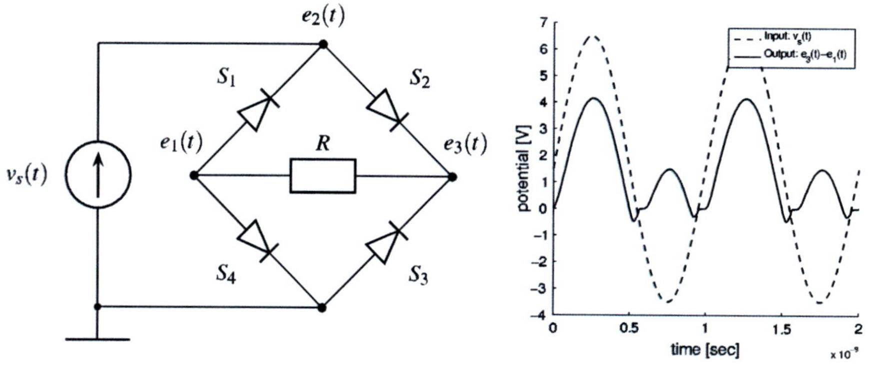

Fig. 10.7 Rectifier network and simulation results. The input v_s is sinusoidal with offset $+1.5$ [V]

Finally we note that the presented reduction method does not create reduced models for semiconductors as such, but accounts for the position of the semiconductors in a given network. The modes or POD basis functions of two identical semiconductors may be different due to their different operating states. To demonstrate this fact, we consider the rectifier network in Fig. 10.7 and measure the distance between the spaces U^1 and U^2 which are spanned, e.g., by the POD-functions U_ψ^1 of the diode S_1 and U_ψ^2 of the diode S_2 respectively, by

$$d(U^1, U^2) := \max_{\substack{u \in U^1 \\ \|u\|_2 = 1}} \min_{\substack{v \in U^2 \\ \|v\|_2 = 1}} \|u - v\|_2.$$

Exploiting the orthonormality of the bases U_ψ^1 and U_ψ^2 and using a Lagrange framework, we find

Table 10.2 Distances between reduced models in the rectifier network

Δ	$d(U^1, U^2)$	$d(U^1, U^3)$
10^{-4}	0.61288	5.373×10^{-8}
10^{-5}	0.50766	4.712×10^{-8}
10^{-6}	0.45492	2.767×10^{-7}
10^{-7}	0.54834	1.211×10^{-6}

$$d(U^1, U^2) = \sqrt{2 - 2\sqrt{\lambda}},$$

where λ is the smallest eigenvalue of the positive definite matrix $SS^\top$ with $S_{ij} = \langle u^1_{\psi,i}, u^2_{\psi,j} \rangle_2$. The distances for the rectifier network are given in Table 10.2. While the reduced model for the diodes S_1 and S_3 are almost equal, the models for the diodes S_1 and S_2 are significantly different. Similar results are obtained for the reduction of n, p, etc.

Acknowledgments The work reported in this paper was supported by the German Federal Ministry of Education and Research (BMBF), grant no. 03HIPAE5. Responsibility for the contents of this publication rests with the authors.

Appendix: Proper Orthogonal Decomposition

Let $y : \Omega \times [0, T] \to \mathbb{R}$ be a function, which in the situation of Sect. 10.4 represents the time continuous finite element approximation of ψ, n, p, g_ψ, J_n and J_p, respectively. For $t \in [0, T]$ let $y(\cdot, t) \in Y_N := \operatorname{span}\{\phi_1, \ldots, \phi_N\}$, where ϕ_i $(i = 1, \ldots, N)$ denote linearly independent elements of a Hilbert space X.

The snapshot variant of (POD) introduced in [18] works as follows; let $Y = [y^1, \ldots, y^l] \in \mathbb{R}^{N \times l}$ contain as columns the coefficient vectors of l time-snapshots $y(\cdot, t_i)$ taken at time instances $t_i \in [0, T]$, i.e.

$$y(\cdot, t_i) = \sum_{j=1}^{N} y^i_j \phi_j.$$

Furthermore, let $M := (\langle \phi_m, \phi_n \rangle_X)_{m,n=1,\ldots,N}$ with its Cholesky factorization $M = LL^\top$. Let $(\tilde{U}, \Sigma, \tilde{V})$ denote the singular value decomposition of $\tilde{Y} := L^\top Y$, i.e. $\tilde{Y} = \tilde{U}\Sigma\tilde{V}^\top$ with $\tilde{U} \in \mathbb{R}^{N \times N}$, $\Sigma \in \mathbb{R}^{N \times l}$, and $\tilde{V} \in \mathbb{R}^{l \times l}$ and Σ a diagonal matrix containing the singular values $0 \le \sigma_i$ in decreasing order for $1 \le i \le l$, where we assume $l \le N$. We set

$$U := L^{-\top} \tilde{U}_{(:,\,1:k)} \equiv [u^1, \ldots, u^k]. \tag{10.21}$$

Then, the k-dimensional POD basis of $\operatorname{span}\{\sum_{j=1}^N y^i_j \phi_j,\ i = 1, \ldots, l\} \subseteq Y_N$ is given by

$$\text{span}\left\{\sum_{j=1}^{N} u_j^i \phi_j,\ i = 1, \ldots, k\right\}.$$

Note that one chooses $k \leq m$, where $\sigma_m > 0$ denotes the smallest non-vanishing singular value. The expression $1 - \Delta(k)^2$ is a measure of the information content of the subspace spanned by the first k POD basis functions, where

$$\Delta(k) = \sqrt{1 - \frac{\sum_{i=1}^{k} \sigma_i^2}{\sum_{i=1}^{l} \sigma_i^2}}. \tag{10.22}$$

An extended introduction to POD can be found in [14].

References

1. Anile, A., Mascali, G., Romano, V.: Mathematical problems in semiconductor physics. Lectures given at the C. I. M. E. summer school, Cetraro, Italy, July 15–22, 1998. Lecture Notes in Mathematics. Springer, Berlin (2003)
2. Anile, A., Nikiforakis, N., Romano, V., Russo, G.: Discretization of semiconductor device problems. II. In: Schilders, W.H.A. et al. (eds.) Handbook of Numerical Analysis, vol. 13. Special volume: Numerical Methods in Electromagnetics, pp. 443–522. Elsevier/North Holland, Amsterdam (2005)
3. Bodestedt, M., Tischendorf, C.: PDAE models of integrated circuits and index analysis. Math. Comput. Model. Dyn. Syst. **13**(1), 1–17 (2007)
4. Brezzi, F., Marini, L., Micheletti, S., Pietra, P., Sacco, R., Wang, S.: Discretization of semiconductor device problems. I. In: Schilders, W.H.A. et al. (eds.) Handbook of Numerical Analysis, vol 13. Special volume: Numerical Methods in Electromagnetics, pp. 317–441. Elsevier/North-Holland, Amsterdam (2005)
5. Estévez Schwarz, D., Feldmann, U., März, R., Sturtzel, S., Tischendorf, C.: Finding beneficial DAE structures in circuit simulation. In: Jäger, W. et al. (eds.) Mathematics—Key Technology for the Future. Joint Projects Between Universities and Industry, pp. 413-428. Springer, Berlin (2003)
6. Estévez Schwarz, D., Tischendorf, C.: Structural analysis of electric circuits and consequences. Int. J. Circuit Theory Appl. **28**(2), 131–162 (2000)
7. Gummel, H.: A selfconsistent iterative scheme for onedimensional steady state transistor calculations. IEEE Trans. Electron Devices **11**(10), 455–465 (1964)
8. Günther, M.: Partielle differential-algebraische Systeme in der numerischen Zeitbereichs-analyse elektrischer Schaltungen. VDI Fortschritts-Berichte, Reihe 20, Rechnerunterstützte Verfahren, Nr. 343 (2001)
9. Günther, M., Feldmann, U., ter Maten, J.: Modelling and discretization of circuit problems. In: Schilders, W.H.A. et al. (eds.) Handbook of Numerical Analysis, vol 13. Special volume: Numerical Methods in Electromagnetics, pp. 523–629. Elsevier/North Holland, Amsterdam (2005)
10. Kahlbacher, M., Volkwein, S.: Galerkin proper orthogonal decomposition methods for parameter dependent elliptic systems. Discuss. Math., Differ. Incl. Control Optim. **27**(1), 95–117 (2007)
11. Markowich, P.: The Stationary Semiconductor Device Equations (Computational Microelectotrnics). Springer-Verlag, Wien-New York (1986)

12. Patera, A., Rozza, G.: Reduced Basis Approximation and A Posteriori Error Estimation for Parametrized Partial Differential Equations. Version 1.0. Copyright MIT 2006–2007, to appear in (tentative rubric) MIT Pappalardo Graduate Monographs in Mechanical Engineering (2007)
13. Petzold, L.R.: A description of DASSL A differential/algebraic system solver. IMACS Trans. Sci. Comput. **1**, 65–68 (1993)
14. Pinnau, R.: Model reduction via proper orthogonal decomposition. Schilders, Wilhelmus H. A. (ed.) et al., Model Order Reduction: Theory, Research Aspects and Applications. Selected papers based on the presentations at the workshop 'Model order reduction, coupled problems and optimization', Leiden, The Netherlands, September 19–23, 2005. Berlin: Springer. Mathematics in Industry 13, 95-109. (2008)
15. Selberherr, S.: Analysis and Simulation of Semiconductor Devices. Springer Verlag Wien New York (1984)
16. Selva Soto, M.: A coupled system for electrical circuits. Numerical simulations. Proc. Appl. Math. Mech. **6**(1), 51–54 (2006)
17. Selva Soto, M., Tischendorf, C.: Numerical analysis of DAEs from coupled circuit and semiconductor simulation. Appl. Numer. Math. **53**(2-4), 471–488 (2005)
18. Sirovich, L.: Turbulence and the dynamics of coherent structures I: Coherent structures. II: symmetries and transformations. III: Dynamics and scaling. Q. Appl. Math. **45**, 561–590 (1987)
19. Tischendorf, C.: Coupled Systems of Differential Algebraic and Partial Differential Equations in Circuit and Device Simulation. Habilitation thesis, Humboldt-University of Berlin (2003)

Chapter 11
Model Reduction of Periodic Descriptor Systems Using Balanced Truncation

Peter Benner, Mohammad-Sahadet Hossain and Tatjana Stykel

Abstract Linear periodic descriptor systems represent a broad class of time evolutionary processes in micro-electronics and circuit simulation. In this paper, we consider discrete-time linear periodic descriptor systems and study the concepts of periodic reachability and observability Gramians. We also discuss a lifted representation of periodic descriptor systems and propose a balanced truncation model reduction method for such systems.The behaviour of the suggested model reduction technique is illustrated using a numerical example.

Keywords Periodic descriptor systems · Lifted state space representation · Periodic projected Lyapunov equations · Balanced realization · Model reduction

11.1 Introduction

Linear discrete-time periodic descriptor systems have received a lot of attention over the last twenty years. They are suitable models for several natural as well as man-made phenomena, and have applications in modeling of periodic time-

---

P. Benner · M.-S. Hossain (✉)
Fakultät für Mathematik, TU Chemnitz, 09107 Chemnitz, Germany
e-mail: mohh@hrz.tu-chemnitz.de

P. Benner
e-mail: benner@mathematik.tu-chemnitz.de

T. Stykel
Institut für Mathematik, MA 4-5, TU Berlin, Straße des 17.Juni 136, 10623 Berlin, Germany
e-mail: stykel@math.tu-berlin.de

P. Benner et al. (eds.), *Model Reduction for Circuit Simulation*,
Lecture Notes in Electrical Engineering, 74, DOI: 10.1007/978-94-007-0089-5_11,

varying filters and networks [16, 21], multirate sampled-data systems [16, 18], circuit simulation [3, 8, 10, 18], micro-electronics [19, 20], aerospace realm [34, 35], control of industrial processes and communication systems [1, 20].

A linear discrete-time periodic descriptor system with time-varying dimensions has the form

$$E_k x_{k+1} = A_k x_k + B_k u_k, \quad y_k = C_k x_k, \quad k \in \mathbb{Z}, \tag{11.1}$$

where $E_k \in \mathbb{R}^{\mu_{k+1}\times n_{k+1}}$, $A_k \in \mathbb{R}^{\mu_{k+1}\times n_k}$, $B_k \in \mathbb{R}^{\mu_{k+1}\times m_k}$, $C_k \in \mathbb{R}^{p_k\times n_k}$ are periodic with a period $K \geq 1$ and $\sum_{k=0}^{K-1}\mu_k = \sum_{k=0}^{K-1} n_k = n$. The matrices E_k are allowed to be singular for all k. For system (11.1), a reduced-order model of dimension r would be a system of the form

$$\tilde{E}_k \tilde{x}_{k+1} = \tilde{A}_k \tilde{x}_k + \tilde{B}_k u_k, \quad \tilde{y}_k = \tilde{C}_k \tilde{x}_k, \quad k \in \mathbb{Z}, \tag{11.2}$$

where $\tilde{E}_k \in \mathbb{R}^{\gamma_{k+1}\times r_{k+1}}$, $\tilde{A}_k \in \mathbb{R}^{\gamma_{k+1}\times r_k}$, $\tilde{B}_k \in \mathbb{R}^{\gamma_{k+1}\times m_k}$, $\tilde{C}_k \in \mathbb{R}^{p_k\times r_k}$ are K-periodic matrices, $\sum_{k=0}^{K-1}\gamma_k = \sum_{k=0}^{K-1} r_k = r$ and $r \ll n$. Apart from having a much smaller state-space dimension, it is also important that the reduced-order model preserves physical properties of the original system such as regularity, stability and passivity, and that the approximation error is small.

The dynamics of the discrete-time periodic descriptor system (11.1) are often addressed by the regularity and the eigenstructure of the periodic matrix pairs $\{E_k, A_k\}_{k=0}^{K-1}$. If all E_k are nonsingular, the eigenvalues (also called characteristic multipliers) of system (11.1) are given by the eigenvalues of the matrix product

$$E_{K-1}^{-1}A_{K-1}E_{K-2}^{-1}A_{K-2}\cdots E_0^{-1}A_0 \tag{11.3}$$

associated with the periodic matrix pairs $\{E_k, A_k\}_{k=0}^{K-1}$. This product only yields a well-defined matrix if all E_k are nonsingular. Even if they are, the formulation of that matrix should be avoided for reasons of numerical stability. Note that even for some E_k being singular, we use (11.3) in a formal way to denote a generalization of matrix pencils to this periodic case (see [2] for details of this formal matrix product calculus). We compute the eigenvalues of (11.3) via the generalized periodic Schur decomposition [11, 32].

There exist unitary matrices $P_k \in \mathbb{C}^{\mu_{k+1}\times\mu_{k+1}}$ and $Q_k \in \mathbb{C}^{n_k\times n_k}$, with $Q_{k+K} = Q_k$ such that the transformed matrices

$$N_k = P_k^* E_k Q_{k+1}, \quad M_k = P_k^* A_k Q_k, \quad k = 0, \ldots, K-1,$$

are all upper triangular, where for the ease of notation we allow complex arithmetic (in practice, however, computations can be performed in real arithmetic leading to quasi-triangular structure of one of the M_k).

Then the formal matrix product

$$N_{K-1}^{-1}M_{K-1}N_{K-2}^{-1}M_{K-2}\cdots N_0^{-1}M_0$$

has the same eigenvalues as (11.3). The blocks on the diagonals of the transformed matrices M_k and N_k are used to define the eigenvalues of the periodic matrix pairs $\{E_k, A_k\}_{k=0}^{K-1}$. A *finite eigenvalue* is given by

$$\lambda_l = \prod_{k=0}^{K-1} \frac{m_{ll}^{(k)}}{n_{ll}^{(k)}},$$

provided $n_{ll}^{(k)} \neq 0$ for $k = 0, \ldots, K-1$. Here $m_{ll}^{(k)} \in \Lambda(M_k)$ and $n_{ll}^{(k)} \in \Lambda(N_k)$, where Λ denotes the eigenspectrum of the corresponding matrix. An eigenvalue is called infinite if $\prod_{k=0}^{K-1} n_{ll}^{(k)} = 0$, but $\prod_{k=0}^{K-1} m_{ll}^{(k)} \neq 0$.

In this paper, we briefly review some basic concepts of discrete-time periodic descriptor systems (Sect. 11.2). In Sect. 11.3, we study the periodic reachability and observability Gramians from [6] using a lifted representation. A balanced truncation model reduction method for periodic descriptor systems is presented in Sect. 11.4. Section 11.5 contains a numerical example that illustrates the properties of the suggested model reduction technique.

11.2 Periodic Descriptor Systems

Lifted representations of discrete-time periodic systems play an important role in extending many theoretical results and numerical algorithms for time-invariant systems to the periodic setting [4, 7, 33]. We consider here the cyclic lifted representation which was introduced first for standard periodic systems in [17]. The *cyclic lifted representation* of the periodic descriptor system (11.1) is given by

$$\mathscr{E}\mathscr{X}_{k+1} = \mathscr{A}\mathscr{X}_k + \mathscr{B}\mathscr{U}_k, \quad \mathscr{Y}_k = \mathscr{C}\mathscr{X}_k, \tag{11.4}$$

where

$$\mathscr{E} = \operatorname{diag}(E_0, E_1, \ldots, E_{K-1}), \quad \mathscr{B} = \operatorname{diag}(B_0, B_1, \ldots, B_{K-1}),$$
$$\mathscr{A} = \begin{bmatrix} 0 & \cdots & 0 & A_0 \\ A_1 & & & 0 \\ & \ddots & & \vdots \\ 0 & & A_{K-1} & 0 \end{bmatrix}, \quad \mathscr{C} = \begin{bmatrix} 0 & \cdots & 0 & C_0 \\ C_1 & & & 0 \\ & \ddots & & \vdots \\ 0 & & C_{K-1} & 0 \end{bmatrix}, \tag{11.5}$$

The descriptor vector, system input and output of (11.4) are related to those of (11.1) via

$$\mathscr{X}_k = [x_1^T, \ldots, x_{K-1}^T, x_0^T]^T, \quad \mathscr{U}_k = [u_0^T, u_1^T, \ldots, u_{K-1}^T]^T, \quad \mathscr{Y}_k = [y_0^T, y_1^T, \ldots, y_{K-1}^T]^T,$$

respectively.

A set of periodic matrix pairs $\{E_k, A_k\}_{k=0}^{K-1}$ is called *regular* if the pencil $z\mathscr{E} - \mathscr{A}$ is regular, i.e., $\det(z\mathscr{E} - \mathscr{A}) \not\equiv 0$. In this case, we can define a transfer function of the lifted system (11.4) as $\mathscr{H}(z) = \mathscr{C}(z\mathscr{E} - \mathscr{A})^{-1}\mathscr{B}$.

The regular set of periodic matrix pairs $\{E_k, A_k\}_{k=0}^{K-1}$ can be transformed into a periodic Kronecker canonical form [6, 32]. For $k = 0, 1, \ldots, K-1$, there exist nonsingular matrices $W_k \in \mathbb{R}^{\mu_{k+1} \times \mu_{k+1}}$ and $Z_k \in \mathbb{R}^{n_k \times n_k}$ such that

$$W_k E_k Z_{k+1} = \begin{bmatrix} I_{n_{k+1}^f} & 0 \\ 0 & E_k^b \end{bmatrix}, \quad W_k A_k Z_k = \begin{bmatrix} A_k^f & 0 \\ 0 & I_{n_k^\infty} \end{bmatrix}, \tag{11.6}$$

where $Z_K = Z_0$, $A_{k+K-1}^f A_{k+K-2}^f \cdots A_k^f = J_k$ is an $n_k^f \times n_k^f$ matrix corresponding to the finite eigenvalues, $E_k^b E_{k+1}^b \cdots E_{k+K-1}^b = N_k$ is an $n_k^\infty \times n_k^\infty$ nilpotent matrix corresponding to an eigenvalue at infinity, $n_k = n_k^f + n_k^\infty$ and $\mu_{k+1} = n_{k+1}^f + n_k^\infty$. The index ν of the periodic descriptor system (11.1) is defined as $\nu = \max(\nu_0, \nu_1, \ldots, \nu_{K-1})$, where ν_k is the nilpotency index of N_k. Note that the finite eigenvalues of $\{E_k, A_k\}_{k=0}^{K-1}$ coincide with the finite eigenvalues of the lifted pencil $z\mathscr{E} - \mathscr{A}$.

For $k = 0, 1, \ldots, K-1$, the matrices

$$P_r(k) = Z_k \begin{bmatrix} I_{n_{k+1}^f} & 0 \\ 0 & 0 \end{bmatrix} Z_k^{-1}, \quad P_l(k) = W_k^{-1} \begin{bmatrix} I_{n_{k+1}^f} & 0 \\ 0 & 0 \end{bmatrix} W_k,$$

are the *spectral projectors* onto the kth right and left deflating subspaces of the periodic matrix pairs $\{E_k, A_k\}_{k=0}^{K-1}$ corresponding to the finite eigenvalues, and $Q_r(k) = I - P_r(k)$ and $Q_l(k) = I - P_l(k)$ are the complementary projectors. Let for every $k = 0, 1, \ldots, K-1$, the vector $Z_k^{-1} x_k = [\,(x_k^f)^T,\ (x_k^b)^T\,]^T$ and the matrices

$$W_k B_k = \begin{bmatrix} B_k^f \\ B_k^b \end{bmatrix}, \quad C_k Z_k = \begin{bmatrix} C_k^f, & C_k^b \end{bmatrix},$$

be partitioned in blocks conformally to the periodic matrix pairs $\{E_k, A_k\}_{k=0}^{K-1}$ in (11.6). Under this transformation, system (11.1) can be decoupled into forward and backward periodic subsystems

$$x_{k+1}^f = A_k^f x_k^f + B_k^f u_k, \quad y_k^f = C_k^f x_k^f, \tag{11.7}$$

$$E_k^b x_{k+1}^b = x_k^b + B_k^b u_k, \quad y_k^b = C_k^b x_k^b, \tag{11.8}$$

respectively, with $y_k = y_k^f + y_k^b$, $k = 0, 1, \ldots, K-1$. The state transition matrix for the forward subsystem (11.7) is given by $\Phi_f(i,j) = A_{i-1}^f A_{i-2}^f \cdots A_j^f$ for $i > j$ and $\Phi_f(i,i) = I_{n_i^f}$. For the backward subsystem (11.8), the state transition matrix is defined as $\Phi_b(i,j) = E_i^b E_{i+1}^b \cdots E_{j-1}^b$ for $i < j$ and $\Phi_b(i,i) = I_{n_i^\infty}$. Using these matrices we can now define the forward and backward fundamental matrices of the periodic descriptor system (11.1) as

$$\Psi_{i,j} = \begin{cases} Z_i \begin{bmatrix} \Phi_f(i,j+1) & 0 \\ 0 & 0 \end{bmatrix} W_j, & i > j, \\ Z_i \begin{bmatrix} 0 & 0 \\ 0 & -\Phi_b(i,j) \end{bmatrix} W_j, & i \leq j. \end{cases}$$

These fundamental matrices play an important role in the definition of the reachability and observability Gramians of the periodic descriptor system (11.1) that we will consider in the next section.

11.3 Periodic Gramians and Matrix Equations

The dynamics of the periodic descriptor system (11.1) are often addressed by the eigenstructure of the periodic matrix pairs $\{E_k, A_k\}_{k=0}^{K-1}$.

Definition 1 The periodic matrix pairs $\{E_k, A_k\}_{k=0}^{K-1}$ are said to be *periodic stable* (*pd-stable*) if all the finite eigenvalues of $\{E_k, A_k\}_{k=0}^{K-1}$ lie inside the unit circle.

In balanced truncation model reduction, Gramians play a fundamental role [15, 25, 28, 31]. For periodic descriptor system (11.1), the reachability and observability Gramians have been first introduced in [6].

Definition 2 Suppose that the periodic matrix pairs $\{E_k, A_k\}_{k=0}^{K-1}$ are pd-stable. For $k = 0, 1, \ldots, K-1$, we define the *causal* and *noncausal reachability Gramians* G_k^{cr} and G_k^{ncr} of system (11.1) as

$$G_k^{cr} = \sum_{j=-\infty}^{k-1} \Psi_{k,j} B_j B_j^T \Psi_{k,j}^T, \quad G_k^{ncr} = \sum_{j=k}^{k+vK-1} \Psi_{k,j} B_j B_j^T \Psi_{k,j}^T.$$

The *complete reachability Gramian* G_k^r is the sum of the causal and noncausal Gramians, i.e., $G_k^r = G_k^{cr} + G_k^{ncr}$ for $k = 0, 1, \ldots, K-1$.

Definition 3 For the pd-stable matrix pairs $\{E_k, A_k\}_{k=0}^{K-1}$ and $k = 0, 1, \ldots, K-1$, the *causal* and *noncausal observability Gramians* G_k^{co} and G_k^{nco} of system (11.1) are defined as

$$G_k^{co} = \sum_{j=k}^{\infty} \Psi_{j,k-1}^T C_j^T C_j \Psi_{j,k-1}, \quad G_k^{nco} = \sum_{j=k-vK}^{k-1} \Psi_{j,k-1}^T C_j^T C_j \Psi_{j,k-1}.$$

The *complete observability Gramian* G_k^o is the sum of the causal and noncausal Gramians, i.e., $G_k^o = G_k^{co} + G_k^{nco}$ for $k = 0, 1, \ldots, K-1$.

Note that the causal and noncausal Gramians correspond to the forward and backward subsystems (11.7) and (11.8), respectively.

11.3.1 Periodic Projected Lyapunov Equations

It has been shown in [26] that the Gramians of discrete-time descriptor systems satisfy projected generalized discrete-time Lyapunov equations with special right-hand sides. A similar result also holds for periodic descriptor systems.

Theorem 1 [6] *Consider a periodic discrete-time descriptor system (11.1), where the periodic matrix pairs* $\{E_k, A_k\}_{k=0}^{K-1}$ *are pd-stable.*

1. *For* $k = 0, 1, \ldots, K-1$, *the causal reachability and observability Gramians* $\{G_k^{cr}\}_{k=0}^{K-1}$ *and* $\{G_k^{co}\}_{k=0}^{K-1}$ *are the unique symmetric, positive semidefinite solutions of the projected generalized discrete-time periodic Lyapunov equations (PGDPLEs)*

$$\begin{aligned} A_k G_k^{cr} A_k^T - E_k G_{k+1}^{cr} E_k^T &= -P_l(k) B_k B_k^T P_l(k)^T, \\ G_k^{cr} &= P_r(k) G_k^{cr} P_r(k)^T, \end{aligned} \tag{11.9}$$

and

$$\begin{aligned} A_k^T G_{k+1}^{co} A_k - E_{k-1}^T G_k^{co} E_{k-1} &= -\, P_r(k)^T C_k^T C_k P_r(k), \\ G_k^{co} &= P_l(k-1)^T G_k^{co} P_l(k-1), \end{aligned}$$

respectively, where $G_K^{cr} = G_0^{cr}, G_K^{co} = G_0^{co}, E_{-1} = E_{K-1},$ *and* $P_l(-1) = P_l(K-1)$.
2. *For* $k = 0, 1, \ldots, K-1$, *the noncausal reachability and observability Gramians* $\{G_k^{ncr}\}_{k=0}^{K-1}$ *and* $\{G_k^{nco}\}_{k=0}^{K-1}$ *are the unique symmetric, positive semidefinite solutions of the PGDPLEs*

$$\begin{aligned} A_k G_k^{ncr} A_k^T - E_k G_{k+1}^{ncr} E_k^T &= Q_l(k) B_k B_k^T Q_l(k)^T, \\ G_k^{ncr} &= Q_r(k) G_k^{ncr} Q_r(k)^T, \end{aligned}$$

and

$$\begin{aligned} A_k^T G_{k+1}^{nco} A_k - E_{k-1}^T G_k^{nco} E_{k-1} &= Q_r(k)^T C_k^T C_k Q_r(k) \\ G_k^{nco} &= Q_l(k-1)^T G_k^{nco} Q_l(k-1), \end{aligned}$$

respectively, where $G_K^{ncr} = G_0^{ncr}$, $G_K^{nco} = G_0^{nco}$ *and* $Q_l(-1) = Q_l(K-1)$.

Numerical solution of the PGDPLEs has been considered in [6]. The method proposed there extends the periodic Schur method [5, 29, 30] and the generalized Schur-Hammarling method [23] developed for periodic standard and projected generalized Lyapunov equations, respectively. This method is based on an initial reduction of the periodic matrix pairs $\{E_k, A_k\}_{k=0}^{K-1}$ to the generalized periodic Schur form [11, 32] and solving the resulting generalized periodic Sylvester and Lyapunov equations of (quasi)-triangular structure using the recursive blocked algorithms [9]. Due to the computational complexity, the periodic generalized Schur-Hammarling method is restricted to problems of small and medium size. In [12, 27, 29], iterative methods based on Smith iterations [22] have been developed for periodic standard Lyapunov equations and also for projected generalized Lyapunov equations. These methods can also be extended to periodic projected Lyapunov equations by using the lifted representation of these equations. Therefore, we will discuss these lifted representations in the following subsection.

11.3.2 Lifted Representation of Periodic Lyapunov Equations

It is known that the Gramians of standard periodic systems satisfy the lifted form of the periodic Lyapunov equations and the solutions of these equations are diagonal matrices [12, 29].

The following theorem describes the block structures of the solutions of periodic Lyapunov equations in lifted form and their relations to the corresponding solutions of PGDPLEs in Theorem 1.

Theorem 2 *Consider the periodic discrete-time descriptor system* (11.1) *and its lifted representation* (11.4)*, where the periodic matrix pairs* $\{E_k, A_k\}_{k=0}^{K-1}$ *are pd-stable. The causal and noncausal reachability Gramians* $\mathscr{G}^{cr}$ *and* $\mathscr{G}^{ncr}$ *satisfy the lifted projected Lyapunov equations*

$$\begin{aligned} \mathscr{A}\mathscr{G}^{cr}\mathscr{A}^T - \mathscr{E}\mathscr{G}^{cr}\mathscr{E}^T &= -\mathscr{P}_l\mathscr{B}\mathscr{B}^T\mathscr{P}_l^T, \quad \mathscr{G}^{cr} = \mathscr{P}_r\mathscr{G}^{cr}\mathscr{P}_r^T, \\ \mathscr{A}\mathscr{G}^{ncr}\mathscr{A}^T - \mathscr{E}\mathscr{G}^{ncr}\mathscr{E}^T &= \mathscr{Q}_l\mathscr{B}\mathscr{B}^T\mathscr{Q}_l^T, \quad \mathscr{G}^{ncr} = \mathscr{Q}_r\mathscr{G}^{ncr}\mathscr{Q}_r^T, \end{aligned} \tag{11.10}$$

respectively, where $\mathscr{E}$, $\mathscr{A}$ *and* $\mathscr{B}$ *are as in* (11.5) *and*

$$\mathscr{G}^{cr} = \mathrm{diag}(G_1^{cr}, \ldots, G_{K-1}^{cr}, G_0^{cr}), \quad \mathscr{G}^{ncr} = \mathrm{diag}(G_1^{ncr}, \ldots, G_{K-1}^{ncr}, G_0^{ncr}),$$

$$\begin{aligned} \mathscr{P}_l &= \mathrm{diag}(P_l(0), P_l(1), \ldots, P_l(K-1)), \quad \mathscr{Q}_l = I - \mathscr{P}_l, \\ \mathscr{P}_r &= \mathrm{diag}(P_r(1), \ldots, P_r(K-1), P_r(0)), \quad \mathscr{Q}_r = I - \mathscr{P}_r. \end{aligned} \tag{11.11}$$

Proof We will only sketch the proof due to space limitation of the paper. Using the block structure of matrix coefficients, a straightforward computation shows that the projected Lyapunov Equation (11.10) is equivalent to the periodic projected Lyapunov Equation (11.9). Since the periodic matrix pairs $\{E_k, A_k\}_{k=0}^{K-1}$ are pd-stable, the pencil $z\mathscr{E} - \mathscr{A}$ is regular and all its eigenvalues lie inside the unit circle. Then (11.10) has a unique solution [24]. The proof for $\mathscr{G}^{ncr}$ can be treated similarly. □

For the observability Gramians, the situation becomes a bit more complex. The reason is that we do not want to destroy the block diagonal structure of the lifted solutions and we would like to use the lifted solution to find a balanced realization of the original system.

Theorem 3 *Consider the periodic discrete-time descriptor system* (11.1) *and its lifted representation* (11.4). *The causal and noncausal observability Gramians* $\mathcal{G}^{co}$ *and* $\mathcal{G}^{nco}$ *satisfy the lifted projected Lyapunov equations*

$$\mathcal{A}^T\mathcal{G}^{co}\mathcal{A} - \mathcal{E}^T\mathcal{G}^{co}\mathcal{E} = -\mathcal{P}_r^T\hat{\mathcal{C}}^T\hat{\mathcal{C}}\mathcal{P}_r, \quad \mathcal{G}^{co} = \mathcal{P}_l^T\mathcal{G}^{co}\mathcal{P}_l,$$
$$\mathcal{A}^T\mathcal{G}^{nco}\mathcal{A} - \mathcal{E}^T\mathcal{G}^{nco}\mathcal{E} = \mathcal{Q}_r^T\hat{\mathcal{C}}^T\hat{\mathcal{C}}\mathcal{Q}_r, \quad \mathcal{G}^{nco} = \mathcal{Q}_l^T\mathcal{G}^{nco}\mathcal{Q}_l,$$

respectively, where $\mathcal{E}$ *and* $\mathcal{A}$ *are as in* (11.5), *the projectors* $\mathcal{P}_l$, $\mathcal{P}_r$, $\mathcal{Q}_l$ *and* $\mathcal{Q}_r$ *are as in* (11.11), $\hat{\mathcal{C}} = \text{diag}(C_1, \ldots, C_{K-1}, C_0)$ *and*

$$\mathcal{G}^{co} = \text{diag}(G_1^{co}, \ldots, G_{K-1}^{co}, G_0^{co}), \quad \mathcal{G}^{nco} = \text{diag}(G_1^{nco}, \ldots, G_{K-1}^{nco}, G_0^{nco}).$$

Proof The proof is analogous to the previous proof of Theorem 2. □

11.3.3 Hankel Singular Values

Let the periodic matrix pairs $\{E_k, A_k\}_{k=0}^{K-1}$ be pd-stable. Then the causal and noncausal matrices $M_k^c = G_k^{cr}E_{k-1}^TG_k^{co}E_{k-1}$ and $M_k^{nc} = G_k^{ncr}A_k^TG_{k+1}^{nco}A_k$, $k = 0, 1, \ldots, K-1$, have real and nonnegative eigenvalues. These eigenvalues are used to define the causal and noncausal Hankel singular values of system (11.1).

Definition 4 Let the set of periodic matrix pairs $\{E_k, A_k\}_{k=0}^{K-1}$ be pd-stable. For $k = 0, 1, \ldots, K-1$, the square roots of the largest n_k^f eigenvalues of the matrix M_k^c, denoted by $\sigma_{k,j}$, are called the *causal Hankel singular values* and the square roots of the largest n_k^∞ eigenvalues of M_k^{nc}, denoted by $\theta_{k,j}$, are called the *noncausal Hankel singular values* of the periodic descriptor system (11.1).

Since the causal and noncausal reachability and observability Gramians are symmetric and positive semidefinite, there exist the Cholesky factorizations

$$G_k^{cr} = R_kR_k^T, \quad G_k^{co} = L_k^TL_k, \quad G_k^{ncr} = \check{R}_k\check{R}_k^T, \quad G_k^{nco} = \check{L}_k^T\check{L}_k, \tag{11.12}$$

Simple calculations show that $\sigma_{k,j} = \zeta_j(L_kE_{k-1}R_k)$ and $\theta_{k,j} = \zeta_j(\check{L}_{k+1}A_k\check{R}_k)$, where $\zeta_j(.)$ denotes the singular values of the corresponding matrices.

11.4 Balanced Truncation Model Reduction

In this section, we present a generalization of a balanced truncation model reduction method to periodic descriptor systems. For a balanced system, the reachability and observability Gramians are both equal to a diagonal matrix

[15, 25]. Balanced realizations for periodic descriptor system have been considered in [6].

Definition 5 A realization (E_k, A_k, B_k, C_k) of a periodic descriptor system (11.1) is called *balanced* if

$$G_k^{cr} = G_k^{co} = \begin{bmatrix} \Sigma_k & 0 \\ 0 & 0 \end{bmatrix}, \quad G_k^{ncr} = G_{k+1}^{nco} = \begin{bmatrix} 0 & 0 \\ 0 & \Theta_k \end{bmatrix},$$

where $\Sigma_k = \mathrm{diag}(\sigma_{k,1}, \ldots, \sigma_{k,n_k^f})$ and $\Theta_k = \mathrm{diag}(\theta_{k,1}, \ldots, \theta_{k,n_k^\infty})$, $k = 0, 1, \ldots, K-1$.

Consider the Cholesky factorizations (11.12) of the reachability and observability Gramians and let

$$L_k E_{k-1} R_k = U_k \Sigma_k V_k^T, \quad \check{L}_{k+1} A_k \check{R}_k = \check{U}_k \Theta_k \check{V}_k^T \tag{11.13}$$

be the singular value decompositions of the matrices $L_k E_{k-1} R_k$ and $\check{L}_{k+1} A_k \check{R}_k$ for $k = 0, 1, \ldots, K-1$. Here $U_k, V_k, \check{U}_k, \check{V}_k$ are orthogonal, and Σ_k and Θ_k are diagonal. If a realization (E_k, A_k, B_k, C_k) with pd-stable matrix pairs $\{E_k, A_k\}_{k=0}^{K-1}$ is minimal, i.e., Σ_k and Θ_k are nonsingular, then there exist nonsingular periodic matrices

$$S_k = [L_{k+1}^T U_{k+1} \Sigma_{k+1}^{-1/2}, \ \check{L}_{k+1}^T \check{U}_k \Theta_k^{-1/2}], \quad T_k = [R_k V_k \Sigma_k^{-1/2}, \ \check{R}_k \check{V}_k \Theta_k^{-1/2}],$$

such that the transformed realization $(S_k^T E_k T_{k+1}, S_k^T A_k T_k, S_k^T B_k, C_k T_k)$ is balanced [6]. Note that as in the case of standard state space systems, the balancing transformation matrices for periodic discrete-time descriptor system (11.1) are not unique.

Model reduction via balanced truncation is discussed very widely for standard discrete-time periodic systems [12, 31] and also for continuous-time descriptor systems [14, 25]. For a balanced system, truncation of states related to the small causal Hankel singular values does not change system properties essentially. Unfortunately, we can not do the same for the noncausal Hankel singular values. If we truncate the states that correspond to the small non-zero noncausal Hankel singular values, then the pencil for the reduced-order system may get finite eigenvalues outside the unit circle that will lead to additional errors in the system approximation.

Assume that the periodic matrix pairs $\{E_k, A_k\}_{k=0}^{K-1}$ are pd-stable. Consider the Cholesky factorizations in (11.12). Let

$$L_k E_{k-1} R_k = [U_{k,1}, U_{k,2}] \begin{bmatrix} \Sigma_{k,1} & \\ & \Sigma_{k,2} \end{bmatrix} [V_{k,1}, V_{k,2}]^T, \quad \check{L}_{k+1} A_k \check{R}_k = \check{U}_k \Theta_k \check{V}_k^T,$$

be singular value decompositions of $L_k E_{k-1} R_k$ and $\check{L}_{k+1} A_k \check{R}_k$, where

$$\Sigma_{k,1} = \operatorname{diag}(\sigma_{k,1}, \ldots, \sigma_{k,r_k^f}), \quad \Sigma_{k,2} = \operatorname{diag}(\sigma_{k,r_{k+1}^f}, \ldots, \sigma_{n_k^f}),$$

with $\sigma_{k,1} \geq \cdots \geq \sigma_{k,r_k^f} > \sigma_{k,r_{k+1}^f} \geq \ldots \geq \sigma_{k,n_k^f} > 0$, and $\Theta_k = \operatorname{diag}(\theta_{k,1}, \ldots, \theta_{k,r_k^\infty})$ is nonsingular for $k = 0, 1, \ldots, K-1$. Then the reduced-order system can be computed as

$$\tilde{E}_k = S_{k,r}^T E_k T_{k+1,r}, \quad \tilde{A}_k = S_{k,r}^T A_k T_{k,r}, \quad \tilde{B}_k = S_{k,r}^T B_k, \quad \tilde{C}_k = C_k T_{k,r}, \tag{11.14}$$

where

$$\begin{aligned} S_{k,r} &= [L_{k+1}^T U_{k+1,1} \Sigma_{k+1,1}^{-1/2}, \check{L}_{k+1}^T \check{U}_k \Theta_k^{-1/2}] \in \mathbb{R}^{\mu_{k+1}, r_{k+1}}, \\ T_{k,r} &= [R_k V_{k,1} \Sigma_{k,1}^{-1/2}, \quad \check{R}_k \check{V}_k \Theta_k^{-1/2}] \in \mathbb{R}^{n_k, r_k}, \end{aligned}$$

with $r_k = r_k^f + r_k^\infty$. Let $\tilde{\mathscr{H}}(z)$ be the transfer function of the reduced-order lifted system formed from the reduced-order subsystems in (11.14). Then we have the following $\mathbb{H}_\infty$-norm error bound

$$\|\mathscr{H} - \tilde{\mathscr{H}}\|_{\mathbb{H}_\infty} = \sup_{\omega \in [0, 2\pi]} \|\mathscr{H}(e^{i\omega}) - \tilde{\mathscr{H}}(e^{i\omega})\|_2 \leq 2 \sum_{k=0}^{K-1} \operatorname{trace}(\Sigma_{k,2}), \tag{11.15}$$

where $\|.\|_2$ denotes the matrix spectral norm and $\Sigma_{k,2}$ contains the truncated causal Hankel singular values. This error bound can be obtained similarly to the standard state space case [13, 31].

11.5 Example

We consider a periodic discrete-time descriptor system with $\mu_k = n_k = 10$, $n = 10$, $m = 2$, $p_k = 3$, and period $K = 3$ as presented in [6, Example 1]. The periodic matrix pairs $\{E_k, A_k\}_{k=0}^{K-1}$ are pd-stable with $n_k^f = 8$ and $n_k^\infty = 2$ for $k = 0, 1, 2$. The norms of the computed solutions of the periodic Lyapunov equations and the corresponding residuals, e.g.,

$$\rho_k^{cr} = \|A_k G_k^{cr} A_k^T - E_k G_{k+1}^{cr} E_k^T + P_l(k) B_k B_k^T P_l(k)^T\|_2,$$

are shown in Tables 11.1 and 11.2.

Table 11.1 Norms and relative residuals for the reachability Gramians

k	$\lVert G_k^{cr} \rVert_2$	ρ_k^{cr}	$\lVert G_k^{ncr} \rVert_2$	ρ_k^{ncr}
0	5.8182×10^2	6.1727×10^{-12}	1.3946×10^1	1.5444×10^{-14}
1	8.2981×10^4	8.2172×10^{-12}	1.3660×10^1	1.7508×10^{-14}
2	7.1107×10^3	3.0961×10^{-12}	1.4308×10^1	3.3847×10^{-14}

Table 11.2 Norms and relative residuals for the observability Gramians

k	$\| G_k^{co} \|_2$	ρ_k^{co}	$\| G_k^{nco} \|_2$	ρ_k^{nco}
0	9.7353×10^1	2.7678×10^{-13}	1.6866×10^0	1.3372×10^{-15}
1	1.1373×10^3	7.7003×10^{-14}	1.7406×10^0	2.1113×10^{-15}
2	9.6984×10^0	1.7859×10^{-14}	1.6866×10^0	1.1626×10^{-15}

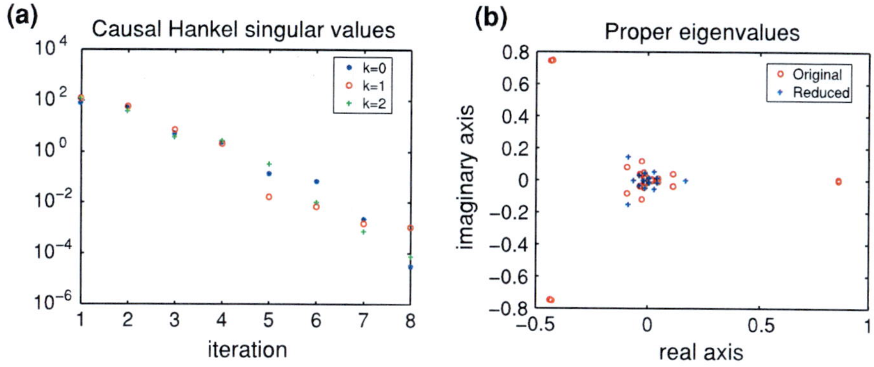

Fig. 11.1 **a** Causal Hankel singular values of different subsystems, **b** finite eigenvalues of the original and the reduced-order lifted systems

The original lifted system has order $n = 30$. Figure 11.1a shows the causal Hankel singular values of the different subsystems for $k = 0, 1, 2$. We see that they decay fast, and, hence system (11.1) can be well approximated by a reduced-order model. We have 24 causal Hankel singular values for the original lifted system and the remaining 6 are noncausal Hankel singular values which are positive. We approximate system (11.1) to the tolerance 10^{-2} by truncating the states corresponding to the smallest 7 causal Hankel singular values.

Figure 11.1b shows the finite eigenvalues of the original and reduced-order lifted systems. We observe that stability is preserved for the reduced-order system. In Fig. 11.2a, we present the norms of the frequency responses $\mathcal{H}(e^{i\omega})$ and $\tilde{\mathcal{H}}(e^{i\omega})$ of the original and reduced-order lifted systems for a frequency range $[0, 2\pi]$. We observe nice match of the system norms.

In Fig. 11.2b, we display the absolute error $\|\mathcal{H}(e^{i\omega}) - \tilde{\mathcal{H}}(e^{i\omega})\|_2$ and the error bound (11.15). One can see that the absolute error is smaller than the error bound.

11.6 Conclusion

In this paper, we have considered the reachability and observability Gramians as well as Hankel singular values for periodic discrete-time descriptor systems. For such systems, a balanced truncation model reduction method has been presented.

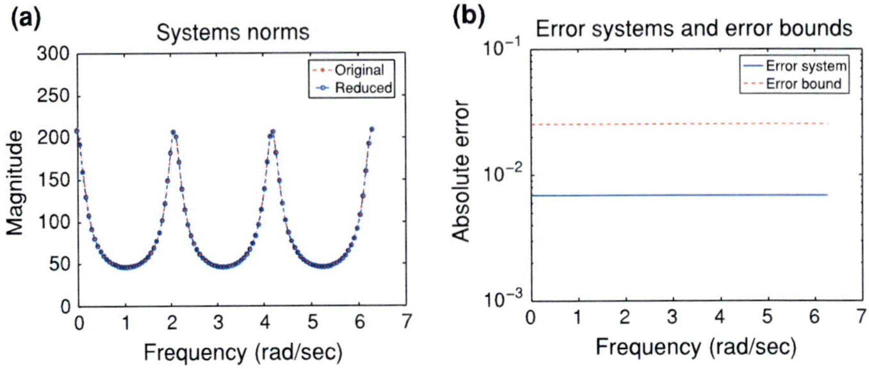

Fig. 11.2 **a** The frequency responses of the original and the reduced-order lifted systems; **b** absolute error and error bound

The proposed method delivers a reduced-order model that preserves the regularity and stability properties of the original system. A computable global error bound for the approximate system is also available.

Acknowledgements This work was supported by the Research Network SyreNe—*System Reduction for Nanoscale IC Design* funded by the German Federal Ministry of Education and Science (BMBF), grants 03BEPAE1 and 03STPAE3. Responsibility for the contents of this publication rests with the authors.

References

1. Arcara, P., Bittanti, S.: Periodic control of helicopter rotors for attenuation of vibrations in forward flight. IEEE Trans. on Control Syst. Tech. **8**(6), 883–894 (2000)
2. Benner, P., Mehrmann, V., Xu, H.: Perturbation analysis for the eigenvalue problem of a formal product of matrices. BIT **42**(1), 1–43 (2002)
3. Benner, P., Quintana-Ortí, E., Quintana-Ortí, G.: Computing passive reduced-order models for circuit simulation. In: Proceedings of International Conference on Parallel Computing in Electrical Engineering. PARELEC 2004, pp. 146–151. IEEE Computer Society, Los Alamitos (2004)
4. Bittanti, S., Colaneri, P.: Invariant representations of discrete-time periodic systems. Automatica **36**, 1777–1793 (2000)
5. Byers, R., Rhee, N.: Cyclic Schur and Hessenberg-Schur numerical methods for solving periodic Lyapunov and Sylvester equations. Technical report, Department of Mathematics, University of Missouri, Kansas (1995)
6. Chu, E.W., Fan, H.Y., Lin, W.W.: Projected generalized discrete-time periodic Lyapunov equations and balanced realization of periodic descriptor systems. SIAM J. Matrix Anal. Appl. **29**(3), 982–1006 (2007)
7. Farhood, M., Beck, C., Dullerud, G.: Model reduction of periodic systems: a lifting approach. Automatica, **41**, 1085–1090 (2005)
8. Freund, R.W.: Padé-type model reduction of second-order and higher-order linear dynamical systems. In: Benner, P., Mehrmann, V., Sorensen, D. (eds.) Dimension Reduction of Large-

Scale Systems, Lecture Notes in Computational Science and Engineering, vol. 45, pp. 191–223. Springer-Verlag, Berlin (2005)
9. Granat, R., Jonsson, I., Kågström, B.: Recursive blocked algorithms for solving periodic triangular Sylvester-type matrix equations. In: Kågström, B., Elmroth, E., Dongarra, J., Waśniewski, J. (eds.) Applied Parallel Computing. State of the Art in Scientific Computing, Lecture Notes in Computer Science, vol. 4699, pp. 531–539. Springer-Verlag (2007)
10. Günther, M., Feldmann, U.: CAD-based electric-circuit modeling in industry. II. Impact of circuit configurations and parameters. Surv. Math. Ind. **8**(2), 131–157 (1999)
11. Kressner, D.: An efficient and reliable implementation of the periodic QZ algorithm, In: S. Bittanti, P. Colaneri (eds.) Periodic Control Systems 2001. A Proceedings volume of the IFAC Workshop, Cernobbio-Como, Italy, 27–28 August 2001, Elsevier Science, Oxford, UK (2001)
12. Kressner, D.: Large periodic Lyapunov equations: Algorithms and applications. In: Proceedings of ECC03, Cambridge (2003)
13. Lall, S., Beck, C., Dullerud, G.: Guaranteed error bounds for model reduction of linear time-varying systems. In: Proceedings of the Americal Control Conference, Philadelphia, pp. 634–638 (1998)
14. Mehrmann, V., Stykel, T.: Balanced truncation model reduction for large-scale systems in descriptor form. In: Benner, P., Mehrmann, V., Sorensen, D. (eds.) Dimension Reduction of Large-Scale Systems, Lecture Notes in Computational Science and Engineering, vol. 45, pp. 83–115. Springer-Verlag, Berlin (2005)
15. Moore, B.: Principal component analysis in linear systems: controllability, observability, and model reduction. IEEE Trans. Automat. Contr. **AC-26**(1):17–32 (1981)
16. Nakhla, M., Gad, E.: Efficient model reduction of linear periodically time-varying systems via compressed transient system function. IEEE Trans. Automat. Contr. **52**(6) (2005)
17. Park, B., Verriest, E.: Canonical forms of discrete linear periodically time-varying systems and a control application. In: Proceedings of the 28th Conference on Decision and Control, Tampa, pp. 1220–1225 (1989)
18. Phillips, J., Daniel, L., Miguel Silveira, L.: Guaranteed passive balancing transformations for model order reduction. IEEE Trans. Comput. Aided Des. Integr. Circuits Syst. **22**, 1027–1041 (2003)
19. Roychowdhury, J.: Reduced-order modelling of linear time-varying systems. Design automation conference, 1999. In: Proceedings of the ASP-DAC '99. Asia and South Pacific **1**, 53–56 (18-21 Jan.1999)
20. Roychowdhury, J.: Reduced-order modeling of time-varying systems. IEEE Control Syst. Mag. **46**, 1273–1288 (1999)
21. Shokoohi, S., Silverman, L., Van Dooren, P.: Linear time-variable systems: Balancing and model reduction. IEEE Trans. Automat. Contr. **AC-28**(8), 810–822 (1983)
22. Smith, R.: Matrix equation $XA + BX = C$. SIAM J. Appl. Math. **16**, 198–201 (1968)
23. Stykel, T.: Numerical solution and perturbation theory for generalized Lyapunov equations. Linear Algebra Appl. **349**, 155–185 (2002)
24. Stykel, T.: Stability and inertia theorems for generalized Lyapunov equations. Linear Algebra Appl. **355**, 297–314 (2002)
25. Stykel, T.: Gramian-based model reduction for descriptor systems. Math. Control Signals Syst. **16**, 297–319 (2004)
26. Stykel, T.: On some norms for descriptor systems. IEEE Trans. Automat. Control **51**(5), 842–847 (2006)
27. Stykel, T.: Low-rank iterative methods for projected generalized Lyapunov equations. Electron. Trans. Numer. Anal. **30**, 187–202 (2008)
28. Van Dooren, P.: Gramian based model reduction of large-scale dynamical systems. In: Numerical Analysis 1999, Chapman & Hall/CRC Research Notes in Mathematics, 420, pp. 231–247. Chapman & Hall/CRC, Boca Raton, FL (2000)
29. Varga, A.: Periodic Lyapunov equations: some applications and new algorithms. Int. J. Control **67**(1), 69–87 (1997)

30. Varga, A.: Balancing related methods for minimal realization of periodic systems. Syst. Control Lett. **36**, 339–349 (1999)
31. Varga, A.: Balanced truncation model reduction of periodic systems. In: Proceedings of CDC'2000, Sydney (2000)
32. Varga, A.: Computation of Kronecker-like forms of periodic matrix pairs. In: Proceedings of Mathematical Theory of Networks and Systems, MTNS 2004, Leuven, 5-9 July 2004
33. Varga, A.: A PERIODIC SYSTEMS toolbox for MATLAB. In: Proceedings of IFAC'05 World Congress, Prague, 3-8 July 2005
34. Wisniewski, R., Blanke, M.: Fully magnetic attitude control for spacecraft subject to gravity gradient. Automatica **35**(7), 1201–1214 (1999)
35. Yang, X., Kawamata, M., Higuchi, T.: Balanced realisations and model reduction of periodically time-varying state-space digital filters, In: IEE Proc. Vision, Image & Singal Proc., vol. 143, pp. 370–376 (1996)

Chapter 12
On Synthesis of Reduced Order Models

Roxana Ionutiu and Joost Rommes

Abstract A framework for model reduction and synthesis is presented, which enables the re-use of reduced order models in circuit simulation. Two synthesis techniques are considered for obtaining the circuit representation (netlist) of the reduced model: (1) by means of realizing the reduced transfer function and (2) by unstamping the reduced system matrices. For both methods, advantages and limitations are discussed. Especially when model reduction exploits structure preservation, we show that using the model as a current-driven element is possible, and allows for synthesis without controlled sources. The presented framework serves as a basis for reduction of large parasitic R/RC/RCL networks.

Keywords Circuit synthesis · Reduced models · Metlist representation · Passivity · Terminals · Impedance · Modified model analysis

PACS 02.60.C · 02.60.Dc · 02.60.Lj · 84.32.Ff · 84.32.Hh · 84.32 Tt

---

R. Ionutiu: Marie Curie Fellowship Programme COMSON (Contract Number MRTN-CT-2005-019417). J. Rommes: Marie-Curie Fellowship Programme O-MOORE-NICE! (FP6 MTKI-CT-2006-042477).

R. Ionutiu (✉)
Department of Mathematics and Computer Science, Eindhoven University of Technology, P.O. Box 513, 5600 MB Eindhoven, The Netherlands
e-mail: r.ionutiu@tue.nl

J. Rommes
NXP Semiconductors, HTC46-2.08, 5656 AE Eindhoven, The Netherlands
e-mail: joost.rommes@nxp.com

P. Benner et al. (eds.), *Model Reduction for Circuit Simulation*, Lecture Notes in Electrical Engineering, 74, DOI: 10.1007/978-94-007-0089-5_12,

12.1 Introduction

The main motivation for this chapter comes from the need for a general framework for the (re)use of reduced order models in circuit simulation. Although many model order reduction methods have been developed and evolved since the 1990s (see for instance [1] for an overview), it is usually less clear how to use these methods efficiently in industrial practice, e.g., in a circuit simulator. One reason can be that the reduced order model does not satisfy certain physical properties, for instance, it may not be stable or passive while the original system is. Failing to preserve these properties is typically inherent to the reduced order method used (or its implementation). Passivity (and stability implicitly) can nowadays be preserved via several methods [2, 10, 16, 19, 21, 26, 27], but none address the practical aspect of (re)using the reduced order models with circuit simulation software (e.g., SPICE [9]). This brings forward another reason of concern within the circuit simulation industry. The linear circuit to be reduced is represented by a *netlist*, which is a description of the circuit element values (R, L, C) and their connections to the circuit nodes (see also Fig. 12.1). However, reduced order models (as a result of model reduction applied on the dynamical system describing the original circuit) are usually represented in terms of *system matrices* or of the *input–output transfer function*. Typically, circuit simulators are not prepared for inputs in a mathematical representation, and would require additional software architecture to handle them. In contrast, a reduced model in *netlist* representation could be easily coupled to bigger systems and directly simulated.

Synthesis is the realization step needed to map the reduced order model from the mathematical representation (in terms of system matrices or transfer function) into a netlist consisting of electrical circuit components [13]. In [7] it was shown that passive systems (with positive real transfer functions) can be synthesized with positive R, L, C elements and transformers (see also [23]). Later developments [6] propose a method to circumvent the introduction of transformers, however the resulting realization is non-minimal (i.e., the number of electrical components generated during synthesis is too large). Allowing for possibly negative R, L, C values, other methods have been proposed via e.g. direct stamping [18, 19] or full realization [14, 20]. These mostly model the input/output connections of the reduced model with controlled sources.

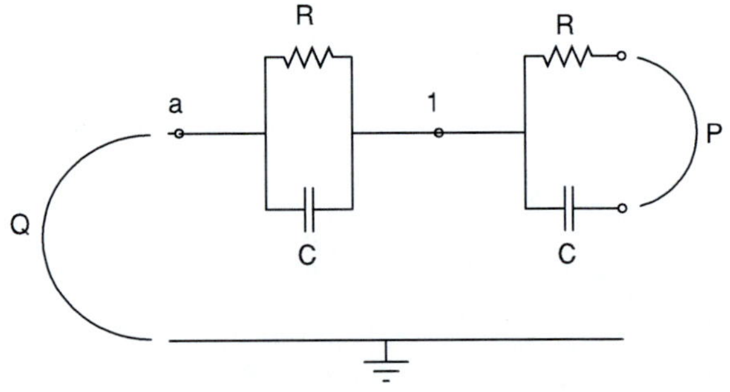

Fig. 12.1 Circuit with terminal a, internal node 1, port P, and port Q(a,0)

In this paper we consider two synthesis methods that do not involve controlled sources: (1) *Foster synthesis* [13], where the realization is done via the system's transfer function and (2) *RLCSYN synthesis by unstamping* [29], which exploits input–output structure preservation in the reduced system matrices (provided that the original system matrices are written in *modified nodal analysis (MNA)* representation). The focus of this paper is on structure preservation and RLCSYN, especially because synthesis by unstamping is simple to implement for both SISO and MIMO systems. Strengthening the result of [29], we give a simple procedure to reduce either current- or voltage-driven circuits directly in impedance form, enabling synthesis without controlled sources. The reduced order model is available as a netlist, making it suitable for simulation and reuse in other designs. Similar software [8] is commercially available.

The material in this chapter is organized as follows. The remainder of this section introduces terminology for the different nodes pertaining to a circuit topology (Sect. 12.1.1). A brief mathematical formulation of model order reduction is given in Sect. 12.1.2. The Foster synthesis is presented in Sect. 12.2. In Sect. 12.3 we focus on reduction and synthesis with structure (and input/output) preservation. Section 12.3.1 describes the procedure to convert admittance models to impedance form, so that synthesized models are easily (re)used in simulation. Based on [29], Sect. 12.3.2 is an outline of SPRIM/IOPOR reduction and RLCSYN synthesis. Numerical considerations on SPRIM/IOPOR are covered in Sect. 12.3.3. Experiments follow in Sect. 12.4, and Sect. 12.5 concludes.

12.1.1 Internal Nodes, Terminals, and Ports

The terms internal nodes, terminals (or external nodes), and ports often occur in electronic engineering related papers. An *internal node* is a node in a circuit that is not visible on the outside of a circuit, i.e., no currents can be injected in an internal node (cf. node 1 in Fig. 12.1). A *terminal (external node)* is a node that is visible on the outside, i.e., a node in which currents can be injected (cf. node a in Fig. 12.1). A port consists of two terminals that can be connected, for instance, by a source or another (sub)circuit (cf. port P in Fig. 12.1). Sometimes terminals are referred to as ports and vice versa: from the context it should then be clear which terminal(s) complete the ports; usually it is implicitly assumed that the ground node completes the ports. In Fig. 12.1, for instance, terminal a can be seen as a port (Q) by including the ground node.

12.1.2 Problem Formulation

In this paper the dynamical systems $\Sigma(\mathbf{A}, \mathbf{E}, \mathbf{B}, \mathbf{C}, \mathbf{D})$ are of the form $\mathbf{E}\dot{\mathbf{x}}(t) = \mathbf{A}\mathbf{x}(t) + \mathbf{B}\mathbf{u}(t)$, $\mathbf{y}(t) = \mathbf{C}\mathbf{x}(t) + \mathbf{D}\mathbf{u}(t)$, where $\mathbf{A}, \mathbf{E} \in \mathbb{R}^{n \times n}$, $\mathbf{E}$ may be singular but the pencil $(\mathbf{A}, \mathbf{E})$ is regular, $\mathbf{B} \in \mathbb{R}^{n \times m}$, $\mathbf{C} \in \mathbb{R}^{p \times n}$, $\mathbf{x}(t) \in \mathbb{R}^n$, and $\mathbf{u}(t) \in \mathbb{R}^m$, $\mathbf{y}(t) \in \mathbb{R}^p$, $\mathbf{D} \in \mathbb{R}^{p \times m}$. If $m, p > 1$, the system is called multiple-input multiple-output (MIMO), otherwise it is called single-input single-output (SISO). The frequency domain transfer function is defined as $\mathbf{H}(s) = \mathbf{C}(s\mathbf{E} - \mathbf{A})^{-1}\mathbf{B} + \mathbf{D}$. For systems in MNA form arising in circuit simulation see Sect. 12.3

The model order reduction problem is to find, given an n-th order (descriptor) dynamical system, a k-th order system, $k<n$: $\widetilde{\mathbf{E}}\dot{\tilde{\mathbf{x}}}(t) = \widetilde{\mathbf{A}}\tilde{\mathbf{x}}(t) + \widetilde{\mathbf{B}}\mathbf{u}(t)$, $\tilde{\mathbf{y}}(t) = \widetilde{\mathbf{C}}\tilde{\mathbf{x}}(t) + \mathbf{D}\mathbf{u}(t)$, such that the output approximation error $\|\mathbf{y}(t) - \tilde{\mathbf{y}}(t)\|$ (in appropriate norm) is small. The corresponding dimensions are: $\widetilde{\mathbf{E}}, \widetilde{\mathbf{A}} \in \mathbb{R}^{k \times k}$, $\widetilde{\mathbf{B}} \in \mathbb{R}^{k \times m}$, $\widetilde{\mathbf{C}} \in \mathbb{R}^{p \times k}$, $\tilde{\mathbf{x}}(t) \in \mathbb{R}^k$, $\mathbf{u}(t) \in \mathbb{R}^m$, $\tilde{\mathbf{y}}(t) \in \mathbb{R}^p$, and $\mathbf{D} \in \mathbb{R}^{p \times m}$. The number of inputs and outputs is the same as for the original system, and the corresponding transfer function becomes: $\widetilde{\mathbf{H}}(s) = \widetilde{\mathbf{C}}(s\widetilde{\mathbf{E}} - \widetilde{\mathbf{A}})^{-1}\widetilde{\mathbf{B}} + \mathbf{D}$. For an overview of model order reduction methods, see [1, 5, 25]. Projection based model order reduction methods construct a reduced order model via the Petrov–Galerkin projection:

$$\widetilde{\Sigma}(\widetilde{\mathbf{E}}, \widetilde{\mathbf{A}}, \widetilde{\mathbf{B}}, \widetilde{\mathbf{C}}, \mathbf{D}) \equiv (\mathbf{W}^*\mathbf{E}\mathbf{V}, \mathbf{W}^*\mathbf{A}\mathbf{V}, \mathbf{W}^*\mathbf{B}, \mathbf{C}\mathbf{V}, \mathbf{D}), \tag{12.1}$$

where $\mathbf{V}, \mathbf{W} \in \mathbb{R}^{n \times m}$ are matrices whose $k \ll n$ columns form bases for relevant subspaces of the state-space. There are several projection methods, that differ in the way the matrices $\mathbf{V}$ and $\mathbf{W}$ are chosen. These also determine which properties are preserved after reduction. Some stability preserving methods are: *modal approximation* [24], *poor man's TBR* [22]. Among *moment matching* [11] methods, the following preserve passivity *PRIMA* [19], *SPRIM* [10], *spectral zero interpolation* [2, 16, 26]. From the balancing methods, *balanced truncation* [4] preserves stability, and *positive real balanced truncation* [21, 27] preserves passivity.

12.2 Foster Synthesis of Rational Transfer Functions

This section describes the Foster synthesis method, which was developed in the 1930s by Foster and Cauer [13] and involves realization based on the system's transfer function. The Foster approach can be used to realize any reduced order model that is computed by standard projection based model order reduction techniques. Realizations will be described in terms of SISO impedances (Z-parameters). For equivalent realizations in terms of admittances (Y-parameters), see for instance [13, 28]. Given the reduced, proper system (12.1) and assuming that all its finite poles are simple [i.e., the matrix pencil $(\widetilde{\mathbf{A}}, \widetilde{\mathbf{E}})$ is non-defective], consider the partial fraction expansion [17] of its transfer function:

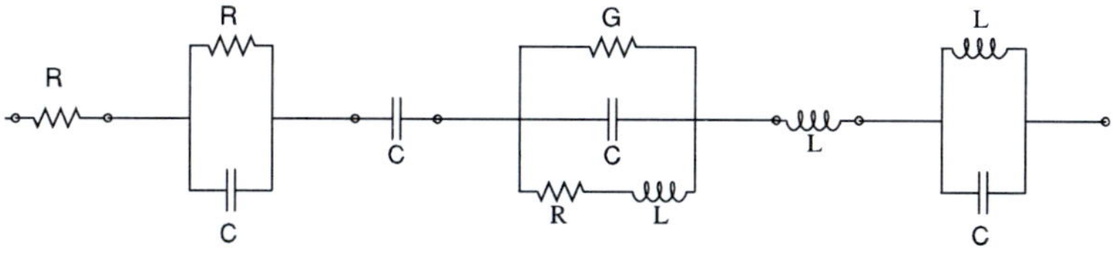

Fig. 12.2 Realization of a general impedance transfer function as a series *RLC* circuit

Table 12.1 Circuit element values for Fig. 12.2 for the Foster impedance realization of (12.3)

pole	residue	$R(\Omega)$	$C(\mathrm{F})$	$L(\mathrm{H})$	$G(\Omega^{-1})$
$p_1 = \infty$	$r_1 \in \mathbb{R}$	r_1			
$p_2 \in \mathbb{R}$	$r_2 \in \mathbb{R}$	$-\frac{r_2}{p_2}$	$\frac{1}{r_2}$		
$p_3 = 0$	$r_3 \in \mathbb{R}$		$\frac{1}{r_3}$		
$p_4 = \sigma + i\omega \in \mathbb{C}$	$r_4 = \alpha + i\beta \in \mathbb{C}$	$\frac{a_0}{a_1} L$	$\frac{1}{a_1}$	$\frac{a_1^3}{a_1^2 b_0 - a_0(a_1 b_1 - a_0)}$	$\frac{a_1 b_1 - a_0}{a_1^2}$
$p_5 \equiv \bar{p}_4$	$r_5 \equiv \bar{r}_4$				
	$a_0 = -2(\alpha\sigma + \beta\omega)$,	$a_1 = 2\alpha$,	$b_0 = \sigma^2 + \omega^2$,	$b_1 = -2\sigma$	
$p_6 = \infty$	$r_6 \in \mathbb{R}$			r_6	
$p_7 \in i\mathbb{R}$	$r_7 \in \mathbb{R}$		$\frac{1}{r_7}$	$\frac{2r_7}{p_7 \bar{p}_7}$	
$p_8 \equiv \bar{p}_7$	$r_8 \equiv \bar{r}_7$				

$$\widetilde{\mathbf{H}}(s) = \sum_{i=1}^{k} \frac{\tilde{r}_i}{s - \tilde{p}_i} + \mathbf{D}, \tag{12.2}$$

The residues are $\tilde{r}_i = \frac{\widetilde{(\mathbf{C}\tilde{\mathbf{x}}_i)}(\tilde{\mathbf{y}}_i^* \mathbf{B})}{\tilde{\mathbf{y}}_i^* \widetilde{\mathbf{E}}\tilde{\mathbf{x}}_i}$, the poles are $\tilde{p}_i$ and, if non-zero, $\mathbf{D}$ gives additional contribution from poles at ∞. An eigentriplet $(\tilde{p}_i, \tilde{\mathbf{x}}_i, \tilde{\mathbf{y}}_i)$ is composed of an eigenvalue $\tilde{p}_i$ of $(\widetilde{\mathbf{A}}, \widetilde{\mathbf{E}})$ and the corresponding right and left eigenvectors $\tilde{\mathbf{x}}_i, \tilde{\mathbf{y}}_i \in \mathbb{C}^k$. The proper expansion (12.2) belongs to the more general class of transfer functions (not necessarily proper) which can be expressed using basic summands of the form:

$$Z(s) = r_1 + \frac{r_2}{s - p_2} + \frac{r_3}{s} + \left(\frac{r_4}{s - p_4} + \frac{\bar{r}_4}{s - \bar{p}_4}\right) + sr_6 + \left(\frac{r_7}{s - p_7} + \frac{r_7}{s - \bar{p}_7}\right). \tag{12.3}$$

For completeness, in (12.3) we can assume that any kind of poles may appear, i.e., either purely real, purely imaginary, in complex conjugate pairs, at ∞ or at 0 (see also Table 12.1). The Foster realization converts each term in (12.3) into the corresponding circuit block with *R*, *L*, *C* components, and connects these blocks in series in the final netlist. This is shown in Fig. 12.2. Note that any reordering of the circuit blocks in the realization of (12.3) in Fig. 12.2 still is a realization of

(12.3). The values for the circuit components in Fig. 12.2 are determined according to Table 12.1.

The realization in netlist form can be implemented in any language such as SPICE [9], so that it can be reused and combined with other circuits as well. The advantages of Foster synthesis are: (1) its straightforward implementation for single-input-single-output (SISO) transfer functions, via either the impedance or the admittance transfer function, (2) after reducing purely *RC* or *RL* circuits via modal approximation [24], the reduced netlists obtained from Foster synthesis are guaranteed to have positive *RC* or *RL* values respectively (see [15] for a proof). Note however that Foster synthesis does not guarantee positive circuit elements in general (e.g., when used to synthesize reduced models originating from *RLC* circuits, or reduced models of *RC* and *RL* circuits that were obtained with methods different than modal approximation). The main disadvantage is that for multi-input-multi-output transfer functions, finding the Foster realization (see for instance [28]) is cumbersome and may also give dense reduced netlists (i.e., all nodes are interconnected). This is because the Foster synthesis of a k-dimensional reduced system with p terminals, will generally yield $O(p^2k)$ circuit elements.

12.3 Structure Preservation and Synthesis by Unstamping

This section describes the second synthesis approach, which is based on *unstamping* the reduced matrix data into an *RLC* netlist and is denoted by RLCSYN [29]. It is suitable for obtaining netlist representations for models reduced via methods that preserve the MNA structure and the input–output connectivity at the circuit terminals. Such a method is the *input–output structure preserving* method SPRIM/IOPOR [29]. The strength of the result in [29] is that the input/output connectivity is synthesized after reduction without controlled sources, provided that the system is in *impedance form* (i.e., inputs are currents injected into the circuit terminals, and outputs are voltages measured at the terminals). For ease of understanding, the input–output structure preserving reduction from [29] can be interpreted as model reduction with *preservation of external nodes* [e.g., after reducing the circuit in Fig. 12.1, the nodes forming ports Q and P are external (terminals) and will also appear in the synthesized reduced model]. This way the reduced netlist can be easily coupled to other circuitry in place of the original netlist, and (re)using the reduced model in simulation becomes straightforward. The main drawback is that, when the reduced system matrices are dense and the number of terminals is large [$O(10^3)$], the RLCSYN unstamping procedure yields dense netlists. For a k dimensional reduced network with p terminals, the RLC-SYN synthesized netlist will generally have $O(p^2k^2)$ circuit elements.

First, we motivate reduction and synthesis in impedance form, and show how it also applies for systems that are originally in admittance form. Then we explain model reduction via SPRIM/IOPOR, followed by RLCSYN synthesis. Finally, a note on numerical aspects concerning the SPRIM/IOPOR projection is given.

12.3.1 A Simple Admittance to Impedance Conversion

In [29] it was shown how SPRIM/IOPOR preserves the structure of the input/output connectivity when the original model is in impedance form, allowing for synthesis via RLCSYN without controlled sources. The emerging question is: how to ensure synthesis without controlled sources, if the original model is in admittance form (i.e., it is voltage driven)? We show that reduction and synthesis via the input impedance transfer function can be performed, also when the original circuit is initially voltage driven.

Consider the modified nodal analysis (MNA) description of an input *admittance*[1] type *RLC* circuit, driven by n_s voltage sources:

$$\underbrace{\begin{pmatrix} \mathscr{C} & \mathbf{0} & \mathbf{0} \\ \mathbf{0} & \mathbf{0} & \mathbf{0} \\ \mathbf{0} & \mathbf{0} & \mathscr{L} \end{pmatrix}}_{\mathbf{E}_Y} \underbrace{\frac{d}{dt}\begin{pmatrix} \mathbf{v}(t) \\ \mathbf{i}_S(t) \\ \mathbf{i}_L(t) \end{pmatrix}}_{\dot{\mathbf{x}}_Y} + \underbrace{\begin{pmatrix} \mathscr{G} & \mathscr{E}_v & \mathscr{E}_l \\ -\mathscr{E}_v{}^* & \mathbf{0} & \mathbf{0} \\ -\mathscr{E}_l{}^* & \mathbf{0} & \mathbf{0} \end{pmatrix}}_{-\mathbf{A}_Y} \underbrace{\begin{pmatrix} \mathbf{v}(t) \\ \mathbf{i}_S(t) \\ \mathbf{i}_L(t) \end{pmatrix}}_{\mathbf{x}_Y} = \underbrace{\begin{pmatrix} \mathbf{0} \\ \mathscr{B} \\ \mathbf{0} \end{pmatrix}}_{\mathbf{B}_Y} \mathbf{u}(t), \tag{12.4}$$

where $\mathbf{u}(t) \in \mathbb{R}^{n_s}$ are input voltages and $\mathbf{y}(t) = \mathbf{C}_Y\mathbf{x}(t) \in \mathbb{R}^{n_s}$ are output currents, $\mathbf{C}_Y = \mathbf{B}_Y^*$. The states are $\mathbf{x}_Y(t)^T = [\mathbf{v}(t), \mathbf{i}_S(t), \mathbf{i}_L(t)]^T$, with $\mathbf{v}(t) \in \mathbb{R}^{n_v}$ the node voltages, $\mathbf{i}_S(t) \in \mathbb{R}^{n_s}$ the currents through the voltage sources, and $\mathbf{i}_L(t) \in \mathbb{R}^{n_l}$ the currents through the inductors, $n_v + n_s + n_l = n$. The $n_v = n_1 + n_2$ node voltages are formed by the n_1 external nodes/terminals[2] and the n_2 internal nodes (assuming that the voltage sources may be ungrounded, n_1 satisfies: $n_s < n_1 \leq 2n_s + 1$). The dimensions of the underlying matrices are: $\mathscr{C} \in \mathbb{C}^{n_v \times n_v}, \mathscr{G} \in \mathbb{C}^{n_v \times n_v}, \mathscr{E}_v \in \mathbb{C}^{n_v \times n_s}$, $\mathscr{L} \in \mathbb{C}^{n_l \times n_l}, \mathscr{E}_l \in \mathbb{C}^{n_v \times n_l}, \mathscr{B} \in \mathbb{C}^{n_1 \times n_s}$. Assuming without loss of generality that (12.4) is permuted such that the *first* n_1 nodes are the external nodes, the input voltages are determined by a linear combination of $\mathbf{v}_{1:n_1}(t)$ only. Thus the following holds:

$$\mathscr{E}_v = \begin{pmatrix} \mathscr{B}_v \\ \mathbf{0}_{n_2 \times n_s} \end{pmatrix} \in \mathbb{C}^{n_v \times n_s}, \quad \mathscr{B}_v \in \mathbb{C}^{n_1 \times n_s}, \quad \mathscr{B} = -\mathscr{B}_v. \tag{12.5}$$

We derive the first order impedance-type system associated with (12.4). Note that by definition, $\mathbf{i}_S(t)$ flows *out of* the circuit terminals into the voltage source (i.e., from the + to the − terminal of the voltage source, see also [19, Fig. 3], [15]).

---

[1] The subscript Y refers to quantities associated with a system in admittance form.

[2] The MNA form (12.4) corresponds to the ungrounded circuit (i.e., the reference node is counted within the n_1 external nodes), resulting in a defective matrix pencil $(\mathbf{A}_Y, \mathbf{E}_Y)$. For subsequent computations such as the construction of a Krylov subspace, the pencil $(\mathbf{A}_Y, \mathbf{E}_Y)$ must be regular. Thus in (12.4), one node must be chosen as a ground (reference) node by removing the row/column corresponding to that node; this ensures that the regularity conditions (i) and (ii) from [23, page 5, Assumption 4] are satisfied. The positive definiteness of $\mathscr{C}, \mathscr{L}, \mathscr{G}$ is also a necessary condition to ensure the circuit's passivity.

We can define new input currents as the currents flowing *into* the circuit terminals: $\mathbf{i}_{in}(t) = -\mathbf{i}_S(t)$. Since $\mathbf{v}_{1:n_1}(t)$ are the terminal voltages, they describe the new output equations, and it is straightforward to rewrite (12.4) in the impedance form:

$$\begin{cases} \underbrace{\begin{pmatrix} \mathscr{C} & \mathbf{0} \\ \mathbf{0} & \mathscr{L} \end{pmatrix}}_{\mathbf{E}} \underbrace{\frac{d}{dt}\begin{pmatrix} \mathbf{v}(t) \\ \mathbf{i}_L(t) \end{pmatrix}}_{\dot{\mathbf{x}}} + \underbrace{\begin{pmatrix} \mathscr{G} & \mathscr{E}_l \\ -\mathscr{E}_l^* & \mathbf{0} \end{pmatrix}}_{-\mathbf{A}} \underbrace{\begin{pmatrix} \mathbf{v}(t) \\ \mathbf{i}_L(t) \end{pmatrix}}_{\mathbf{x}} = \underbrace{\begin{pmatrix} \mathscr{E}_v \\ \mathbf{0} \end{pmatrix}}_{\mathbf{B}} \mathbf{i}_{in}(t) \\ \underbrace{\begin{pmatrix} \mathscr{E}_v^* & \mathbf{0} \end{pmatrix}}_{\mathbf{C}} \underbrace{\begin{pmatrix} \mathbf{v}(t) \\ \mathbf{i}_L(t) \end{pmatrix}}_{\mathbf{x}} = \mathbf{y}(t) = \mathscr{B}_v \mathbf{v}_{1:n_1}(t), \quad \mathscr{E}_v^* = \begin{pmatrix} \mathscr{B}_v^* \mathbf{0}_{n_s \times n_2} \end{pmatrix} \end{cases} \tag{12.6}$$

where $\mathbf{B}$ describes the new *input incidence matrix* corresponding the input currents, $\mathbf{i}_{in}$. The new *output incidence matrix* is $\mathbf{C}$, corresponding to the voltage drops over the circuit terminals. We emphasize that (12.6) has fewer unknowns than (12.4), since the currents $\mathbf{i}_S$ have been eliminated. The transfer function associated to (12.6) is an input impedance: $\mathbf{H}(s) = \frac{\mathbf{y}(s)}{\mathbf{i}_{in}(s)}$. In Sect. 12.3.2 we explain how to obtain an impedance type reduced order model in the input/output structure preserved form:

$$\begin{cases} \underbrace{\begin{pmatrix} \widetilde{\mathscr{C}} & \mathbf{0} \\ \mathbf{0} & \widetilde{\mathscr{L}} \end{pmatrix}}_{\widetilde{\mathbf{E}}} \underbrace{\frac{d}{dt}\begin{pmatrix} \widetilde{\mathbf{v}}(t) \\ \widetilde{\mathbf{i}}_L(t) \end{pmatrix}}_{\dot{\widetilde{\mathbf{x}}}} + \underbrace{\begin{pmatrix} \widetilde{\mathscr{G}} & \widetilde{\mathscr{E}}_l \\ -\widetilde{\mathscr{E}}_l^* & \mathbf{0} \end{pmatrix}}_{-\widetilde{\mathbf{A}}} \underbrace{\begin{pmatrix} \widetilde{\mathbf{v}}(t) \\ \widetilde{\imath}_L(t) \end{pmatrix}}_{\widetilde{\mathbf{x}}} = \underbrace{\begin{pmatrix} \widetilde{\mathscr{E}}_v \\ \mathbf{0} \end{pmatrix}}_{\widetilde{\mathbf{B}}} \mathbf{i}_{in}(t) \\ \underbrace{\begin{pmatrix} \widetilde{\mathscr{E}}_v^* & \mathbf{0} \end{pmatrix}}_{\widetilde{\mathbf{C}}} \underbrace{\begin{pmatrix} \widetilde{\mathbf{v}}(t) \\ \widetilde{\mathbf{i}}_L(t) \end{pmatrix}}_{\widetilde{\mathbf{x}}} = \mathbf{y}(t) = \mathscr{B}_{v1:n_1}(t), \quad \widetilde{\mathscr{E}}_v^* = \begin{pmatrix} \mathscr{B}_v^* \mathbf{0}_{n_s \times k_2} \end{pmatrix} \end{cases} \tag{12.7}$$

where $\widetilde{\mathscr{C}}$, $\widetilde{\mathscr{L}}$, $\widetilde{\mathscr{G}}$, $\widetilde{\mathscr{E}}_v$ are the reduced MNA matrices, and the reduced input impedance transfer function is: $\widetilde{\mathbf{H}}(s) = \frac{\widetilde{\mathbf{y}}(s)}{\mathbf{i}_{in}(s)}$. Due to the input/output preservation, the circuit terminals are easily preserved in the reduced model (12.7).

It turns out that after reduction and synthesis, the reduced model (12.7) can still be used as a voltage driven admittance block in simulation. This is shown next. We can rewrite the second equation in (12.7) as: $\begin{pmatrix} -\widetilde{\mathscr{E}}_v^* & \mathbf{0} & \mathbf{0} \end{pmatrix} \begin{pmatrix} \widetilde{\mathbf{v}}(t)^T & \widetilde{\mathbf{i}}_S(t)^T & \widetilde{\mathbf{i}}_L(t)^T \end{pmatrix}^T = \mathscr{B}\mathbf{u}(t)$. This result together with $\mathbf{i}_{in}(t) = -\mathbf{i}_S(t)$, reveals that (12.7) can be rewritten as:

$$\underbrace{\begin{pmatrix} \widetilde{\mathscr{C}} & \mathbf{0} & \mathbf{0} \\ \mathbf{0} & \mathbf{0} & \mathbf{0} \\ \mathbf{0} & \mathbf{0} & \widetilde{\mathscr{L}} \end{pmatrix}}_{\widetilde{\mathbf{E}}_Y} \underbrace{\frac{d}{dt}\begin{pmatrix} \widetilde{\mathbf{v}}(t) \\ \mathbf{i}_S(t) \\ \widetilde{\mathbf{i}}_L(t) \end{pmatrix}}_{\dot{\widetilde{\mathbf{x}}}_Y} + \underbrace{\begin{pmatrix} \widetilde{\mathscr{G}} & \widetilde{\mathscr{E}}_v & \widetilde{\mathscr{E}}_l \\ -\widetilde{\mathscr{E}}_v^* & \mathbf{0} & \mathbf{0} \\ -\widetilde{\mathscr{E}}_l^* & \mathbf{0} & \mathbf{0} \end{pmatrix}}_{-\widetilde{\mathbf{A}}_Y} \underbrace{\begin{pmatrix} \widetilde{\mathbf{v}}(t) \\ \mathbf{i}_S(t) \\ \widetilde{\mathbf{i}}_L(t) \end{pmatrix}}_{\widetilde{\mathbf{x}}_Y} = \underbrace{\begin{pmatrix} \mathbf{0} \\ \mathscr{B} \\ \mathbf{0} \end{pmatrix}}_{\widetilde{\mathbf{B}}_Y} \mathbf{u}(t), \tag{12.8}$$

which has the same structure as the original admittance model (12.4). Conceptually one could have reduced system (12.4) directly via the input admittance. In that case, synthesis by unstamping via RLCSYN [29] would have required controlled sources [14]. As shown above, this is avoided by: applying the simple admittance-to-impedance conversion (12.4)–(12.6), reducing (12.6)–(12.7), and finally reinserting voltage sources after synthesis (if the input–output structure preserved admittance reduced admittance (12.8) is needed).

12.3.2 I/O Structure Preserving Reduction and RLCSYN Synthesis

The reduced input impedance model (12.7) is obtained via the input–output structure preserving SPRIM/IOPOR projection [29] as follows. Let $\mathbf{V} = \left(\mathbf{V}_1^T, \mathbf{V}_2^T, \mathbf{V}_3^T\right)^T \in \mathbb{C}^{((n_1+n_2+n_l)\times k)}$ be the projection matrix obtained with PRIMA [19], where $\mathbf{V}_1 \in \mathbb{C}^{(n_1\times k)}$, $\mathbf{V}_2 \in \mathbb{C}^{(n_2\times k)}$, $\mathbf{V}_3 \in \mathbb{C}^{(n_l\times k)}$, $k \geq n_1$, $i = 1\ldots3$. After appropriate orthonormalization (e.g., via Modified Gram–Schmidt [24, Chapter. 1]), we obtain: $\widetilde{\mathbf{V}}_i = \text{orth}(\mathbf{V}_i) \in \mathbb{C}^{n_i\times k_i}$, $k_i \leq k$. The SPRIM [10] block structure preserving projection is: $\widetilde{\mathbf{V}} = \text{blkdiag}\left(\widetilde{\mathbf{V}}_1, \widetilde{\mathbf{V}}_2, \widetilde{\mathbf{V}}_3\right) \in \mathbb{C}^{\mathrm{n}\times(\mathrm{k}_1+\mathrm{k}_2+\mathrm{k}_3)}$, which does not yet preserve the structure of the input and output matrices. The input–output structure preserving SPRIM/IOPOR [29] projection is $\widetilde{\mathbf{W}} = \begin{pmatrix} \mathbf{W} & \mathbf{0} \\ \mathbf{0} & \widetilde{\mathbf{V}}_3 \end{pmatrix} \in \mathbb{C}^{n\times(n_1+k_2+k_3)}$ where:

$$\mathbf{W} = \begin{pmatrix} \mathbf{I}_{n_1\times n_1} & \mathbf{0} \\ \mathbf{0} & \widetilde{\mathbf{V}}_2 \end{pmatrix} \in \mathbb{C}^{(n_1+n_2)\times(n_1+k_2)}. \tag{12.9}$$

Recalling (12.5), we obtain the reduced system matrices in (12.7): $\widetilde{\mathscr{C}} = \mathbf{W}^*\mathscr{C}\mathbf{W}$, $\widetilde{\mathscr{G}} = \mathbf{W}^*\mathscr{G}\mathbf{W}$, $\widetilde{\mathscr{L}} = \widetilde{\mathbf{V}}_3^*\mathscr{L}\widetilde{\mathbf{V}}_3$, $\widetilde{\mathscr{E}}_l = \mathbf{W}^*\mathscr{E}_l\widetilde{\mathbf{V}}_3$, $\widetilde{\mathscr{E}}_v = \mathbf{W}^*\mathscr{E}_v = (\,\mathscr{B}_v^* \quad \mathbf{0}_{n_s\times k_2}\,)^*$, which compared to (12.5) clearly preserve input–output structure. Therefore a netlist representation for the reduced impedance-type model can be obtained, that is driven by injected currents just as the original circuit. This is done via the RLCSYN [29] unstamping procedure. To this end, we use the Laplace transform and convert (12.7) to the second order form:

$$\begin{cases} [s\widetilde{\mathscr{C}} + \widetilde{\mathscr{G}} + \frac{1}{s}\widetilde{\Gamma}]\widetilde{\mathbf{v}}(s) = \widetilde{\mathscr{E}}_v\mathbf{i}_{in}(s) \\ \widetilde{\mathbf{y}}(s) = \widetilde{\mathscr{E}}_v^*\widetilde{\mathbf{v}}(s), \end{cases} \tag{12.10}$$

where $\widetilde{\mathbf{i}}_L(s) = \frac{1}{s}\widetilde{\mathscr{L}}^{-1}\left(\widetilde{\mathscr{E}_l}^*\right)\widetilde{\mathbf{v}}(s)$ and $\widetilde{\Gamma} = \widetilde{\mathscr{E}}_l\widetilde{\mathscr{L}}^{-1}\widetilde{\mathscr{E}}_l^*$.

The presentation of RLCSYN follows [29, Sect. 4], [15] and is only summarized here. In circuit simulation, the process of forming the $\mathscr{C}, \mathscr{G}, \mathscr{L}$ system matrices from the individual branch element values is called "stamping".

The reverse operation of "unstamping" involves decomposing entry-wise the values of the reduced system matrices in (12.10) into the corresponding R, L, and C values. The resulting Rs, Ls and Cs are connected in the reduced netlist according to the MNA topology. The reduced input/output matrices of (12.10) directly reveal the input connections in the reduced model via injected currents, without any controlling elements. The prerequisites for an unstamping realization procedure therefore are:

1. The original system is in MNA impedance form (12.6). If the system is of admittance type (12.4), apply the admittance-to-impedance conversion from Sect. 12.3.1.
2. In (12.6), no Ls are directly connected to the input terminals so that, after reduction, diagonalization and regularization preserve the input/output structure.
3. System (12.6) is reduced with SPRIM/IOPOR [28] to (12.7) and converted to second order form (12.10). The alternative is to obtain the second order form of the original system first, and reduce it directly with SAPOR/IOPOR [3, 29].
4. The reduced system (12.10) must be diagonalized and regularized according to [29]. Diagonalization ensures that all inductors in the synthesized model are connected to ground (i.e., there are no inductor loops). Regularization eliminates spurious over-large inductors. These steps however are not needed for purely RC circuits.

12.3.3 On SPRIM/IOPOR and Rank Loss of $\widetilde{\mathbf{A}}$

In some cases it was observed (and shown by results in Sect. 12.4.2) that models reduced with SPRIM or SPRIM/IOPOR exhibit poles and zeros at 0. This section explains when this happens and supports theoretically the interpretation of the results in Sect. 12.4.2

Proposition 1 *Let* $\mathbf{W}$ *and* $\widetilde{\mathbf{V}}_3$ *be the SPRIM/IOPOR projection matrices* (12.9), *with full column rank. If* $\widetilde{\mathscr{E}}_l = \mathbf{W}^* \mathscr{E}_l \widetilde{\mathbf{V}}_3$ *in* (12.3.2) *has deficient column rank, then the SPRIM/IOPOR reduced model* (12.7) *has at least a pole-zero pair at* 0.

Proof Recalling that $\widetilde{\mathbf{V}}_2$ has full column rank, it is clear from (12.9) that $\mathbf{W}$ also has full column rank. Nonetheless $\mathbf{W}^* \in \mathbb{C}^{(n_1+k_2)\times(n_1+n_2)}$ has more columns than rows (usually $k_2 \ll n_2$), thus $\mathbf{W}^*$ is column rank deficient. Hence $\widetilde{\mathscr{E}}_l = \mathbf{W}^* \mathscr{E}_l \widetilde{\mathbf{V}}_3$ will have deficient column rank (even if $\mathscr{E}_l$ and $\widetilde{\mathbf{V}}_3$ have full column rank). Then in (12.7) we have: $\text{rank}(\widetilde{\mathbf{A}}) < \#\text{cols}(\widetilde{\mathbf{A}})$. Thus $\widetilde{\mathbf{A}}$ has deficient column rank $\Rightarrow \widetilde{\mathbf{A}}$ has at least an eigenvalue at $0 \Rightarrow 0$ is also an eigenvalue of the pencil $(\widetilde{\mathbf{A}}, \widetilde{\mathbf{E}})$ and this is a pole at 0. The zeros of the reduced system (12.7) are determined via the

Rosenbrock matrix [23, Chapter 5]: $\widetilde{\mathbf{A}}_z = \begin{bmatrix} \widetilde{\mathbf{A}} & \widetilde{\mathbf{B}} \\ -\widetilde{\mathbf{C}} & \mathbf{0} \end{bmatrix} = \begin{bmatrix} -\widetilde{\mathscr{G}} & -\widetilde{\mathscr{E}}_l & \widetilde{\mathscr{E}}_v \\ \widetilde{\mathscr{E}}_l^* & \mathbf{0} & \mathbf{0} \\ -\widetilde{\mathscr{E}}_v^* & \mathbf{0} & \mathbf{0} \end{bmatrix}$,

which immediately reveals that if $\widetilde{\mathscr{E}}_l$ is column rank deficient, then $\widetilde{\mathbf{A}}_z$ also loses rank and will have a 0 eigenvalue. Therefore the SPRIM/IOPOR reduced system (12.7) will also have a zero at 0. □

Analytically, pole-zero pairs at 0 are harmless since they theoretically cancel. Numerically this may not be the case, altering the approximation for low frequencies (as seen for instance in the result in Sect. 12.4.2).

12.4 Numerical Examples

Two circuits are chosen to demonstrate the applicability of the model reduction and synthesis procedures treated in this paper. For the single-input-single-output (SISO) example, one can easily provide synthesized models via both Foster and RLCSYN. For the multi-input-multi-output (MIMO) example, a synthesized model can be obtained straightforwardly with RLCSYN, thus RLCSYN synthesis is preferred over Foster synthesis.

12.4.1 SISO RLC *Network*

We reduce the SISO *RLC* transmission line in Fig. 12.3. Note that the circuit is driven by the voltage **u**, thus it is of admittance type (12.4). The admittance simulation of the model reduced with the *dominant spectral zero method (Dominant SZM)* [16], synthesized with the Foster approach, is shown in Fig. 12.5. The behavior of the original model is well approximated for the entire frequency range, and can also reproduce oscillations at dominant frequency points.

The benefit of the admittance-to-impedance transformation described in Sect. 12.3.1 is seen in Fig. 12.6. By reducing the system in impedance form with SPRIM/IOPOR and synthesizing (12.7) (using the second order form (12.10)) with RLCSYN [29], we are able to recover the reduced admittance (12.8) as well. The approximation is good for the entire frequency range.

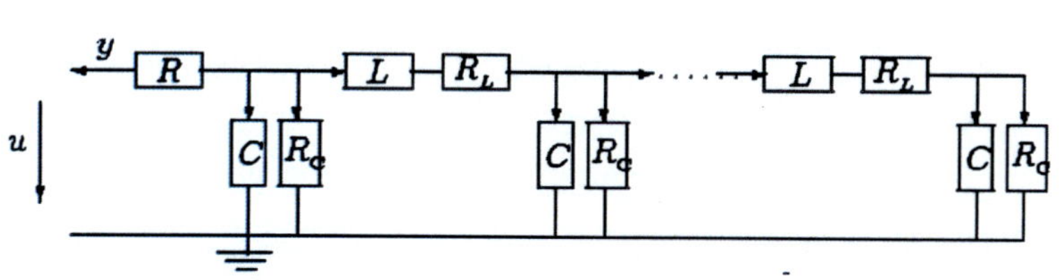

Fig. 12.3 Transmission line from Sect.12.4.1

12.4.2 MIMO RLC *Network*

We reduce the MIMO *RLC* netlist resulting from the parasitic extraction [12] of the coil structure in Fig. 12.4. The model has 4 pins (external nodes). Pin 4 is connected to other circuit nodes only via *C*'s, which causes the original model (12.6) to have a pole at 0. The example shows that: (1) the SPRIM/IOPOR model preserves the terminals and is synthesizable with RLCSYN without controlled sources, and (2) structure preserving projections may affect numerical cancellations of pole-zero pairs at 0.

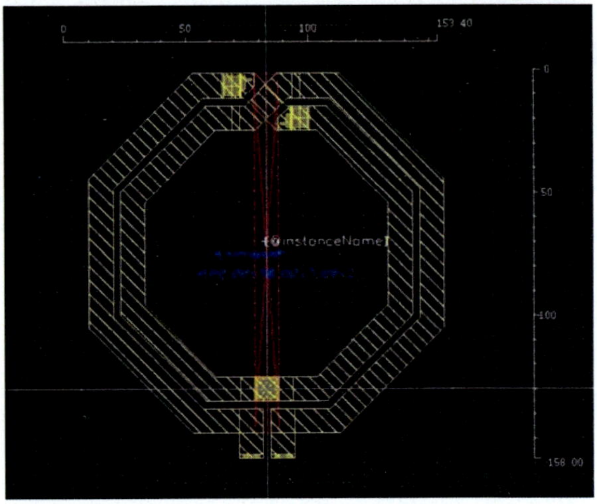

Fig. 12.4 Coil structure from Sect.12.4.2

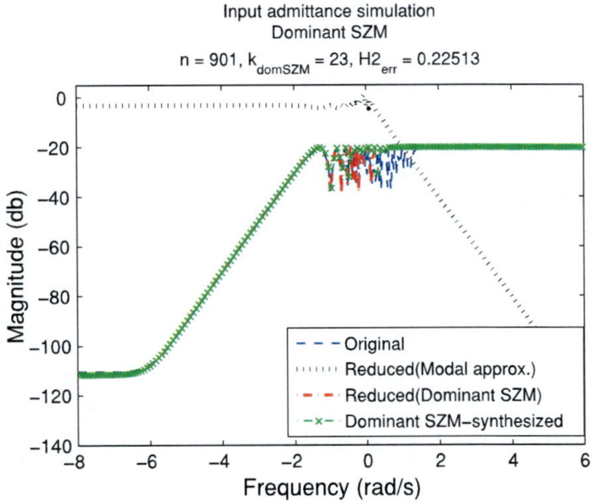

Fig. 12.5 Input admittance transfer function: original, reduced with Dominant SZM in admittance form and synthesized with Foster admittance

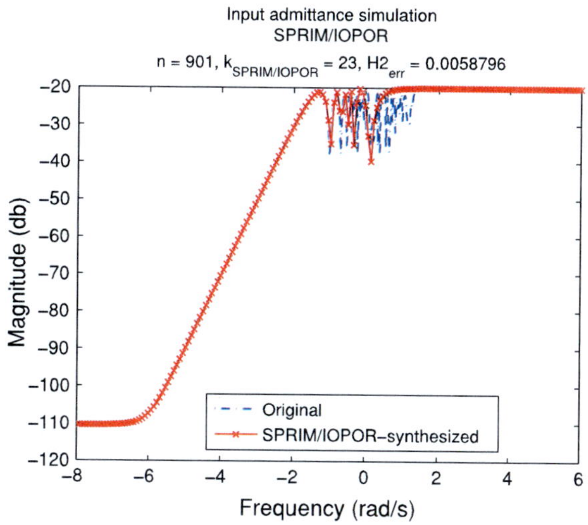

Fig. 12.6 Input admittance transfer function: original and synthesized SPRIM/IOPOR model (via impedance), after reconnecting the voltage source at the input terminal

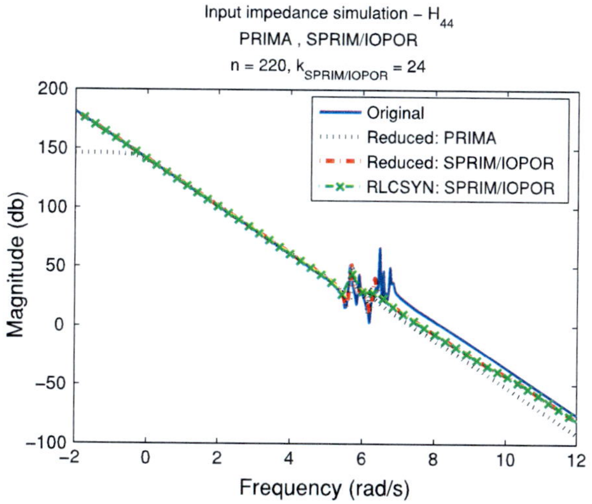

Fig. 12.7 Input impedance transfer function with "v_4" kept: $\mathbf{H}_{44}$ for PRIMA, SPRIM/IOPOR and RLCSYN realization

Fig. 12.7 shows the simulation of the transfer function from input 4 to output 4, which clearly reflects the presence of the pole at 0 due to the large response magnitude for low frequencies. By inspecting the response from input 3 to output 3 however, we notice in Fig. 12.8 that the SPRIM/IOPOR model is less accurate around DC than PRIMA. The SPRIM/IOPOR (12.9) projection approximates the original pole at 0 by a small pole (not at 0), but still places zeros at 0 due to the dependencies created in the Rosenbrock matrix $\widetilde{\mathbf{A}}_z$ (see Sect. 12.3.3). Such pole-zero pairs can no longer cancel numerically, affecting the approximation quality for low frequencies.

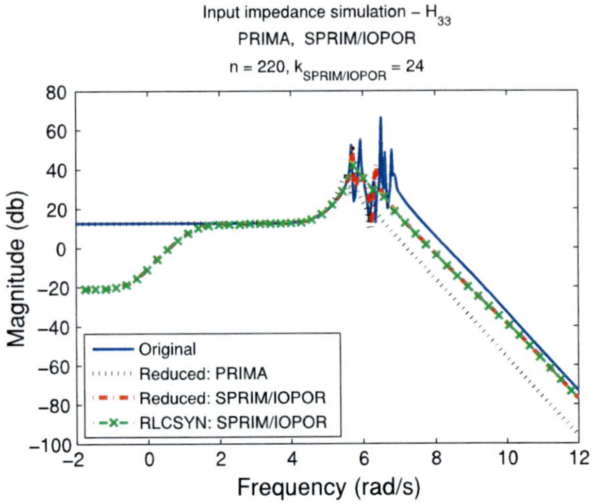

Fig. 12.8 Input impedance transfer function with "v_4" kept: $\mathbf{H}_{33}$ for PRIMA, SPRIM/IOPOR and RLCSYN realization

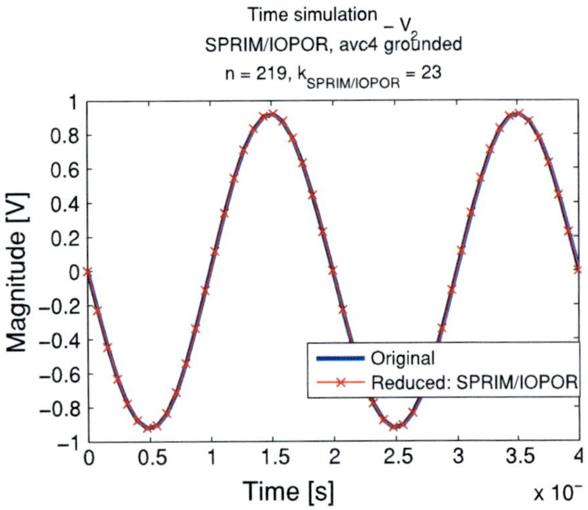

Fig. 12.9 Transient simulation with "v_4" grounded: voltage measured at node 2 for SPRIM/IOPOR (from RLCSYN realization)

The alternative is to ground pin 4 prior to reduction. This will remove the pole at 0 from the original model, resolve the corresponding numerical cancellation effects, and improve approximation around DC. As seen from Fig. 12.10, SPRIM/IOPOR applied on the remaining 3-terminal system gives a better approximation than PRIMA for the entire frequency range. The transient simulation in Fig. 12.9 confirms that the SPRIM/IOPOR model is both accurate and stable. With pin 4 grounded however, we loose the ability to (re)connect the synthesized model in simulation via all the terminals .

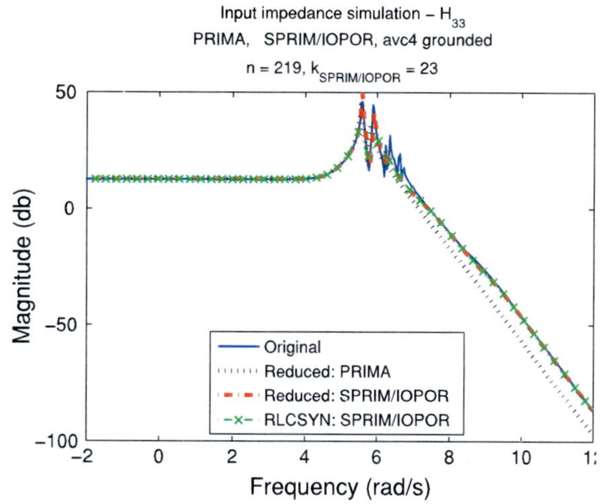

Fig. 12.10 Input impedance transfer function with "v_4" grounded: $\mathbf{H}_{33}$ for PRIMA, SPRIM/IOPOR and RLCSYN realization

Table 12.2 Summary and comparison of synthesis methods

To summarize	Projection and Foster realization of $\hat{\mathbf{H}}(s)$	I/O pres. SPRIM and RLCSYN unstamping
Properties	Passivity guaranteed by positive realness of $\widehat{\mathbf{H}}(s)$ from appropriate $\mathbf{V}$, $\mathbf{W}$	Passivity ensured via congruence transformation $\tilde{\mathbf{W}}$, $\tilde{\mathbf{W}}^T\tilde{\mathbf{W}} = \mathbf{I}$ on Σ in MNA form
Advantages	Σ need not be in MNA form	Preserved I/O and MNA structure
	RC, *RL* reduced with modal approximation $\Rightarrow$ positive elements	Unstamping easy for MIMO via impedance $\hat{\Sigma}$
Main hurdles	MIMO realization for >2 terminals only for **admittance** $\mathbf{H}(s)$ and $\hat{\mathbf{H}}(s)$	$\tilde{\mathcal{G}}$ may loose rank
	Dense MIMO netlists	Reduced $\tilde{\mathcal{G}}$,$\tilde{\mathcal{C}}$, are dense
	For p ports, k poles $O(p^2k)$ circuit elements	For p ports, k moments $O(p^2k^2)$ circuit elements
	Positive R, L, C not guaranteed	Positive R, L, C not guaranteed

12.5 Conclusions and Outlook

A framework for realizing reduced mathematical models into *RLC* netlists was developed. Model reduction by projection for *RLC* circuits was described and associated with two synthesis approaches: Foster realization (for SISO transfer functions) and RLCSYN [29] synthesis by unstamping (for MIMO systems). The approaches were tested on several examples, and a comparison is shown in Table 12.2. Future research will investigate reduction and synthesis methods for *RCLK* circuits with many terminals.

References

1. Antoulas, A.C.: Approximation of Large-Scale Dynamical Systems. SIAM, Phildelphia (2005)
2. Antoulas, A.C.: A new result on passivity preserving model reduction. Syst. Control Lett. **54**, 361–374 (2005)
3. Bai, Z., Li, R., Su, Y.: A unified Krylov projection framework for structure-preserving model reduction. In: Schilders, W., van der Vorst, H., Rommes, J. (eds.) Model order reduction, Mathematics in Industry, vol. 11. Springer, Berlin(2008)
4. Benner, P.: Advances in balancing-related model reduction for circuit simulation. In: Roos, J., Costa, L.R.J. (Eds.) Scientific Computing in Electrical Engineering, SCEE 2008, Mathematics in Industry. vol.14, Springer, Berlin, pp. 469–482 (2010)
5. Benner, P., Mehrmann, V., Sorensen, D. (eds.): Dimension reduction of large-scale systems, Lecture Notes in Computational Science and Engineering, vol.45. Springer, Berlin (2005)
6. Bott, R., Duffin, R.: Impedance synthesis without the use of transformers. J. Appl. Phys. **20**, 816 (1949)
7. Brune, O.: Synthesis of a finite two-terminal network whose driving point impedance is a prescribed function of frequency. J. Math. Phys. **10**, 191–236 (1931)
8. Edxact: Jivaro. URL http://www.edxact.com
9. EECS Department, University of California at Berkeley: Spice. URL http://bwrc.eecs.berkeley.edu/classes/icbook/spice/
10. Freund, R.: Sprim: Structure-preserving reduced-order interconnect macromodeling. In: Proceedings of IEEE/ACM International Conference on Computer Aided Design, pp. 80–87. Los Alamitos, CA (2004)
11. Grimme, E.J.: Krylov projection methods for model reduction. Ph.D. thesis, University of Illinois (1997)
12. Guille, A., Hanssen, M., Niehof, J.: Comparison between RLCk extraction and EM simulation on RF circuits. Technical report, NXP Semiconductors (2008)
13. Guillemin, E.A.: Synthesis of Passive Networks, 2nd edn. Wiley, New York (1959)
14. Heres, P.J.: Robust and efficient Krylov subspace methods for model order reduction. Ph.D. thesis, Eindhoven University of Technology (2005)
15. Ionutiu, R., Rommes, J.: Circuit synthesis of reduced order models. Technical Note 2008/00316, NXP Semiconductors (2009)
16. Ionutiu, R., Rommes, J., Antoulas, A.: Passivity preserving model reduction using dominant spectral zero interpolation. IEEE Trans.CAD Circ.Syst. **27**(12), 2250–2263 (2008)
17. Kailath, T.: Linear Systems. Prentice-Hall, New Jersey (1980)
18. Kerns, K.J., Yang, A.T.: Preservation of passivity during RLC network reduction via split congruence transformations. IEEE Trans. Comput-Aided Des. Integr. Circuits Syst. **17**(7), 582–591 (1998)
19. Odabasioglu, A., Celik, M., Pileggi, T.: PRIMA: Passive reduced-order interconnect macromodeling algorithm. IEEE Trans. Comput-Aided Des. Integr. Circuits Syst. **17**(8), 645–654 (1998)
20. Palenius, T., Roos, J.: Comparison of reduced-order interconnect macromodels for time-domain simulation. IEEE Trans. Microw. Theory Tech. **52**(9),191–236 (2004)
21. Phillips, J., Daniel, L., Silveira, L.: Guaranteed passive balancing transformations for model order reduction. IEEE Trans. Comput-Aided Des. Integr. Circuits Syst. **22**(8), 1027–1041 (2003). DOI 10.1109/TCAD.2003.814949
22. Phillips, J.R., Silveira, L.M.: Poor man's TBR: A simple model reduction scheme. IEEE Trans.CAD Circ.Syst. **24**(1), 283–288 (2005)
23. Reis, T.: Circuit synthesis of passive descriptor systems—a modified nodal approach. Int. J. Circuit Theory Appl. (2008) (to appear)
24. Rommes, J.: Methods for eigenvalue problems with applications in model order reduction. Ph.D. thesis, Utrecht University (2007)

25. Schilders, W.H.A., van der Vorst, H.A., Rommes, J. (eds.): Model order reduction: theory, research aspects and applications, Mathematics in Industry, vol. 11. Springer, Berlin (2008)
26. Sorensen, D.: Passivity preserving model reduction via interpolation of spectral zeros. Syst. Control Lett. **54**, 347–360 (2005)
27. Stykel, T., Reis, T.: Passivity-preserving model reduction of differential-algebraic equations in circuit simulation. In: Proceedings in Applied Mathematics and Mechanics (ICIAM 2007, Zurich, Switzerland), pp. 1021, 601–1021, 602 (2007)
28. Tan, S.X.D., He, L.: Advanced Model Order Reduction Techniques in VLSI Design. Cambridge University Press, Cambridge (2007)
29. Yang, F., Zeng, X., Su, Y., Zhou, D.: RLC equivalent circuit synthesis method for structure-preserved reduced-order model of interconnect in VLSI. Commun. Comput. Phys. **3**(2), 376–396 (2008)

Chapter 13
Model Reduction Methods for Linear Network Models of Distributed Systems with Sources

Stefan Ludwig and Wolfgang Mathis

Abstract Distributed systems with distributed sources are modeled as large electrical networks with linear RLC-elements, independent sources and pins for the connection with other network models. Often these networks are too large to be simulated efficiently. Model reduction can be used to reduce these networks while approximating the behavior at the pins for the connection with other nonlinear and linear network models. In the standard model reduction the independent sources are extracted and connected by ports with the RLC-part of the network which is to be reduced. This extraction only enables a weak reduction for networks with a large number of independent sources, as the number of ports is very high. In this article an efficient reduction of networks with a large number of sources is proposed by taking the waveforms of the sources into account. The method is based on the reduction of the dimension of the function space of the waveforms with the help of function approximation. The basis functions of the function approximation are used in the network, which results a lower number of independent sources. Extracting this lower number of independent sources and reducing the network enables a higher model reduction with state of the art techniques. With the proposed method a smaller size and higher accuracy of the reduced model can be achieved. The validity and efficiency of the proposed method is shown by reducing example models.

Keywords Reduced order modeling · Linear RLC networks with sources · Generalized state space systems · Terminals · Impedance · Modified model analysis

S. Ludwig (✉) · W. Mathis
Institute of Electromagnetic Theory, Leibniz University of Hannover, Appelstr. 9A, 30167 Hannover, Germany
e-mail: ludwig@tet.uni-hannover.de

W. Mathis
e-mail: mathis@tet.uni-hannover.de

P. Benner et al. (eds.), *Model Reduction for Circuit Simulation*,
Lecture Notes in Electrical Engineering, 74, DOI: 10.1007/978-94-007-0089-5_13,

13.1 Introduction

In the modeling of distributed systems very large linear RLC-networks are created. For these very large networks a smaller network with approximated behavior at some nodes of interest, typically the pins for the connection with other network models, is required to enable fast simulations. With the help of model order reduction (MOR) and network synthesis it is possible to find such a reduced network [3, 7, 10, 24]. Replacing the original network with the reduced network enables investigations with a lower effort for simulations.

For distributed systems with distributed sources the modeled networks also contain, next to passive RLC-elements a high number of independent sources. These models of distributed systems with distributed sources typically occur in the IC design and verification as for example in the modeling of power grids, substrate couplings and conducted emission behavior [14, 19, 23, 33]. In the standard approach the independent sources are extracted from the reducible RLC-part of the network and connected by ports. With this extraction only the passive RLC-part is reduced and passivity can be preserved during a MOR [2, 24, 28, 32]. For networks with a large number of independent sources this extraction produces a large number of ports of the network to be reduced. A large number of ports is a strong limitation for the MOR [20, 30]. In this article we will propose an additional step prior to the MOR to overcome this limitation. In this step the dimension of the function space of the waveforms of the independent sources is lowered. Realizing this smaller function space as network elements and replacing the original sources by controlled sources results in a lower number of independent sources. Extracting this lower number of independent sources leads to a lower number of ports of the reducible network. In addition, network properties like passivity and reciprocity are preserved in the reducible part. The behavior at the pins for the connection with other nonlinear elements like drivers and loads is preserved in this first port reduction step but the behavior of the distributed sources is approximated. The resulting reducible network with a lower number of ports can now more efficiently reduced with state of the art MOR techniques by approximating the behavior at the pins.

This article is structured as follows. In Sect. 13.2 the basics of model reduction are shown. In Sect. 13.3 a description of a large number of independent sources as additional step before an efficient model reduction is proposed. The efficiency of the complete reduction process is shown by reducing two example networks in Sect. 13.4. The article is concluded in Sect. 13.5.

13.2 Background for Model Reduction of Linear Networks

This article will focus on distributed systems modeled as linear positive real valued RLC-networks with independent current sources. The impedance description of the network is considered. However, all presented methods are also valid for the admittance or hybrid case and for networks with independent voltage sources.

For MOR the electrical circuit is described by differential algebraic equations. With the help of modified nodal analysis, circuit equations in the frequency domain of order N in the form of

$$(s\mathbf{C} + \mathbf{G})\mathbf{x} = \mathbf{Bu}, \quad \mathbf{y} = \mathbf{L}^T\mathbf{X} \tag{13.1}$$

are generated. The real positive definite matrices $\mathbf{C}, \mathbf{G} \in \mathbb{R}^{N\times N}$ contain the stamps for the capacitors, resistors and inductors. $\mathbf{u}, \mathbf{y}$ are the Laplace transformed p input currents and output voltages respectively, which are connected with the system vector $\mathbf{x}$ by the system matrices $\mathbf{B} \in \mathbb{R}^{N\times p}$ and $\mathbf{L} \in \mathbb{R}^{p\times N}$. The resulting impedance transfer function is given with

$$\mathbf{Z}(s) = \mathbf{L}^{\mathrm{T}}(s\mathbf{C} + \mathbf{G})^{-1}\mathbf{B}. \tag{13.2}$$

With the help of MOR a reduced system in the form of Eq. 13.1 is built by approximating the transfer function behavior. MOR algorithms based on the projection are commonly used for the reduction of large models. The advantage of the projection methods is the simple implementation and the low computational effort for the construction of the reduced model. For the reduction a projection matrix $\mathbf{T} \in \mathbb{R}^{N\times n}$ has to be generated. By changing the system vector $\widetilde{\mathbf{x}} = \mathbf{T}^T\mathbf{x}$ the reduced system matrices of order n

$$\widetilde{\mathbf{C}} = \mathbf{T}^T\mathbf{CT}; \quad \widetilde{\mathbf{G}} = \mathbf{T}^T\mathbf{GT}; \quad \widetilde{\mathbf{B}} = \mathbf{T}^T\mathbf{B}; \quad \widetilde{\mathbf{L}} = \mathbf{T}^T\mathbf{L} \tag{13.3}$$

are calculated [24]. If the projection matrix is real, this method preserves the passivity of suitably described systems [24]. For the generation of the projection matrix different methods are used.

The most common method is that the projection matrix is generated by using a single-point Krylov subspace [24]. It can be shown, that the moments of the reduced system match the moments of the original system to some order [24]. As an extension the union of multiple Krylov subspaces at different expansion points is used to generate the projection matrix [7], which is in the following named MP-Krylov. With this type of projection matrices the order of the reduced model is proportional to the number of ports. For models with a large number of ports, this represents a strong limitation for the possible order reduction [30].

Another approach for the generation of the projection matrix is proposed in [27] with Poor Man's Truncated Balanced Realization (PMTBR) using congruence transformation (Section V.E in [27]). An approximation to the controllability Gramian is generated and an approximation to the Hankel singular values is calculated. As shown in [28, 29] the approximated Hankel singular values are used for order and error control. With this projection matrix the order does not directly depend on the number of ports. However, there is an indirect dependence on the number of ports and the possible order reduction [27].

For investigations with electrical simulators the order reduced system is synthesized with network-synthesis algorithms as described for example in [20, 25, 35].

As these methods are only applicable to networks with a relatively low number of ports some advanced MOR methods dealing with networks with a large number of ports are developed. The underlying principles can be divided into two methods: using properties of the network structure and using input information. The methods presented in [9, 18] belong to the first class and use the correlation of the ports which is efficient if the models are regularly structured. The same class includes methods of dividing the network into substructures with a lower number of ports, as for example presented in [5]. For the class of methods where input information is used, for example the input correlated TBR method where the dependencies of the input excitations are taken into account in the model reduction process is presented in [31]. The reduced order analysis methods presented in [16, 17, 34] rely on determined inputs at all ports of the network, which is more likely for a simulation method as all inputs have to be determined and no connection with other nonlinear network elements is possible. The method presented in [21] uses the information of parameterized loads connected to the reducible network for a more efficient reduction with parametric MOR methods. In [20] a method based on the determined behavior of independent sources for reducing the number of ports is presented, which allows for the connection with other network models at some specified nodes but is limited to groups of internal sources with equal or proportional waveforms.

13.3 Description of Distributed Sources

In this article a generalization of the method of [20] is presented which is not restricted to groups of independent sources with proportional waveforms and includes waveforms that can be decomposed into or approximated by a small set of basis functions. Another advantage of this method is that unlike in [9, 18] no limitations for the structure of the linear network are necessary, Unlike in [16, 17, 34] not all inputs of the model have to be determined during the reduction which allows for the connection with other network models in simulations. The disadvantage is that only the number of ports where the input is determined is reduced and thereby the method is designated for networks with a large number of independent sources and a low number of pins for the connection with other network models. The method is an additional prior step to enhance the MOR.

In the standard description the independent sources are extracted from the RLC-network and connected through ports with the remaining network (Fig. 13.1). In this way the reducible part of the network contains only positive real valued RLC-elements and thereby the network is passive and reciprocal. Passivity can be preserved during MOR using adequate reduction algorithms [2, 24, 28, 32] as well as reciprocity [10]. For networks with a large number of independent sources, the disadvantage of this extraction is the large number of ports which limits the order reduction [8, 20, 30].

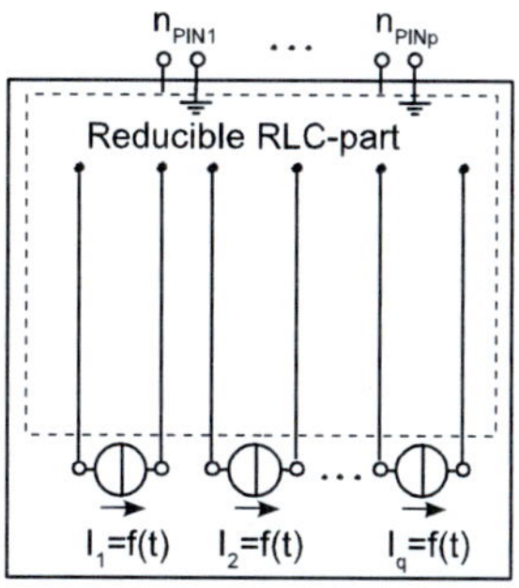

Fig. 13.1 Reducible network with extracted independent sources and a large number of ports

In this article a description of the sources in the network, which results in a lower number of independent sources while approximating the behavior at the pins is presented. The description of the sources is a prior step to the MOR and meets the following requirements: Firstly, preserve reciprocity and passivity of the reducible network in order to enhance network synthesis and guarantee stable time-domain simulations of the reduced model. Secondly, preserve properties of matrices $\mathbf{C}, \mathbf{G}$ like (semi-) definiteness for passivity-preserving reduction [10, 24] and (J-) symmetry and typical block structure to enhance network synthesis after the reduction [10, 25, 35]. The general structure of the network is shown in Fig. 13.2 for an RLC-network with q independent current sources.

Firstly we will have a look at the waveforms of the independent sources. A reduction of the dimension of the function space based on decomposition and approximation of the waveform functions is used. The q waveforms of the independent sources of the network are described by superposition with the weights $w_{i,l}$ of b basis functions

$$I_l \approx \sum_{i=1}^{b} w_{i,l} I_i(t) \quad 1 \leq l \leq q. \tag{13.4}$$

For this function approximation several methods are known in the fields of approximation theory, black box modeling and artificial neural networks. In the following some of the most often used methods are presented.

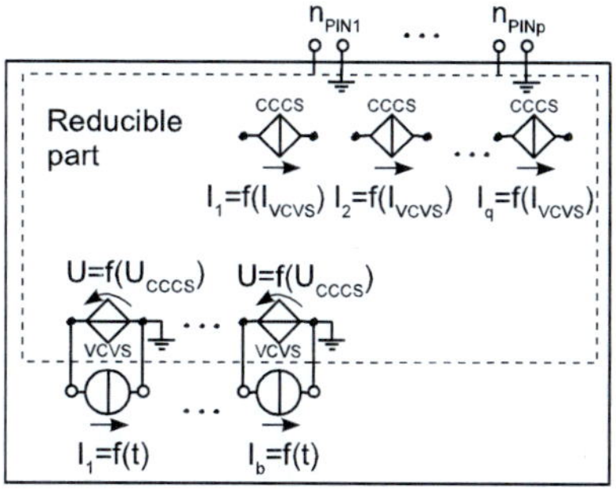

Fig. 13.2 Passive and reciprocal reducible network with replaced sources and a small number of ports

A special and most simple case is the grouping of equal or proportional waveforms as presented in [20], where one representative waveform of every group is used as basis function. The number of groups specifies the number of necessary basis functions.

For sources with waveforms described by continuous piecewise-linear (PWL) functions an exact decomposition into (finite) ramp functions is possible, e.g. described in [1]. The number of basis functions is now given by the number of points which are used for the waveform description. If the number of points describing the waveform is lower than the number of sources a reduction of the function space is achieved. If the PWL functions are periodic, the basis functions can also chosen as periodic ramp functions and thereby the number of basis functions is given with the number of data points in one period of the waveform. For discontinuous PWL functions the representations of [6] can be used. If the waveforms are piecewise constant functions, shifted bar functions can be used as basis functions.

If waveforms are specified in a limited range of time, for example if the waveforms are generated from measurements or modeling in a finite time range, several classes of basis functions used in the field of approximation theory and artificial neural network modeling can be used. For every waveform the approximating basis functions can be obtained as algebraic polynomials to any degree of accuracy as stated in the Weierstrass approximation theorem. An example is the well known Taylor series expansion. Radial basis functions which are used in the black box modeling are also suitable as basis functions, as they are universal approximators [11, 26]. Another class of basis functions are the sigmoid functions, which are almost always capable of universal approximation [13]. For periodic or time- or band-limited waveforms the approximation with a fourier series expansion can be used. The basis functions are then trigonometric functions. The series converges almost everywhere to the waveform functions which is stated by the Carleson–Hunt theorem and thereby a finite number of basis functions is sufficient.

The discrete wavelet transform, which is widely used in signal- and data processing can also be used for the reduction of the dimension of the function space. With this transformation one dimensional wavelets are used as basis function for the approximation as for example presented in [22].

For this approximative methods the number of necessary basis functions depends on the desired accuracy of the approximation of the waveforms. For the number of basis functions a tradeoff between the number of basis functions and the desired accuracy has to be found. For a higher accuracy a higher number of basis functions is necessary and the reduction of the dimension of the function space is lower.

Any other method of decomposition into or approximation by a finite number of basis functions is feasible, as long as the number of basis functions b is smaller than the number of independent sources q in the network. As for most approximation methods the number of basis functions depends on the desired accuracy and the number of basis functions has to be lower than the number of independent sources waveforms this method is most suitable for networks with a large number of independent sources.

The inclusion of the reduced dimension function space in the network model in the form of Fig. 13.2 is divided into the following four steps. Independently on the decomposition or approximation method for every basis function an independent current source with the basis waveform is added in the first step in the network model, resulting in b additional independent sources. In the second step all q original independent current sources are replaced by current controlled current sources (CCCS), controlled by the b additional independent sources. The gains of the CCCS are given by the weights $w_{i,l}$ in Eq. 13.4. In the third step, voltage controlled voltage sources (VCVS), connected with the additional independent sources, are used for the preservation of passivity and reciprocity if the gains are chosen according to the gains of the CCCS, as is shown later on in this section. In the last step, only the b additional independent current sources are extracted and connected through ports with the reducible part of the network. In this way a reducible network with a reduced number of ports is generated. The resulting reducible network also contains controlled sources in addition to the RLC-elements and is called network with replaced sources throughout the article.

In the following the description of the reducible network with replaced sources and a lowered number of ports is given, whereas the system matrices $\mathbf{C}, \mathbf{G}$ have the typical properties of pure RLC-networks like (J-)symmetry, (semi-)definiteness and block structure. An RLC-network with N nodes, p pins and q independent current sources, which waveforms are the superposition of b basis waveforms, is considered. For the generation of the system matrices of the network with replaced sources the incidence matrices $\mathbf{K}_C, \mathbf{K}_R, \mathbf{K}_L, \mathbf{K}_{\text{CCCS}}, \mathbf{K}_{\text{pin}}$ for capacitors, resistors, inductors, CCCSs and pins are used. The matrices $\widehat{\mathbf{P}}_{\text{CCCS}} \in \mathbb{R}^{q \times b}, \widehat{\mathbf{P}}_{\text{VCVS}} \in \mathbb{R}^{b \times q}$ contain the parameters of the gains of the controlled sources. The branch constitutive relations are given with

$$\begin{aligned} \mathbf{i}_C = s\widehat{\mathbf{C}}\mathbf{u}_C, \quad \mathbf{i}_R = \widehat{\mathbf{G}}\mathbf{u}_R, \quad \mathbf{u}_L = s\widehat{\mathbf{L}}\mathbf{i}_L \\ \mathbf{i}_{\text{CCCS}} = \widehat{\mathbf{P}}_{\text{CCCS}}\mathbf{i}_{\text{VCVS}}, \quad \mathbf{u}_{\text{VCVS}} = \widehat{\mathbf{P}}_{\text{VCVS}}\mathbf{u}_{\text{CCCS}}. \end{aligned} \tag{13.5}$$

With Kirchhoff's Current Law $\mathbf{Ki} = \mathbf{0}$ we get

$$\begin{pmatrix} \mathbf{K}_C & \mathbf{K}_R & \mathbf{K}_L & -\mathbf{K}_{\text{CCCS}} & -\mathbf{K}_{\text{pin}} & \mathbf{0} & \mathbf{0} \\ \mathbf{0} & \mathbf{0} & \mathbf{0} & \mathbf{0} & \mathbf{0} & -\mathbf{I} & \mathbf{I} \end{pmatrix} \begin{pmatrix} \mathbf{i}_C \\ \mathbf{i}_R \\ \mathbf{i}_L \\ \mathbf{i}_{\text{CCCS}} \\ \mathbf{i}_{\text{pin}} \\ \mathbf{i}_{\text{VCVS}} \\ \mathbf{i}_b \end{pmatrix} = \mathbf{0} \tag{13.6}$$

with $\mathbf{I}$ as the identity matrix. Notably the network is divided into two parts which are not interchanging charges. The first part contains the RLC-elements, the CCCSs and the pins and the second part the VCVS and the additional independent sources. The voltages are defined as

$$\begin{pmatrix} \mathbf{u}_C \\ \mathbf{u}_R \\ \mathbf{u}_L \\ \mathbf{u}_{\mathrm{CCCS}} \\ \mathbf{u}_{\mathrm{pin}} \\ \mathbf{u}_{\mathrm{VCVS}} \end{pmatrix} = \begin{pmatrix} \mathbf{K}_C^T & \mathbf{0} \\ \mathbf{K}_R^T & \mathbf{0} \\ \mathbf{K}_L^T & \mathbf{0} \\ \mathbf{K}_{\mathrm{CCCS}}^T & \mathbf{0} \\ \mathbf{K}_{\mathrm{pin}}^T & \mathbf{0} \\ \mathbf{0} & \mathbf{I} \end{pmatrix} \begin{pmatrix} \mathbf{u}^\phi \\ \mathbf{u}_{\mathrm{VCVS}}^\phi \end{pmatrix} \tag{13.7}$$

where $\mathbf{u}^\phi$ contains the potentials of the first part of the network and $\mathbf{u}_{\mathrm{VCVS}}^\phi$ contains the potentials of the second part of the network. We define a system of circuit equations with

$$\overbrace{\begin{pmatrix} s\mathbf{K}_C\widehat{\mathbf{C}}\mathbf{K}_C^T + \mathbf{K}_R\widehat{\mathbf{G}}\mathbf{K}_R^T & \mathbf{K}_L \\ -\mathbf{K}_L^T & s\widehat{\mathbf{L}} \end{pmatrix}}^{(s\mathbf{C}+\mathbf{G})} \overbrace{\begin{pmatrix} \mathbf{u}^\phi \\ \mathbf{i}_L \end{pmatrix}}^{\mathbf{x}} = \overbrace{\begin{pmatrix} \mathbf{K}_{\mathrm{CCCS}}\widehat{\mathbf{P}}_{\mathrm{CCCS}} & \mathbf{K}_{\mathrm{pin}} \\ \mathbf{0} & \mathbf{0} \end{pmatrix}}^{\mathbf{B}} \overbrace{\begin{pmatrix} \mathbf{i}_b \\ \mathbf{i}_{\mathrm{pin}} \end{pmatrix}}^{\mathbf{u}} \tag{13.8}$$

$$\underbrace{\begin{pmatrix} \mathbf{u}_{\mathrm{VCVS}} \\ \mathbf{u}_{\mathrm{pin}} \end{pmatrix}}_{\mathbf{y}} = \underbrace{\begin{pmatrix} \widehat{\mathbf{P}}_{\mathrm{VCVS}}\mathbf{K}_{\mathrm{CCCS}}^T & \mathbf{0} \\ \mathbf{K}_{\mathrm{pin}}^T & \mathbf{0} \end{pmatrix}}_{\mathbf{L}^T} \underbrace{\begin{pmatrix} \mathbf{u}^\phi \\ \mathbf{i}_L \end{pmatrix}}_{\mathbf{x}} \tag{13.9}$$

Since the currents $\mathbf{i}_b$ and $\mathbf{i}_{\mathrm{pin}}$ are the inputs $\mathbf{u}$ of the system, the voltages $\mathbf{u}^\phi$ and inductor currents $\mathbf{i}_L$ form the system vector $\mathbf{x}$ and the voltages across the additional independent sources $\mathbf{u}_{\mathrm{VCVS}}$ and at the pins $\mathbf{u}_{\mathrm{pin}}$ are the outputs $\mathbf{y}$, the system has the form of equations (13.1) with $\mathbf{C}, \mathbf{G}$ having the typical MNA properties of pure RLC-networks without controlled sources. The controlled sources only appear in the matrices $\mathbf{B}$ and $\mathbf{L}$. The gains $\widehat{\mathbf{P}}_{\mathrm{VCVS}}$ of the VCVSs are now chosen with

$$\widehat{\mathbf{P}}_{\mathrm{VCVS}} = \widehat{\mathbf{P}}_{\mathrm{CCCS}}^T \tag{13.10}$$

and thereby the condition $\mathbf{B} = \mathbf{L}$ holds. With this description it is shown, that the controlled sources which normally influence the matrix $\mathbf{G}$ can be described in a way, where only $\mathbf{B}, \mathbf{L}$ are changed and properties of $\mathbf{C}, \mathbf{G}$ like definiteness, symmetry and structure of pure RLC-networks are conserved, enhancing the reduction.

This system of the network with replaced sources now has, in addition to the ports for the pins, only b ports for the independent sources which is less than the number of ports q if all sources were extracted. In the following, it is shown that this reducible network is passive and reciprocal.

Theorem 1 *The network with replaced sources is passive.*

Proof For the proof of passivity, the conditions for passivity have to be considered. The necessary and sufficient condition for the passivity of the impedance transfer function $\mathbf{Z}(s)$ of the model is positive-realness [15]:

1. $\mathbf{Z}(s^*) = \mathbf{Z}^*(s)$ where * is the conjugate complex operator
2. $\mathbf{Z}(s)$ is positive, that means $\mathbf{a}^H[\mathbf{Z}(s) + \mathbf{Z}^H(s)]\mathbf{a} \geq 0$, with H as Hermitian operator is satisfied for all complex s with $Re(s) \geq 0$ and for all finite complex vectors $\mathbf{a}$

Since the matrices stay real while replacing the independent current sources, the first condition is still fulfilled.

The second condition of positive-realness can be written as

$$\mathbf{a}^H[\mathbf{Z}(s) + \mathbf{Z}^H(s)]\mathbf{a} = \mathbf{a}^H\mathbf{L}^T[(s\mathbf{C} + \mathbf{G})^{-1} + (s\mathbf{C} + \mathbf{G})^{-H}]\mathbf{B}\mathbf{a}. \qquad (13.11)$$

With $\mathbf{B} = \mathbf{L}$ a finite complex vector $\mathbf{b} = \mathbf{B}a = \mathbf{L}a$ is defined and the transfer function with replaced independent current sources is positive

$$\mathbf{b}^H[(s\mathbf{C} + \mathbf{G})^{-1} + (s\mathbf{C} + \mathbf{G})^{-H}]\mathbf{b} \geq 0. \qquad (13.12)$$

Since the controlled sources are described in a way that does not influence the system matrices they stay positive definite and real while replacing the independent sources. This means that the passivity of the network is conserved, as long as $\mathbf{L} = \mathbf{B}$ holds, which is guaranteed by Eq. 13.10. □

Theorem 2 *The network with replaced sources is reciprocal.*

Proof For the proof of reciprocity, the condition for reciprocity has to be considered. The condition for a model to be reciprocal is that the impedance transfer function is symmetrical [4, 15]:

$$\mathbf{Z}(s) = \mathbf{Z}^T(s). \qquad (13.13)$$

This condition can be written as

$$\mathbf{L}^T(s\mathbf{C} + \mathbf{G})^{-1}\mathbf{B} = \mathbf{B}^T(s\mathbf{C} + \mathbf{G})^{-T}\mathbf{L}, \qquad (13.14)$$

which is fulfilled since system matrices $\mathbf{C}$ and $\mathbf{G}$ can be written in symmetric form by multiplying $\mathbf{C}, \mathbf{G}$ with an appropriate diagonal matrix $\mathbf{J}$ and with Eq. 13.10 the condition $\mathbf{B} = \mathbf{L}$ still holds. □

13.4 Examples

13.4.1 RCI-Grid

As an example, an RCI-grid network as shown in Fig. 13.3a which is widely used in power grid simulations [14, 23] is utilized. All reductions and simulations of this example are performed on an actual personal computer platform.[1] The example

[1] CPU: Core2Duo 2.66 GHz, RAM: 2GB

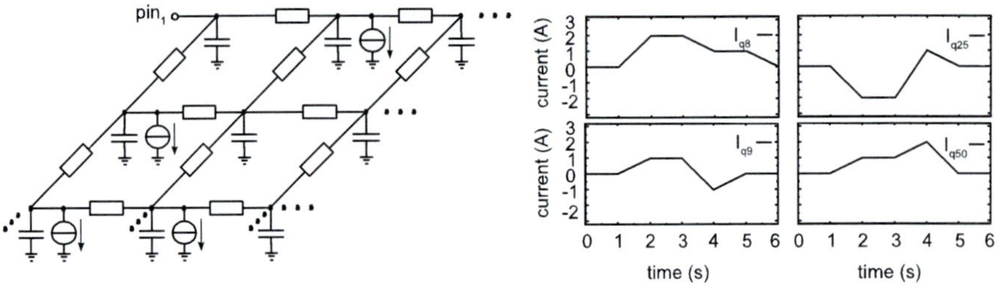

(a) Section of the network structure **(b)** Some of the 55 current sources waveforms

Fig. 13.3 Example 50 × 50 RCI-grid network. **a** Section of the network structure. **b** Some of the 55 current sources waveforms

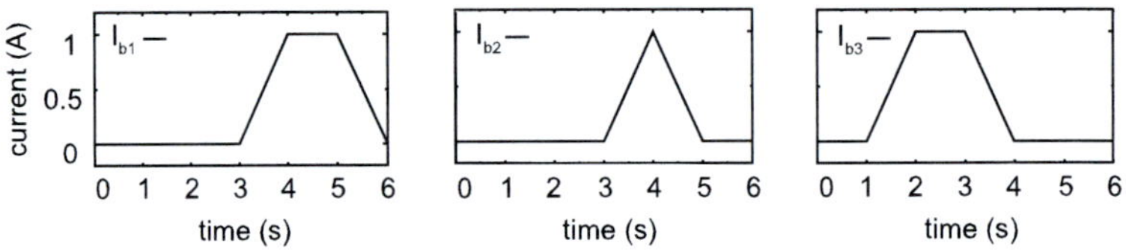

Fig. 13.4 The three basis waveforms of the RCI-grid example using decomposition

grid with 50 × 50 nodes contains 4900 resistors and 2500 capacitors and two pins at two opposite corners for the connection of nonlinear driver models.

The values of the elements are randomly chosen in the ranges $1 \pm 0.5\,\Omega$ and $0.1 \pm 0.05\,\mathrm{F}$ for the resistors and capacitors, respectively. $q = 55$ independent current sources with PWL-waveforms are connected between arbitrary chosen nodes in the network and the ground node. Some PWL-waveforms used in the example are shown in Fig. 13.3b. All PWL-waveforms are decomposed into the three basis-waveforms shown in Fig. 13.4.

With the standard method of extracting all independent sources the system has 57 ports. Only three independent sources are necessary if the waveforms of the independent sources are decomposed into the three basis-waveforms shown in Fig. 13.4 and the network with replaced sources of Sect. 13.3 is used. This results in five ports of the reducible part of the network. By this network alteration the behavior at the two pins of the model is not changed as the decomposition is exact. The transfer function of the system of the network with replaced sources can be calculated almost four times faster before the reduction with MOR due to the lower number of ports (Table 13.1). The network with extracted sources as well as the network with replaced sources are reduced with the PMTBR algorithm as described in Sect. 13.2. The 25 frequency points used in the reduction process are equally distributed in the frequency range $10^{-9} \leq s_i \leq 10^9$. The reduced systems are generated within less than one minute, which is not critical as the model reduction process is done offline before the simulation. In Fig. 13.5 the approximated Hankel singular values of both systems are shown. The singular values of the system with replaced sources decay faster. This means, that for a given

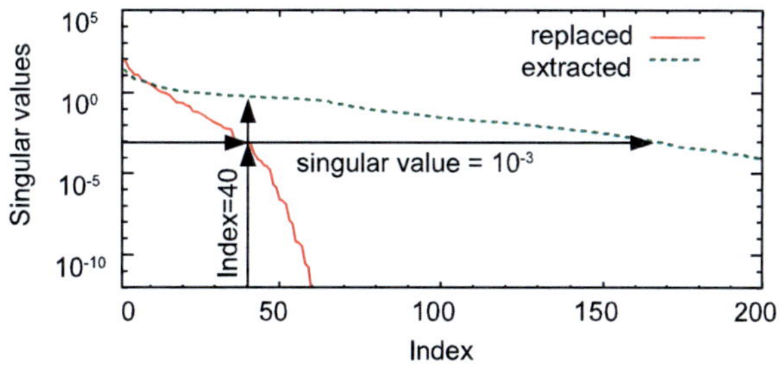

Fig. 13.5 Hankel singular values approximated with PMTBR for the 50 × 50 RCI-grid

Table 13.1 Comparison of reduction results of the 50 × 50 RCI-grid

Original order, $N = 2{,}500$	Ports, p	Reduced order, n	Reduction (%)	Reduction time (s)	MATLAB speed-up	Accuracy
Extracted sources	57	2,500			1×	Exact at the pins
Extracted sources, reduced with PMTBR	57	165	93.4	30	9.4×	Very good
Extracted sources, reduced with PMTBR	57	40	98.4	30	42×	Bad
Replaced sources	5	2,500			3.8×	Exact at the pins
Replaced sources, reduced with PMTBR	5	40	98.4	2	52×	Very good

maximum error the network with replaced sources can be more efficiently reduced than the network with extracted sources. Likewise for a given reduced order the reduced model of the network with replaced sources is more exact than the reduced model of the network with extracted sources. Firstly both systems are reduced to a maximal singular value of 10^{-3}, resulting in an order of $n = 40$ for the system with replaced sources and $n = 165$ for the system with extracted sources. The MATLAB-speed-up of both reduced systems for the calculation of the transfer function in several frequency points is shown in Table 13.1. Both reduced systems with the maximal singular value of 10^{-3} show a good agreement with the original system, as shown for the transfer function Z_{11} in Fig. 13.6. However, since the reduced system of the network with replaced sources is smaller, it can be simulated faster than the reduced system of the network with extracted sources (52× in comparison to 9.4×, Table 13.1). Secondly, both systems are reduced to the same order of $n = 40$. The speed-up of the system of the network with extracted sources is almost as high as for the system of the network with replaced sources (42× and 52×, Table 13.1). Nevertheless, the reduced system of the network with replaced sources is more accurate (Fig. 13.6) than the reduced system of the network with extracted sources.

With this example it is shown that the proposed replacing of independent sources as an additional step before the reduction can lead to smaller and therefore faster models than the standard method of extracting the sources. It was also

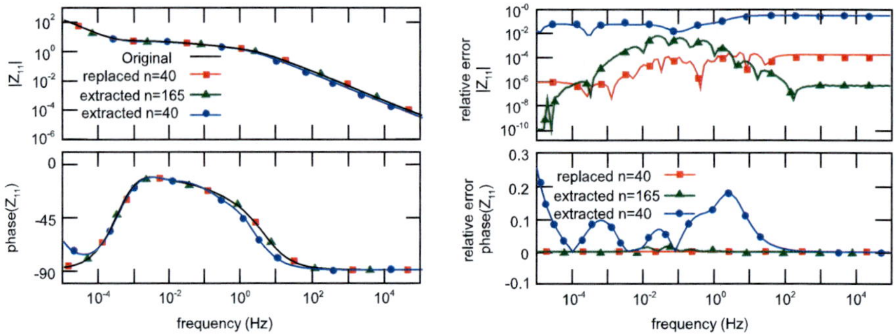

Fig. 13.6 Magnitude and phase of transfer function Z_{11} of 50×50 RCI-grid

Fig. 13.7 Section of the IC conducted emission network

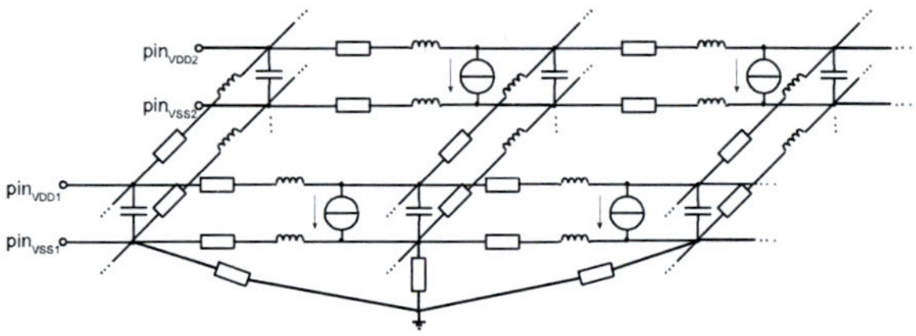

shown that for the same size of the reduced system our proposed method can lead to more accurate reduced models.

13.4.2 IC Conducted Emission Model

As a second example for a network with a large number of independent sources an industrial model describing the IC conducted emission (ICEM) behavior of the Infineon 32 Bit microcontroller TC1796 is utilized [33]. All reductions and simulations of this example are performed on an actual server platform.[2] The network is generated with the Infineon Technologies AG Germany EXPO-Tool [12]. In Fig. 13.7 a small section of the network is shown. The model contains more than 35,000 passive RLC-elements and 328 independent current sources. For the connection with the printed circuit board model the network contains 61 pins. The circuit equations of the complete model have an order of 25,062.

If every independent source is extracted and connected through a port with the reducible part of the network, the circuits equations contain 389 ports including the 61 pins. With respect to their prescribed waveforms the current sources of the network are grouped into eight groups and therefore eight additional independent sources are necessary. Replacing the 328 independent sources with CCCSs and

[2] CPU: 8× DualCore, Intel Harwich Platform, RAM: 64GB

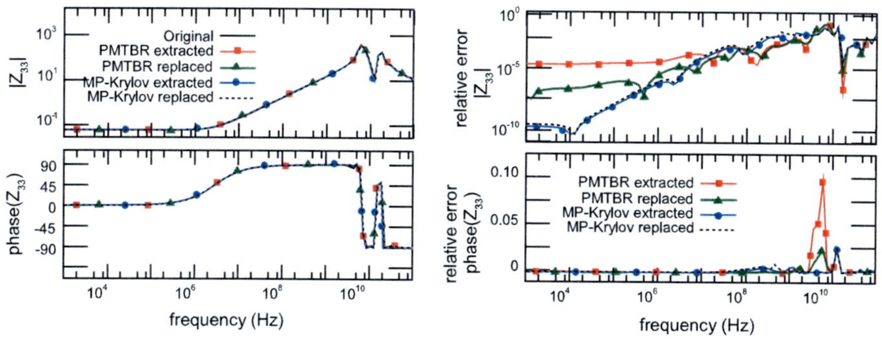

Fig. 13.8 Magnitude and phase of transfer function Z_{33} of ICEM

extracting the eight additional independent sources, as described in Sect. 13.3, results in a reducible network with only 69 ports, including the 61 pins. The network with extracted as well as the network with replaced sources are reduced with the MOR algorithms described in Sect. 13.2. The results for the order reduction are presented in Table 13.2. All models show a close alignment in the frequency range of interest as shown for the arbitrary chosen transfer function Z_{33} in Fig. 13.8.

For the MP-Krylov reduction two expansion points, $s_1 = 2\pi 10^3, s_2 = 2\pi 10^9$ are used where in every point one moment is to be matched. For preservation of reciprocity the projection matrices are split according to [10]. As the reduced models are generated once and used often the necessary time of below one minute for the reduction process is not critical. As there is a direct dependency of the order of the reduced system on the number of ports with MP-Krylov, there is a significant difference in the reduced order for both systems. Since the system of the network with extracted sources has more than five times as many ports as the system of the network with replaced sources, the reduced model is over five times the size of the reduced model of the network with replaced sources. This leads to much faster simulations of the reduced model of our proposed method for the calculation of the transfer function in several frequency points with MATLAB (132× in comparison to 2.8×, Table 13.2).

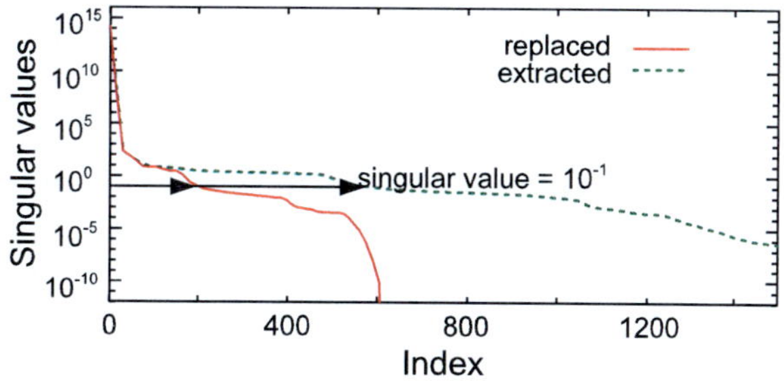

Fig. 13.9 Hankel singular values approximated with PMTBR for the ICEM

Table 13.2 Comparison of the reduction results of the ICEM

Original order, $N = 25{,}062$	Ports, p	Reduced order, n	Reduction (%)	Reduction time	MATLAB speed-up	Accuracy
Extracted sources, reduced with MP-Krylov	389	1,556	93.8	30 s	2.8×	Very good
Replaced sources, reduced with MP-Krylov	69	276	98.9	10 s	132×	Very good
Extracted sources, reduced with PMTBR	389	590	97.6	2 h	33×	Good
Replaced sources, reduced with PMTBR	69	201	99.2	5 m	188×	Good

For PMTBR 20 frequency points are equally distributed in the range $10^1 \leq s_i \leq 10^{10}$. For the reduction prior to the simulation of the system of the network with extracted sources around five minutes and for the system of the network with replaced sources around two hours are necessary. With reduction using PMTBR the reduced order does not directly depend on the number of ports and the reduced order is chosen according to the Hankel singular values. The approximation to the Hankel singular values calculated during PMTBR is shown in Fig. 13.9. It can be seen, that the singular values for the network with replaced sources decay faster than for the network with extracted sources. The largest singular value is set to 10^{-1}, which results in a reduced order of 201 and 590 for the replaced and extracted sources network, respectively. With this MOR algorithm the reduced model of the network with replaced sources is three times smaller than the reduced model of the network with extracted sources and the achieved speed-up in calculating the transfer function behavior with MATLAB is 188× in comparison to 33× (Table 13.2).

With this IC emission model example is has been shown that the proposed network with replaced sources can lead to much smaller and therefore faster reduced models than the standard method of extracting all sources.

13.5 Conclusion

In this article an efficient description of independent sources in electrical networks that can achieve a higher model reduction is presented. The dimension of the function space of the waveforms of the sources is reduced with the help of function decomposition and approximation. A network with a reduced number of independent sources, whereas the behavior at the pins of the network is preserved is generated. This network can be more efficiently reduced with practically all model reduction techniques. The validity and superior efficiency of the proposed method compared to standard methods is shown by reducing two example networks.

References

1. Aliprantis, C., Harris, D., Tourkey, R.: Continous piecewise linear functions. Macroecon. Dyn. **10**, 77–99 (2005)
2. Antoulas, A.: A new result on passivity preserving model reduction. Syst. Control Lett. **54**(4), 361–374 (2005)
3. Antoulas, A.C.: Approximation of Large-Scale Dynamical Systems. Society for Industrial and Applied Mathematics, New York (2005)
4. Belevitch, V.: Classical Network Theory. Holden-Day, San Francisco (1968)
5. Benner, P., Feng, L., Rudnyi, E.: Using the superposition property for model reduction of linear systems with a large number of inputs. In: International Symposium on Mathematical Theory of Networks and Systems (2008)
6. Chua, L., Kang, S.: Section-wise piecewise-linear functions: canonical representation, properties, and applications. Proc. IEEE **65**, 915–929 (1977)
7. Elfadel, I.M., Ling, D.D.: A block rational Arnoldi algorithm for multipoint passive model-order reduction of multiport RLC networks. In: International Conference Computer-Aided Design (1997)
8. Feldmann, P.: Model order reduction techniques for linear systems with large number of terminals. In: Design, Automation and Test in Europe (2004)
9. Feldmann, P., Liu, F.: Sparse and efficient reduced order modeling of linear subcircuits with large number of terminals. In: International Conference on Computer-Aided Design of Integrated Circuits and Systems (2004)
10. Freund, R.W.: SPRIM: structure-preserving reduced order interconnect macromodelling. In: IEEE/ACM International Conference on Computer Aided Design (2004)
11. Hartman, E., Keeler, J., Kowalski, J.: Layered neural networks with gaussian hidden units as universal approximations. Neural Comput. **2**, 210–215 (1990)
12. Hesidenz, D., Steinecke, T.: Chip-package emi modeling and simulation tool 'EXPO'. In: International Workshop on Electromagnetic Compatibility of Integrated Circuits, Munich, Germany (2005)
13. Hornik, K., Stinchcombe, M., White, H.: Multilayer feedforward networks are universal approximators. Neural Netw. **2**, 359–366 (1989)
14. Kozhaya, J., Nassif, S., Najm, F.: A multigrid-like technique for power grid analysis. IEEE Trans. Comput. Aided Design Integr. Circuits Syst. **21**, 1148–1160 (2002)
15. Kuh, E.S., Rohrer, R.A.: Linear Active Networks. Holden-Day, San Francisco (1967)
16. Lee, Y.M., Cao, Y., Chen, T.H., Wang, J., Chen, C.C.-P.: Hiprime: hierarchical and passivity preserved interconnect macromodeling engine for rlkc power delivery. In: International Conference on Computer-Aided Design of Integrated Circuits and Systems (2005)
17. Li, D., Tan, S.D., McGaughy, B.: ETBR: extended truncated balanced realization method for on-chip power grid network analysis. In: Design, Automation and Test in Europe (2008)
18. Liu, P., Tan, S.D., McGaughy, B., Wu, L., He, L.: Termmerg: an efficient terminal-reduction method for interconnect circuits. IEEE Trans. Comput. Aided Design Integr. Circuits Syst. **26**, 1382–1392 (2007)
19. Ludwig, S., Mathis, W.: Reduction of network models of parasitic coupling effects in mixed-signal VLSI circuits. In: International Symposium on Theoretical Electrical Engineering (2009)
20. Ludwig, S., Radić-Weissenfeld, L., Mathis, W., John, W.: Efficient model reduction of passive electrical networks with a large number of independent sources. In: IEEE International Symposium on Circuits and Systems (2008)
21. Ma, M., Khazaka, R.: Model order reduction with parametric port formulation. IEEE Trans. Adv. Packaging **30**, 763–775 (2007)
22. Mallat, S.: A Wavelet Tour of Signal Processing. Academic Press, New York (1999)
23. Nassif, S., Jozhaya, J.: Fast power grid simulation. In: Design Automation Conference (2000)

24. Odabasioglu, A., Celik, M., Pileggi, L.T.: PRIMA: passive reduced-order interconnect macromodeling algorithm. IEEE Trans. Comput. Aided Design Integr. Circuits Syst. **17**, 645–654 (1998)
25. Palenius, T., Roos, J.: Comparison of reduced-order interconnect macromodels for time-domain simulation. IEEE Trans. Microw. Theory Tech. **52**, 2240–2250 (2004)
26. Park, J., Sandberg, I.: Universal approximation using radial-basis-function networks. Neural Comput. **3**, 246–257 (1991)
27. Phillips, J.R., Silveira, L.M.: Poor mans's TBR: a simple model reduction scheme. IEEE Trans. Comput. Aided Design Integr. Circuits Syst. **24**, 43–55 (2005)
28. Phillips, J.R., Daniel, L., Silveira, L.M.: Guaranteed passive balancing transformations for model order reduction. IEEE Trans. Comput. Aided Design Integr. Circuits Syst. **22**(8), 1027–1041 (2003)
29. Phillips, J.R., Zhu, Z., Silveira, L.M.: Model Order Reduction, chap. PMTBR: A Famliy of Approximate Principal-Components-Like Reduction Algorithms. Springer (2008)
30. Silva, J.M.S., Villena, J.F., Flores, P., Silveira, L.M.: Outstanding Issues in Model Order Reduction. Math. Ind. **11**, 139–152 (2007)
31. Silveira, L., Phillips, J.: Exploiting input information in a model reduction algorithm for massively coupled parasitic networks. In: Design Automation Conference (2004)
32. Sorensen, D.: Passivity preserving model reduction via interpolation of spectral zeros. Syst. Control Lett. **54**(4), 347–360 (2005)
33. Steinecke, T., Goekcen, M., Hesidenz, D., Gstoettner, A.: High-accuracy emission simulation models for VLSI chips including package and printed circuit board. In: IEEE International Symposium on Electromagnetic Compatibility, Honululu, Hawaii, USA (2007)
34. Wang, J., Nguyen, T.: Extended krylov subspace method for reduced orderanalysis of linear circuits with multiple sources. In: Design Automation Conference (2000)
35. Yang, F., Zeng, X., Su, Y., Zhou, D.: RLCSYN: RLC equivalent circuit synthesis for structure-preserved reduced-order model of interconnect. In: IEEE International Symposium on Circuits and Systems (2007)

Chapter 14
Structure Preserving Port-Hamiltonian Model Reduction of Electrical Circuits

Rostylav V. Polyuga and Arjan J. van der Schaft

Abstract This paper discusses model reduction of electrical circuits based on a port-Hamiltonian representation. It is shown that by the use of the Kalman decomposition an uncontrollable and/or unobservable port-Hamiltonian system is reduced to a controllable/observable system that inherits the port-Hamiltonian structure. Energy and co-energy variable representations for port-Hamiltonian systems are defined and the reduction procedures are used for both representations. These exact reduction procedures motivate two approximate reduction procedures that are structure preserving for general port-Hamiltonian systems, one reduction procedure is called the effort-constraint reduction methods. The other procedure is structure preserving under a given condition. A numerical example illustrating the model reduction of a ladder network as a port-Hamiltonian system is considered.

14.1 Introduction

Port-based network modeling of physical systems (both in electrical and mechanical domains) leads directly to their representation as port-Hamiltonian systems

R. V. Polyuga (✉)
Centre for Analysis, Scientific Computing and Applications
Department of Mathematics and Computer Science,
Eindhoven University of Technology,
P.O. Box 513, 5600 MB Eindhoven,
The Netherlands
e-mail: R.Polyuga@tue.nl, rostyslav.polyuga@gmail.com.

A. J. van der Schaft
Johann Bernoulli Institute for Mathematics and Computer Science, University of Groningen, P.O. Box 407, 9700 AK Groningen, The Netherlands
e-mail: A.J.van.der.Schaft@math.rug.nl

P. Benner et al. (eds.), *Model Reduction for Circuit Simulation*,
Lecture Notes in Electrical Engineering, 74, DOI: 10.1007/978-94-007-0089-5_14,

which are, if the Hamiltonian is non-negative, an important class of *passive* state-space systems. At the same time network modeling of physical systems often leads to high-dimensional dynamical models. Large state-space dimensions are obtained as well if distributed-parameter models are spatially discretized. Therefore an important issue concerns model reduction of these high-dimensional systems, both for analysis and control. The goal of this work is to show that the specific model reduction techniques of linear port-Hamiltonian systems preserve the port-Hamiltonian structure, and, as a consequence, passivity, and demonstrate the possibility of applying those techniques to electrical circuits as port-Hamiltonian systems.

Port-Hamiltonian systems are endowed with more structure than just passivity. Other important issues like interconnection between port-Hamiltonian systems, the presence of conservation laws and energy dissipation are also reflected by the port-Hamiltonian structure. In Sect. 14.2 we provide a brief overview of linear port-Hamiltonian systems. General theory on port-Hamiltonian systems can be found in [12]. We will show by applying the Kalman decomposition in Sect. 14.3 that the reduction of the dynamics of an uncontrollable/unobservable linear port-Hamiltonian system to a dynamics on the reachability/observability subspace preserves the Port-Hamiltonian structure. This result holds both for energy and co-energy variable representations of linear port-Hamiltonian systems. The co-energy variable representation of port-Hamiltonian systems is considered in Sect. 14.4. It is shown in Sect. 14.4 that the reduced models in the co-energy coordinates take a somewhat "dual" form to the reduced models obtained in the standard energy coordinates.

Within the systems and control literature a popular and elegant tool for model reduction is balancing, going back to [8]. One favorable property of model reduction based on balancing, as compared with other techniques such as modal analysis, is that the approximation of the dynamical system is explicitly based on its input-output properties. Balancing for port-Hamiltonian systems is considered in Sect. 14.6, see also [13]. We will apply two structure preserving model reduction procedures in Sect. 14.6 to linear port-Hamiltonian systems and show that the reduced order models are again port-Hamiltonian. One reduction procedure is called the effort-constraint reduction method. The other procedure is structure preserving under a given condition. Preliminary results of this paper are reported in [11] using scattering coordinates. Similar reduced order port-Hamiltonian models are obtained in [6, 7] employing perturbation analysis. In Sect. 14.7 we consider numerical simulations of a ladder network, and apply the effort-constraint method and the balanced truncation method in order to obtain reduced order models and compare the results.

14.2 Linear Port-Hamiltonian Systems

Port-based network modeling of physical systems leads to their representation as port-Hamiltonian systems (see e.g. [4, 9]). In the linear case, and in the absence of algebraic constraints, port-Hamiltonian systems take the form (see [11–13])

$$\begin{cases} \dot{x} = (J - R)Qx + Bu, \\ y = B^T Qx, \end{cases} \tag{14.1}$$

with $H(x) = \frac{1}{2}x^T Qx$ *the total energy* (Hamiltonian), $Q = Q^T \geq 0$ the *energy* matrix and $R = R^T \geq 0$ the *dissipation* matrix. The matrices $J = -J^T$ and B specify the *interconnection* structure of the system. By skew-symmetry of J and since R is positive semidefinite it immediately follows that

$$\frac{d}{dt}\frac{1}{2}x^T Qx = u^T y - x^T QRQx \leq u^T y. \tag{14.2}$$

Thus if $Q \geq 0$ (and the Hamiltonian is non-negative) any port-Hamiltonian system is *passive* (see [12, 16]). The state variables $x \in \mathbb{R}^n$ are also called *energy* variables, since the total energy $H(x)$ is expressed as a function of these variables. Furthermore, the variables $u \in \mathbb{R}^m, y \in \mathbb{R}^m$ are called *power* variables, since their product $u^T y$ equals the power supplied to the system.

In the sequel we will often abbreviate $J - R$ to $F = J - R$. Clearly

$$F + F^T \leq 0. \tag{14.3}$$

Conversely, any F satisfying (14.3) can be written as $J - R$ as above by decomposing F into its skew-symmetric and symmetric part

$$\begin{aligned} J &= \frac{1}{2}(F - F^T), \\ R &= -\frac{1}{2}(F + F^T). \end{aligned} \tag{14.4}$$

Two special cases of port-Hamiltonian systems correspond to either $R = 0$ or $J = 0$. In fact, if $R = 0$ (no internal energy dissipation) then the dissipation inequality (14.2) reduces to an equality

$$\frac{d}{dt}\frac{1}{2}x^T Qx = u^T y. \tag{14.5}$$

In this case the transfer matrix $G(s) = B^T Q(sI - JQ)^{-1}B$ of the system (for invertible Q) satisfies

$$G(s) = -G^T(-s). \tag{14.6}$$

Conversely, any transfer matrix $G(s)$ satisfying $G(s) = -G^T(-s)$ can be shown to have a minimal realization

$$\begin{cases} \dot{x} = JQx + Bu, \\ y = B^T Qx, \end{cases} \tag{14.7}$$

(with in fact Q being invertible again).

The other special case corresponds to $J = 0$, in which case the system takes the form

$$\begin{cases} \dot{x} = -RQx + Bu, \\ y = B^T Qx, \end{cases} \tag{14.8}$$

with transfer matrix $G(s) = B^T Q(sI + RQ)^{-1}B$ satisfying (for invertible Q)

$$G(s) = G^T(s). \tag{14.9}$$

Conversely, any transfer matrix $G(s)$ satisfying (14.9) is represented by a minimal state-space representation (14.8) with Q invertible, where, however, R need not necessarily be positive semidefinite.

In these two special cases, either $R = 0$ or $J = 0$, there is a direct relationship between controllability and observability properties of the port-Hamiltonian system.

Proposition 1 *Consider a port-Hamiltonian system* (14.7) or (14.8), *and assume* $\det Q \neq 0$. *The system is controllable if and only if it is observable, while the unobservability subspace* $\mathcal{N}$ *is related to the reachability subspace* $\mathcal{R}$ *by*

$$\mathcal{N} = \mathcal{R}^{\perp} \tag{14.10}$$

with $\perp$ *denoting the orthogonal complement with respect to the (possibly indefinite) inner product defined by* Q.

Proof For any port-Hamiltonian system (14.1) with $F = J - R$ we have

$$\begin{bmatrix} B^T Q \\ B^T QFQ \\ B^T QFQFQ \\ \vdots \end{bmatrix} = \begin{bmatrix} B & \vdots & F^T QB & \vdots & F^T QF^T QB & \vdots & \dots \end{bmatrix}^T Q. \tag{14.11}$$

Since the kernel of the matrix on the left-hand side defines the unobservability subspace, while on the right-hand side the image of the matrix preceding Q defines the reachability subspace if $F^T = F$ or $F^T = -F$, the assertion follows. □

Nevertheless, in general controllability and observability for a port-Hamiltonian system are not equivalent, as the following example shows.

Example 1 Consider a port-Hamiltonian system

$$\begin{cases} \begin{bmatrix} \dot{x}_1 \\ \dot{x}_2 \end{bmatrix} = \begin{bmatrix} -1 & 1 \\ -1 & -1 \end{bmatrix} \begin{bmatrix} 1 & -1 \\ -1 & 2 \end{bmatrix} \begin{bmatrix} x_1 \\ x_2 \end{bmatrix} + \begin{bmatrix} 1 \\ 0 \end{bmatrix} u, \\ y = \begin{bmatrix} 1 & 0 \end{bmatrix} \begin{bmatrix} 1 & -1 \\ -1 & 2 \end{bmatrix} \begin{bmatrix} x_1 \\ x_2 \end{bmatrix}, \end{cases} \tag{14.12}$$

corresponding to $J = \begin{pmatrix} 0 & 1 \\ -1 & 0 \end{pmatrix}$ and $R = \begin{pmatrix} 1 & 0 \\ 0 & 1 \end{pmatrix}$. The system is observable but *not* controllable.

14.3 The Kalman Decomposition of Port-Hamiltonian Systems

14.3.1 Reduction to a Controllable Port-Hamiltonian System

Consider a port-Hamiltonian system on a state space $\mathscr{X}$ which is not controllable. Take linear coordinates $x = \begin{bmatrix} x_1 \\ x_2 \end{bmatrix}$ such that vectors of the form $\begin{bmatrix} x_1 \\ 0 \end{bmatrix}$ span the reachability subspace $\mathscr{R} \subset \mathscr{X}$:

$$\begin{cases} \begin{bmatrix} \dot{x}_1 \\ \dot{x}_2 \end{bmatrix} = \begin{bmatrix} F_{11} & F_{12} \\ F_{21} & F_{22} \end{bmatrix} \begin{bmatrix} Q_{11} & Q_{12} \\ Q_{21} & Q_{22} \end{bmatrix} \begin{bmatrix} x_1 \\ x_2 \end{bmatrix} + \begin{bmatrix} B_1 \\ B_2 \end{bmatrix} u, \\ y = \begin{bmatrix} B_1^T & B_2^T \end{bmatrix} \begin{bmatrix} Q_{11} & Q_{12} \\ Q_{21} & Q_{22} \end{bmatrix} \begin{bmatrix} x_1 \\ x_2 \end{bmatrix}. \end{cases} \tag{14.13}$$

By invariance of $\mathscr{R}$ (see e.g. [10]) this implies

$$\begin{aligned} F_{21}Q_{11} + F_{22}Q_{21} &= 0, \\ B_2 &= 0. \end{aligned} \tag{14.14}$$

It follows that the dynamics restricted to $\mathscr{R}$ is given as

$$\begin{cases} \dot{x}_1 = (F_{11}Q_{11} + F_{12}Q_{21})x_1 + B_1 u, \\ y = B_1^T Q_{11} x_1. \end{cases} \tag{14.15}$$

Now let us assume that F_{22} in (14.14) is invertible. Then it follows from (14.14) that $Q_{21} = -F_{22}^{-1}F_{21}Q_{11}$. Substitution in (14.15) yields

$$\begin{cases} \dot{x}_1 = (F_{11} - F_{12}F_{22}^{-1}F_{21})Q_{11}x_1 + B_1 u, \\ y = B_1^T Q_{11} x_1, \end{cases} \tag{14.16}$$

which is again a port-Hamiltonian system. Indeed, $F + F^T \leq 0$ implies that the Schur complement $\bar{F} = F_{11} - F_{12}F_{22}^{-1}F_{21}$ satisfies $\bar{F} + \bar{F}^T \leq 0$.

Remark 1 Note that $\bar{F}$ is skew-symmetric if F is skew-symmetric, and is symmetric if F is symmetric.

Remark 2 The Schur complement of a general F with singular F_{22} is not defined. Nevertheless, it is still possible to extend the definition of the Schur complement of F to the case where F_{22} is singular if F is symmetric, which corresponds to the purely damped port-Hamiltonian systems (14.8). For details see Lemma 1 in the appendix at the end of the paper.

14.3.2 Reduction to an Observable Port-Hamiltonian System

Consider again a port-Hamiltonian system (14.1) and suppose the system is not observable. Then there exist coordinates $x = \begin{bmatrix} x_1 \\ x_2 \end{bmatrix}$ such that the unobservability subspace $\mathcal{N}$ is spanned by vectors of the form $\begin{bmatrix} 0 \\ x_2 \end{bmatrix}$. By invariance of $\mathcal{N}$ (see again [10]) it follows that

$$\begin{aligned} F_{11}Q_{12} + F_{12}Q_{22} &= 0, \\ B_1^T Q_{12} + B_2^T Q_{22} &= 0. \end{aligned} \tag{14.17}$$

The dynamics on the quotient space $\mathcal{X}/\mathcal{N}$ is

$$\begin{cases} \dot{x}_1 = (F_{11}Q_{11} + F_{12}Q_{21})x_1 + B_1 u, \\ y = B_1^T Q_{11} x_1 + B_2^T Q_{21} x_1. \end{cases} \tag{14.18}$$

Assuming invertibility of Q_{22} it follows from (14.17) that $F_{12} = -F_{11}Q_{12}Q_{22}^{-1}$ and $B_2^T = -B_1^T Q_{12} Q_{22}^{-1}$. Substitution in (14.18) yields

$$\begin{cases} \dot{x}_1 = F_{11}(Q_{11} - Q_{12}Q_{22}^{-1}Q_{21})x_1 + B_1 u, \\ y = B_1^T (Q_{11} - Q_{12}Q_{22}^{-1}Q_{21})x_1, \end{cases} \tag{14.19}$$

which is again a port-Hamiltonian system with Hamiltonian $\bar{H}(x_1) = \frac{1}{2}x_1^T(Q_{11} - Q_{12}Q_{22}^{-1}Q_{21})x_1$.

Remark 3 Note that $(Q_{11} - Q_{12}Q_{22}^{-1}Q_{21}) \geq 0$ if $Q \geq 0$.

Remark 4 Since Q is symmetric the definition of the Schur complement of Q can be extended to the case that Q_{22} is singular. For details see Lemma 1 in the appendix at the end of the paper.

14.3.3 The Kalman Decomposition

It is well known that a linear system $\dot{x} = Ax + Bu, y = Cx$ can be represented in a suitable basis as (see [10, 17])

$$A = \begin{bmatrix} A_{11} & A_{12} & 0 & 0 \\ 0 & A_{22} & 0 & 0 \\ A_{31} & A_{32} & A_{33} & A_{34} \\ 0 & A_{42} & 0 & A_{44} \end{bmatrix}, \quad B = \begin{bmatrix} B_1 \\ 0 \\ B_3 \\ 0 \end{bmatrix}, C = \begin{bmatrix} C_1^T \\ C_2^T \\ 0 \\ 0 \end{bmatrix}^T,$$

with $\mathcal{X} = \mathcal{X}_1 \times \mathcal{X}_2 \times \mathcal{X}_3 \times \mathcal{X}_4$, where $\mathcal{X}_1$ is the part of the system that is both controllable and observable, $\mathcal{X}_2$ is uncontrollable but observable, $\mathcal{X}_3$ is controllable but unobservable, while $\mathcal{X}_4$ is uncontrollable and unobservable, that is

$$\begin{aligned} \mathcal{N} &= \mathcal{X}_3 \times \mathcal{X}_4, \\ \mathcal{R} &= \mathcal{X}_1 \times \mathcal{X}_3. \end{aligned} \tag{14.20}$$

Writing out

$$\begin{bmatrix} A_{11} & A_{12} & 0 & 0 \\ 0 & A_{22} & 0 & 0 \\ A_{31} & A_{32} & A_{33} & A_{34} \\ 0 & A_{42} & 0 & A_{44} \end{bmatrix} = \begin{bmatrix} F_{11} & F_{12} & F_{13} & F_{14} \\ F_{21} & F_{22} & F_{23} & F_{24} \\ F_{31} & F_{32} & F_{33} & F_{34} \\ F_{41} & F_{42} & F_{43} & F_{44} \end{bmatrix} \begin{bmatrix} Q_{11} & Q_{12} & Q_{13} & Q_{14} \\ Q_{21} & Q_{22} & Q_{23} & Q_{24} \\ Q_{31} & Q_{32} & Q_{33} & Q_{34} \\ Q_{41} & Q_{42} & Q_{43} & Q_{44} \end{bmatrix},$$

this implies that the blocks of the F and Q matrix satisfy

$$\begin{aligned} &(a) F_{11}Q_{13} + F_{12}Q_{23} + F_{13}Q_{33} + F_{14}Q_{43} = 0, \\ &(b) F_{11}Q_{14} + F_{12}Q_{24} + F_{13}Q_{34} + F_{14}Q_{44} = 0, \\ &(c) F_{21}Q_{11} + F_{22}Q_{21} + F_{23}Q_{31} + F_{24}Q_{41} = 0, \\ &(d) F_{21}Q_{13} + F_{22}Q_{23} + F_{23}Q_{33} + F_{24}Q_{43} = 0, \\ &(e) F_{21}Q_{14} + F_{22}Q_{24} + F_{23}Q_{34} + F_{24}Q_{44} = 0, \\ &(f) F_{41}Q_{11} + F_{42}Q_{21} + F_{43}Q_{31} + F_{44}Q_{41} = 0, \\ &(g) F_{41}Q_{13} + F_{42}Q_{23} + F_{43}Q_{33} + F_{44}Q_{43} = 0, \end{aligned} \tag{14.21}$$

and similarly by writing out

$$\begin{bmatrix} B_1^T & 0 & B_3^T & 0 \end{bmatrix} \begin{bmatrix} Q_{11} & Q_{12} & Q_{13} & Q_{14} \\ Q_{21} & Q_{22} & Q_{23} & Q_{24} \\ Q_{31} & Q_{32} & Q_{33} & Q_{34} \\ Q_{41} & Q_{42} & Q_{43} & Q_{44} \end{bmatrix} = \begin{bmatrix} C_1 & C_2 & 0 & 0 \end{bmatrix},$$

we obtain

$$\begin{aligned} B_1^T Q_{13} + B_3^T Q_{33} &= 0, \\ B_1^T Q_{14} + B_3^T Q_{34} &= 0. \end{aligned} \tag{14.22}$$

The resulting dynamics on $\mathcal{X}_1$ (the part of the system that is both controllable and observable) can be identified in port-Hamiltonian form, by combining the previous two reduction schemes corresponding to controllability and observability. Indeed, application of Sect. 14.3.2 yields the following observable system on $\mathcal{X}_1 \times \mathcal{X}_2$

$$\begin{cases} \begin{bmatrix} \dot{x}_1 \\ \dot{x}_2 \end{bmatrix} = \begin{bmatrix} F_{11} & F_{12} \\ F_{21} & F_{22} \end{bmatrix} \bar{Q} \begin{bmatrix} x_1 \\ x_2 \end{bmatrix} + \begin{bmatrix} B_1 \\ 0 \end{bmatrix} u, \\ y = \begin{bmatrix} B_1^T & 0 \end{bmatrix} \bar{Q} \begin{bmatrix} x_1 \\ x_2 \end{bmatrix}, \end{cases} \tag{14.23}$$

where

$$\bar{Q} = \begin{bmatrix} Q_{11} & Q_{12} \\ Q_{21} & Q_{22} \end{bmatrix} - \begin{bmatrix} Q_{13} & Q_{14} \\ Q_{23} & Q_{24} \end{bmatrix} \begin{bmatrix} Q_{33} & Q_{34} \\ Q_{43} & Q_{44} \end{bmatrix}^{-1} \begin{bmatrix} Q_{31} & Q_{32} \\ Q_{41} & Q_{42} \end{bmatrix}. \tag{14.24}$$

Next, application of Sect. 14.3.1 to (14.23) yields the following port-Hamiltonian description of the dynamics on $\mathscr{X}_1$

$$\begin{cases} \dot{x}_1 &= (F_{11} - F_{12}F_{22}^{-1}F_{21})\bar{Q}_{11}x_1 + B_1 u, \\ y &= B_1^T \bar{Q}_{11}x_1, \end{cases} \tag{14.25}$$

having the same transfer matrix as the original system (14.1).

Further analysis (using the well-known matrix inversion formula) yields

$$\begin{aligned} \bar{Q}_{11} = Q_{11} &- Q_{13}(Q_{33} - Q_{34}Q_{44}^{-1}Q_{43})^{-1}Q_{31} \\ &+ Q_{14}Q_{44}^{-1}Q_{43}(Q_{33} - Q_{34}Q_{44}^{-1}Q_{43})^{-1}Q_{31} \\ &+ Q_{13}Q_{33}^{-1}Q_{34}(Q_{44} - Q_{43}Q_{33}^{-1}Q_{34})^{-1}Q_{41} \\ &- Q_{14}(Q_{44} - Q_{43}Q_{33}^{-1}Q_{34})^{-1}Q_{41}. \end{aligned} \tag{14.26}$$

Remark 5 By first applying the procedure of Sect. 14.3.1 and then applying the procedure of Sect. 14.3.2 for zero initial conditions it can be shown that we obtain the same port-Hamiltonian formulation.

14.4 The Co-Energy Variable Representation

In this section we assume throughout that the matrix Q is *invertible*. This means that

$$e = Qx \tag{14.27}$$

is a valid coordinate transformation, and the port-Hamiltonian system (14.1) in these new coordinates takes the form

$$\begin{cases} \dot{e} &= QFe + QBu, \quad F = J - R, \\ y &= B^T e. \end{cases} \tag{14.28}$$

Since $e = Qx = \frac{\partial H}{\partial x}(x)$, with $H(x) = \frac{1}{2}x^T Qx$ the energy, the variables e are usually called the *co-energy* variables.

Example 2 Consider the LC-circuit in Fig. 14.1, with q the charge on the capacitor and ϕ_1, ϕ_2 the fluxes over the inductors L_1, L_2 correspondingly. The energy (in the case of a linear capacitor and inductors) is given as

$$H(q, \phi_1, \phi_2) = \frac{1}{2C}q^2 + \frac{1}{2L_1}\phi_1^2 + \frac{1}{2L_2}\phi_2^2, \tag{14.29}$$

and $x = [q, \phi_1, \phi_2]^T$ are the energy variables, in which the system takes the port-Hamiltonian form

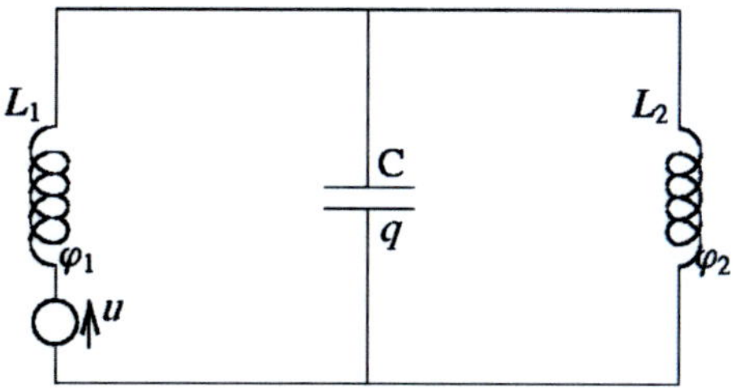

Fig. 14.1 LC-circuit

$$\left\{\begin{array}{l}\begin{bmatrix}\dot{q}\\ \dot{\phi}_1\\ \dot{\phi}_2\end{bmatrix}=\begin{bmatrix}0 & 1 & -1\\ -1 & 0 & 0\\ 1 & 0 & 0\end{bmatrix}\begin{bmatrix}\frac{q}{C}\\ \frac{\phi_1}{L_1}\\ \frac{\phi_2}{L_2}\end{bmatrix}+\begin{bmatrix}0\\ 1\\ 0\end{bmatrix}u,\\ y=\phi_1/L_1,\end{array}\right. \tag{14.30}$$

with u, y being the voltage across and the current through the voltage source. The co-energy variables

$$e=\begin{bmatrix}q/C\\ \phi_1/L_1\\ \phi_2/L_2\end{bmatrix}=\begin{bmatrix}V_C\\ I_{L1}\\ I_{L2}\end{bmatrix}$$

are the voltage over the capacitor and the currents through the inductors, leading to the following form of the dynamics

$$\left\{\begin{array}{l}\begin{bmatrix}\dot{V}_C\\ \dot{I}_{L1}\\ \dot{I}_{L2}\end{bmatrix}=\begin{bmatrix}\frac{1}{C} & 0 & 0\\ 0 & \frac{1}{L_1} & 0\\ 0 & 0 & \frac{1}{L_2}\end{bmatrix}\begin{bmatrix}0 & 1 & -1\\ -1 & 0 & 0\\ 1 & 0 & 0\end{bmatrix}\begin{bmatrix}V_C\\ I_{L1}\\ I_{L2}\end{bmatrix}+\begin{bmatrix}0\\ \frac{1}{L_1}\\ 0\end{bmatrix}u,\\ y=I_{L1}.\end{array}\right. \tag{14.31}$$

Note that

$$\frac{d}{dt}\frac{1}{2}e^TQ^{-1}e=\frac{1}{2}e^T(F+F^T)e+e^TBu=-e^TRe+u^Ty, \tag{14.32}$$

and thus if $Q>0$ then $V(e)=\frac{1}{2}e^TQ^{-1}e$ (the Legendre transform of $H(x)=\frac{1}{2}x^TQx$) is a storage function of (14.28). $V(e)$ is called the *co-energy* of the system, which is in this linear case equal to the energy $(V(Qx)=H(x))$.

A main advantage of the co-energy variable representation of a port-Hamiltonian system is that additional *constraints* on the system are often expressed as constraints on the co-energy variables (see also Sect. 14.6)

The reduction of the port-Hamiltonian system to its controllable and/or observable part takes the following form in the co-energy variable representation. Interestingly enough, the formulas take a somewhat "dual" form to the formulas obtained in the energy variable representation.

Consider the system (14.28) in co-energy variable representation. Take linear coordinates $e = \begin{bmatrix} e_1 \\ e_2 \end{bmatrix}$ such that the reachability subspace $\mathscr{R}$ is spanned by vectors of the form $\begin{bmatrix} e_1 \\ 0 \end{bmatrix}$:

$$\begin{cases} \begin{bmatrix} \dot{e}_1 \\ \dot{e}_2 \end{bmatrix} = \begin{bmatrix} Q_{11} & Q_{12} \\ Q_{21} & Q_{22} \end{bmatrix} \begin{bmatrix} F_{11} & F_{12} \\ F_{21} & F_{22} \end{bmatrix} \begin{bmatrix} e_1 \\ e_2 \end{bmatrix} + \begin{bmatrix} Q_{11} & Q_{12} \\ Q_{21} & Q_{22} \end{bmatrix} \begin{bmatrix} B_1 \\ B_2 \end{bmatrix} u, \\ y = \begin{bmatrix} B_1^T & B_2^T \end{bmatrix} \begin{bmatrix} e_1 \\ e_2 \end{bmatrix}. \end{cases} \tag{14.33}$$

By invariance of $\mathscr{R}$ this implies

$$\begin{aligned} Q_{21}F_{11} + Q_{22}F_{21} &= 0, \\ Q_{21}B_1 + Q_{22}B_2 &= 0. \end{aligned} \tag{14.34}$$

Hence the dynamics restricted to $\mathscr{R}$ equals

$$\begin{cases} \dot{e}_1 &= (Q_{11}F_{11} + Q_{12}F_{21})e_1 + (Q_{11}B_1 + Q_{12}B_2)u \\ &= (Q_{11} - Q_{12}Q_{22}^{-1}Q_{21})F_{11}e_1 + (Q_{11} - Q_{12}Q_{22}^{-1}Q_{21})B_1u, \\ y &= B_1^T e_1, \end{cases} \tag{14.35}$$

which is a port-Hamiltonian system in co-energy variable representation, with energy matrix $\bar{Q} = Q_{11} - Q_{12}Q_{22}^{-1}Q_{21}$, and interconnection/damping matrix F_{11}. Notice that these formulas are dual to the corresponding formulas (14.16) for the controllable part of the system in energy variable representation, where the resulting interconnection/damping matrix is a Schur complement, while the resulting energy matrix is Q_{11}. This duality is associated with the Legendre transform of the Hamiltonian $H(x)$.

Analogously, take linear coordinates $e = \begin{bmatrix} e_1 \\ e_2 \end{bmatrix}$ such that the unobservability subspace $\mathscr{N}$ is spanned by the vectors $\begin{bmatrix} 0 \\ e_2 \end{bmatrix}$. This implies

$$\begin{aligned} Q_{11}F_{12} + Q_{12}F_{22} &= 0, \\ B_2 &= 0, \end{aligned} \tag{14.36}$$

leading to the observable reduced dynamics

$$\begin{cases} \dot{e}_1 = (Q_{11}F_{11} + Q_{12}F_{21})e_1 + Q_{11}B_1u \\ \quad = Q_{11}(F_{11} - F_{12}F_{22}^{-1}F_{21})e_1 + Q_{11}B_1u, \\ y = B_1^T e_1. \end{cases} \tag{14.37}$$

Combination of the above leads to a similar Kalman decomposition as in the energy variable representation.

Remark 6 To extend the definition of the Schur complements of Q, F for singular Q_{22}, F_{22} see Remarks 2, 4.

14.5 Balancing for Port-Hamiltonian Systems

Definition 1 The *controllability* and *observability* function of a linear system are defined as

$$L_c(x_0) = \min_{u \in L_2(-\infty,0)} \frac{1}{2} \int_{-\infty}^{0} \| u(t) \|^2 \, dt, \qquad x(-\infty) = 0, \quad x(0) = x_0 \tag{14.38}$$

and

$$L_o(x_0) = \frac{1}{2} \int_{0}^{\infty} \| y(t) \|^2 \, dt, \qquad u(t) = 0, \quad x(0) = x_0, \tag{14.39}$$

respectively.

The value of the controllability function at x_0 is the minimum amount of input energy required to reach the state x_0 from the zero state, and the value of the observability function at x_0 is the amount of output energy generated by the state x_0.

Theorem 1 *Consider a linear time invariant (LTI) asymptotically stable system* [8]

$$\begin{cases} \dot{x} = Ax + Bu, \\ y = Cx. \end{cases} \tag{14.40}$$

Then $L_c(x_0) = \frac{1}{2} x_0^T W^{-1} x_0$ *and* $L_o(x_0) = \frac{1}{2} x_0^T M x_0$, *where* $W = \int_0^\infty e^{At} BB^T e^{A^T t} dt$ *is the controllability Gramian and* $M = \int_0^\infty e^{A^T t} C^T C e^{At} dt$ is the *observability Gramian. Furthermore W and M are symmetric and positive definite, and are unique solutions of the Lyapunov equations*

$$AW + WA^T = -BB^T \tag{14.41}$$

and

$$A^T M + MA = -C^T C \tag{14.42}$$

respectively.

In the port-Hamiltonian case Eqs. (14.41) and (14.42) specialize to

$$(J-R)QW + WQ(J-R)^T = -BB^T \tag{14.43}$$

and

$$Q(J-R)^T M + M(J-R)Q = -QBB^T Q. \tag{14.44}$$

Sometimes it is useful to proceed using so-called scattering representation for port-Hamiltonian systems (motivated by electrical domain), where controllability and observability Gramians are related to the energy matrix Q as $M \le Q \le W^{-1}$, and Hankel singular values σ_i are all $\Leftarrow 1$ (see [11, 13]).

Nevertheless in this paper we proceed without using scattering coordinates.

The balancing coordinate transformation $S, x = Sx_b$, where x_b denotes balanced coordinates, clearly preserves the port-Hamiltonian structure of the system (14.1):

$$\begin{cases} \dot{x}_b = (J_b - R_b)Q_b x_b + B_b u, \\ y = B_b^T Q_b x_b, \end{cases} . \tag{14.45}$$

where $S^{-1}RS^{-T} = R_b = R_b^T \ge 0$ is the dissipation matrix, $S^{-1}JS^{-T} = J_b = -J_b^T$ is the structure matrix and $S^T QS = Q_b = Q_b^T \ge 0$ is the energy matrix in the balanced coordinates x_b. In this case, $B_b = S^{-1}B$.

Similarly, the port-Hamiltonian structure is preserved applying balancing coordinate transformation $T, e = Te_b$, to the port-Hamiltonian system (14.28) in co-energy coordinates.

Now bringing the system (14.1) into a balanced form where $W = M$ (see [8, 14]) and computing the square roots of the eigenvalues of MW which are equal to the Hankel singular values (see [3]) provides us the information about the number of state components of the system to be reduced. These state components require large amount of the incoming energy to be reached and give small amount of the outgoing energy to be observed. Therefore they are less important from the energy point of view and can be removed from the system (see also [1]).

14.6 Reduction of Port-Hamiltonian Systems in the General Case

For a general port-Hamiltonian system in energy (14.1) or co-energy (14.28) coordinates with no uncontrollable/unobservable but with "hardly" controllable/observable states we may apply balancing as explained in Sect. 14.5 and use one of the following structure-preserving reduction techniques. Since the techniques considered apply to the port-Hamiltonian systems in balanced coordinates, for the sake of simplicity in this section we skip the subscript 'b' writing x, e, J, R, Q, B instead of $x_b, e_b, J_b, R_b, Q_b, B_b$.

14.6.1 Effort-Constraint Reduction

Consider a full order port-Hamiltonian system (14.1). We balance the system (14.1), but now in co-energy coordinates (and thus with another change of coordinates (14.27), obtaining the following balanced representation of our system

$$\begin{cases} \dot{e} = Q(J - R)e + QBu, \\ y = B^T e, \end{cases} \tag{14.46}$$

where the lower part of the state vector $e = \begin{bmatrix} e_1 \\ e_2 \end{bmatrix}$ is the most difficult to reach and to observe.

Consider the system (14.1) again, but now in the coordinates where the system (14.46) is balanced

$$\begin{cases} \dot{x} = (J - R)e + Bu, \\ y = B^T e. \end{cases} . \tag{14.47}$$

A natural choice for the reduced model would be a model which contains only the e_1 dynamics since the lower part of the state vector e_2 is much less relevant from the energy point of view

$$e_2 = Q_{21}x_1 + Q_{22}x_2 \approx 0. \tag{14.48}$$

Therefore the reduced system takes the following form

$$\begin{cases} \dot{x}_1 = (J_{11} - R_{11})e_1 + B_1 u \\ \quad\;\; = (J_{11} - R_{11})(Q_{11}x_1 + Q_{12}x_2) + B_1 u, \\ y = B_1^T e_1 = B_1^T(Q_{11}x_1 + Q_{12}x_2). \end{cases} \tag{14.49}$$

After substituting $x_2 \approx -Q_{22}^{-1}Q_{21}x_1$ from (14.48) into (14.49), assuming that Q_{22}^{-1} exists, the reduced system will take the final form in energy coordinates

$$\begin{cases} \dot{x}_1 = (J_{11} - R_{11})(Q_{11} - Q_{12}Q_{22}^{-1}Q_{21})x_1 + B_1 u, \\ \bar{y} = B_1^T(Q_{11} - Q_{12}Q_{22}^{-1}Q_{21})x_1, \end{cases} . \tag{14.50}$$

which is again a port-Hamiltonian system produced by the effort-constraint method.

14.6.2 An Alternative Reduction Method

Another structure-preserving way of model reduction of port-Hamiltonian systems assumes that we balance the system (14.1) and approximate the lower part of the state vector, but now in energy coordinates, *plus* its dynamics. Using the notation $F := J - R$ we obtain

$$\begin{aligned} &x_2 \approx 0, \\ &\dot{x}_2 = (F_{21}Q_{11} + F_{22}Q_{21})x_1 + B_2u \approx 0, \end{aligned} \tag{14.51}$$

with the reduced port-Hamiltonian system of the form

$$\begin{cases} \dot{x}_1 = (F_{11}Q_{11} + F_{12}Q_{21})x_1 + B_1u, \\ y = (B_1^T Q_{11} + B_2^T Q_{21})x_1. \end{cases} \tag{14.52}$$

From (14.51) it immediately follows that $Q_{21}x_1 \approx -F_{22}^{-1}F_{21}Q_{11}x_1 - F_{22}^{-1}B_2u$, assuming that F_{22}^{-1} exists. Substituting in (14.52) yields

$$\begin{cases} \dot{x}_1 = (F_{11} - F_{12}F_{22}^{-1}F_{21})Q_{11}x_1 + (B_1 - F_{12}F_{22}^{-1}B_2)u, \\ \hat{y} = (B_1^T - B_2^T F_{22}^{-1}F_{21})Q_{11}x_1 - (B_2^T F_{22}^{-1}B_2)u, \end{cases} \tag{14.53}$$

which is if $(F_{12}F_{22}^{-1})^T = F_{22}^{-1}F_{21}$ again a reduced system in the port-Hamiltonian form.

Remark 7 The reduced order port-Hamiltonian systems (14.50) and (14.53) are automatically passive since the preservation of the port-Hamiltonian structure implies the preservation of the passivity property (see [12]).

Remark 8 Although the approximation method with the reduced model (14.53) is similar to the well-known balanced truncation ($x_2 \approx 0$, see e.g. [1] and the references therein) which gives the reduced order model of the form

$$\begin{cases} \dot{x}_1 = A_{11}x_1 + B_1u, \\ y_{bt} = C_1x_1, \end{cases} \tag{14.54}$$

and less well-known singular perturbation method ($\dot{x}_2 \approx 0$, see [2, 5]) with the reduced order model

$$\begin{cases} \dot{x}_1 = (A_{11} - A_{12}A_{22}^{-1}A_{21})x_1 + (B_1 - A_{12}A_{22}^{-1}B_2)u, \\ y_{sp} = (C_1 - C_2A_{22}^{-1}A_{21})x_1 - C_2A_{22}^{-1}B_2u, \end{cases} \tag{14.55}$$

we want to underline that it is different from these reduction methods since it is easy to show that neither of them preserves the port-Hamiltonian structure.

14.7 Example

We consider a ladder network, similar to that of [15]. In our case we take the current I as the input and the voltage of the first capacitor U_{c_1} as the port-Hamiltonian output. The state variables are as follows: x_1 is the charge q_1 over C_1, x_2 is the flux ϕ_1 over L_1, x_3 is the charge q_2 over C_2, x_4 is the flux ϕ_2 over L_2, etc.

In our case, as in [15], the resulting Hankel singular values obtained after balancing are not distinct enough. In order to overcome this difficulty we inject

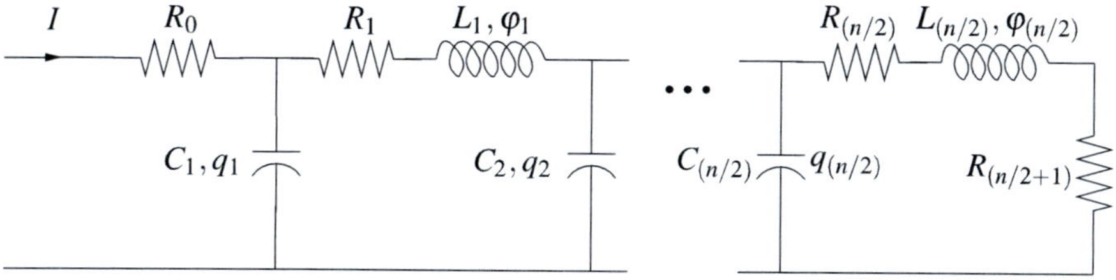

Fig. 14.2 Ladder network

additional dissipative elements, in this case resistors $R_1, \ldots, R_n$, into the model as shown in Fig. 14.2.

We take unit values of the capacitors C_i and inductors L_i, while $R_0 = 0.2$, $R_i = 0.2, i = 1, \ldots, n/2, R_{n/2+1} = 0.4$. A minimal realization of this port-Hamiltonian ladder network for the order $n = 6$ is

$$A = \begin{bmatrix} 0 & -\frac{1}{L_1} & 0 & 0 & 0 & 0 \\ \frac{1}{C_1} & -\frac{R_1}{L_1} & -\frac{1}{C_2} & 0 & 0 & 0 \\ 0 & \frac{1}{L_1} & 0 & -\frac{1}{L_2} & 0 & 0 \\ 0 & 0 & \frac{1}{C_2} & -\frac{R_2}{L_2} & -\frac{1}{C_3} & 0 \\ 0 & 0 & 0 & \frac{1}{L_2} & 0 & -\frac{1}{L_3} \\ 0 & 0 & 0 & 0 & \frac{1}{C_3} & -\frac{R_3+R_4}{L_3} \end{bmatrix}, \quad B = \begin{bmatrix} 1 \\ 0 \\ 0 \\ 0 \\ 0 \\ 0 \end{bmatrix}, \quad C = \begin{bmatrix} \frac{1}{C_1} \\ 0 \\ 0 \\ 0 \\ 0 \\ 0 \end{bmatrix}^T,$$

where $A = (J - R)Q$ with

$$J = \begin{bmatrix} 0 & -1 & 0 & 0 & 0 & 0 \\ 1 & 0 & -1 & 0 & 0 & 0 \\ 0 & 1 & 0 & -1 & 0 & 0 \\ 0 & 0 & 1 & 0 & -1 & 0 \\ 0 & 0 & 0 & 1 & 0 & -1 \\ 0 & 0 & 0 & 0 & 1 & 0 \end{bmatrix}, \quad R = \begin{bmatrix} 0 & 0 & 0 & 0 & 0 & 0 \\ 0 & R_1 & 0 & 0 & 0 & 0 \\ 0 & 0 & 0 & 0 & 0 & 0 \\ 0 & 0 & 0 & R_2 & 0 & 0 \\ 0 & 0 & 0 & 0 & 0 & 0 \\ 0 & 0 & 0 & 0 & 0 & R_3 + R_4 \end{bmatrix},$$

$$Q = \begin{bmatrix} \frac{1}{C_1} & 0 & 0 & 0 & 0 & 0 \\ 0 & \frac{1}{L_1} & 0 & 0 & 0 & 0 \\ 0 & 0 & \frac{1}{C_2} & 0 & 0 & 0 \\ 0 & 0 & 0 & \frac{1}{L_2} & 0 & 0 \\ 0 & 0 & 0 & 0 & \frac{1}{C_3} & 0 \\ 0 & 0 & 0 & 0 & 0 & \frac{1}{L_3} \end{bmatrix}.$$

Adding another LC pair to the network (with appropriate resistors), which would correspond to an increase of the dimension of the model by two, will modify the ABC-model in the following way. The sub-diagonal of the matrix A will contain additionally $L_{n/2-1}^{-1}, C_{n/2}^{-1}$. The super diagonal of A will contain

Table 14.1 Decreasing Hankel singular values of the full order system

Order	1	2	3	4	5	6
Hankel singular values	0.8019	0.3746	0.2766	0.2172	0.2150	0.1537
Order	7	8	9	10	11	12
Hankel singular values	0.1497	0.0966	0.0727	0.0715	0.0125	0.0123

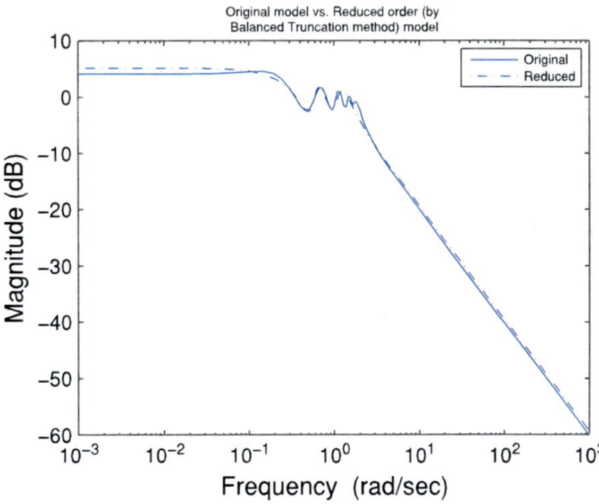

Fig. 14.3 Frequency response

$-C_{n/2}^{-1}, -L_{n/2}^{-1}$. Furthermore, the main diagonal of A will have $-\frac{R_{n/2-1}}{L_{n/2-1}}$ in the $(n-2, n-2)$ position, zero in the $(n-1, n-1)$ position and $-\frac{R_{n/2}+R_{n/2+1}}{L_{n/2}}$ in the (n, n) position.

We considered the 12-dimensional full order minimal port-Hamiltonian network and reduce it to a 5-dimensional one by the effort-constraint method considered in the previous section and the usual balanced truncation method. The non-minimal system can be first reduced to a minimal one as shown in Sect. 14.3. The Hankel singular values of the full order system in decreasing order are shown in Table 14.1.

It is a well-known fact that the transfer functions of reduced order models obtained by the balanced truncation method approximate the full order transfer functions well in the high-frequency region and not that well in the low-frequency one (of course, depending on the application considered). Since the effort-constraint method is similar to the balanced truncation method (with the above explained modification in order to preserve the port-Hamiltonian structure and passivity) we expected approximations of similar nature. In Figs. 14.3 and 14.4 the frequency response of the full order model is shown vs. the frequency responses of the reduced order models, obtained by the balanced truncation method and the effort-constraint method respectively. The figures show that the reduced order transfer functions indeed behave in a similar way.

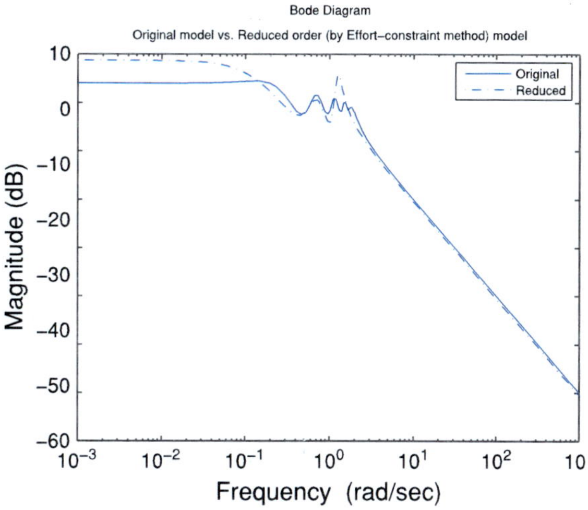

Fig. 14.4 Frequency response

Table 14.2 $\mathscr{H}_\infty$- and $\mathscr{H}_2$-norms for the error systems

Reduced order system by	$\mathscr{H}_\infty$-norm	$\mathscr{H}_2$-norm
Balanced truncation method	0.3411	0.1541
Effort-constraint method	1.1999	0.4491

In Table 14.2 $\mathscr{H}_\infty$- and $\mathscr{H}_2$-norms are shown for the error systems obtained after balanced truncation reduction and effort-constraint reduction. It follows that the error-norms for the effort-constraint method are larger than those for the balanced truncation method.

Port-Hamiltonian systems are somewhere in between general passive systems and electrical circuits. We believe that whenever a transfer function of a full order electrical circuit is approximated well by the balanced truncation method, we can always apply the (port-Hamiltonian) structure preserving effort-constraint method in order to obtain an approximation of a similar quality as the approximation obtained by the balanced truncation method along with the preservation of the port-Hamiltonian structure and passivity.

Important questions concerning general error bounds for the structure preserving port-Hamiltonian model reduction methods and about the physical realization of the obtained port-Hamiltonian reduced order models are currently under investigation.

14.8 Conclusions

We have shown in Sect. 14.3 that a full order uncontrollable/unobservable port-Hamiltonian system can be reduced to a controllable/observable system, which is again port-Hamiltonian, by exploiting the invariance of the reachability/unobservability subspaces of the original system. We discussed energy and co-energy variable representations of port-Hamiltonian systems in Sect. 14.4, illustrated by the example of electrical networks, where the energy variables are charges and fluxes, while the co-energy variables are voltages and currents

Balancing for port-Hamiltonian systems is discussed in Sect. 14.5. The effort-constraint method and the other alternative reduction method are introduced in Sect. 14.6 and applied to a general port-Hamiltonian full order system showing that the proposed approximations preserve the port-Hamiltonian structure for the reduced order systems as well as the passivity property. In Sect. 14.7 we considered a full order ladder network and applied the balanced truncation method and the effort-constraint method in order to obtain reduced order models.

Port-Hamiltonian structure preserving model reduction methods motivate to investigate further important issues about error bounds between full order and reduced order systems, and about the physical realization of the reduced order systems, e.g. as an electrical circuit.

Acknowledgements We gratefully acknowledge the helpful remarks and suggestions of the anonymous referees which significantly improved the presentation of this paper.

Appendix

Lemma 1 *Consider a symmetric negative (positive) semidefinite matrix* $F = F^T$ *partitioned as*

$$F = \begin{bmatrix} F_{11} & F_{12} \\ F_{21} & F_{22} \end{bmatrix}.$$

Then the Schur complement of F *given as*

$$\bar{F} = F_{11} - F_{12}F_{22}^{-1}F_{21}$$

can be defined even for singular F_{22}.

Proof First consider the case when $F_{22} \neq 0$. Since the kernel of F_{22} is nonempty, the symmetric matrix F_{22} takes (after a possible change of coordinates) the following form

$$F_{22} = \begin{bmatrix} F_{22}^a & 0 \\ 0 & 0 \end{bmatrix}$$

with F_{22}^a invertible. Then the properties of a negative (positive) semidefinite symmetric matrix (14.57) shown in Lemma 2 yield

$$F_{12} = \begin{bmatrix} F_{12}^a & 0 \\ F_{12}^c & 0 \end{bmatrix}, \quad F_{21} = \begin{bmatrix} F_{21}^a & F_{21}^b \\ 0 & 0 \end{bmatrix}.$$

Therefore the Schur complement with a perturbed F_{22} reads

$$\begin{aligned}\bar{F}_{\varepsilon} &= F_{11} - \begin{bmatrix} F_{12}^a & 0 \\ F_{12}^c & 0 \end{bmatrix} \begin{bmatrix} (F_{22}^a)^{-1} & 0 \\ 0 & \varepsilon^{-1} \end{bmatrix} \begin{bmatrix} F_{21}^a & F_{21}^b \\ 0 & 0 \end{bmatrix} \\ &= F_{11} - \begin{bmatrix} F_{12}^a (F_{22}^a)^{-1} F_{21}^a & F_{12}^a (F_{22}^a)^{-1} F_{21}^b \\ F_{12}^c (F_{22}^a)^{-1} F_{21}^a & F_{12}^c (F_{22}^a)^{-1} F_{21}^b \end{bmatrix}\end{aligned} \tag{14.56}$$

which is independent of ε. Hence we can let $\varepsilon \to 0$. This shows that the Schur complement can be still defined even for singular F_{22}.

If $F_{22} = 0$ then (14.57) yield $F_{12} = 0, F_{21} = 0$ and $\bar{F} = F_{11}$ which completes the proof. □

Lemma 2 *Consider a negative semidefinite symmetric matrix $F = F^T \leq 0$ partitioned as*

$$F = \begin{bmatrix} F_{11} & F_{12} \\ F_{21} & F_{22} \end{bmatrix}.$$

Then

$$\begin{aligned}\ker F_{22} &\subseteq \ker F_{12}, \\ \operatorname{im} F_{21} &\subseteq \operatorname{im} F_{22}.\end{aligned} \tag{14.57}$$

Proof First we prove that $\ker F_{22} \subset \ker F_{12}$. Since F is negative semidefinite it follows that $x^T F x \leq 0$ for all real vectors $x = \begin{bmatrix} x_1 \\ x_2 \end{bmatrix}$. Take x_2 which is in the kernel of F_{22} and $x_1 = F_{12} x_2$. Then for a small positive constant ε we have

$$\begin{aligned}\begin{bmatrix} \varepsilon x_1^T & x_2^T \end{bmatrix} \begin{bmatrix} F_{11} & F_{12} \\ F_{21} & F_{22} \end{bmatrix} \begin{bmatrix} \varepsilon x_1 \\ x_2 \end{bmatrix} &= \varepsilon^2 x_1^T F_{11} x_1 + \varepsilon x_2^T F_{21} x_1 + \varepsilon x_1^T F_{12} x_2 + x_2^T F_{22} x_2 = \\ &= \varepsilon^2 x_1^T F_{11} x_1 + 2\varepsilon \|x_1\|^2.\end{aligned}$$

Since the term $2\varepsilon\|x_1\|^2$ is strictly positive we can choose ε such that $2\varepsilon\|x_1\|^2$ prevails over $\varepsilon^2 x_1^T F_{11} x_1$ and therefore the expression above is positive. This is a contradiction to the negative semidefiniteness of F. Hence, the above expression is nonpositive if $x_1 = 0 = F_{12} x_2$ with x_2 in the kernel of F_{12}.

Using the fact that the image of a matrix is orthogonal to the kernel of the transpose of the same matrix we write for any z which is in the image of F_{21}

$$\begin{aligned} z \in \operatorname{im} F_{21} &\implies z \perp \ker F_{21}^T \\ &\implies z \perp \ker F_{12} \\ &\implies z \perp \ker F_{22} \\ &\implies z \in \operatorname{im} F_{22}^T = \operatorname{im} F_{22}.\end{aligned}$$

Therefore $\operatorname{im} F_{21} \subseteq \operatorname{im} F_{22}$ which proves the claim. □

Remark 9 To prove expressions (14.57) for a positive semidefinite symmetric matrix take $x = \begin{bmatrix} -\varepsilon x_1 \\ x_2 \end{bmatrix}$.

References

1. Antoulas, A.C.: Approximation of Large-Scale Dynamical Systems. SIAM (2005)
2. Fernando, K., Nicholson, H.: Singular perturbational model reduction of balanced systems. IEEE Trans. Automat. Contr. **27**, 466–468 (1982)
3. Glover, K.: All optimal hankel-norm approximations of linear multivariable systems and their L^{∞}-error bounds. Int. J. Control **39**, 1115–1193 (1984)
4. Golo, G.: Interconnection structures in port-based modelling: tools for analysis and simulation. Ph.D. thesis, University of Twente (2002)
5. Green, M., Limebeer, D.: Linear Robust Control. Prentice-Hall (1995)
6. Hartmann, C.: Balancing of dissipative Hamiltonian systems. In: Troch I., Breitenecker F. (eds.) Proceedings of MathMod 2009, Vienna, February 11–13, 2009, no. 35 in ARGESIM-Reports, pp. 1244–1255. Vienna University of Technology, Vienna (2009)
7. Hartmann, C., Vulcanov, V.M., Schütte, C.: Balanced truncation of linear second-order systems: a Hamiltonian approach. Multiscale Model. Simul. (submitted 2008). Available from URL http://proteomics-berlin.de/28/
8. Moore, B.: Principal component analysis in linear systems. IEEE Trans. Automat. Control **AC-26**, 17–32 (1981)
9. Pasumarthy, R.: On analysis and control of interconnected finite-and infinite-dimensional physical systems. Ph.D. thesis, University of Twente (2006)
10. Polderman, J.W., Willems, J.C.: Introduction to Mathematical Systems Theory. Springer-Verlag (1998)
11. Polyuga, R.V., van der Schaft, A.J.: Structure preserving model reduction of port-Hamiltonian systems. In Proceedings of the 18th International Symposium on Mathematical Theory of Networks and Systems, Blacksburg, July 28– August 1, 2008. http://sites.google.com/site/rostyslavpolyuga/publications'
12. van der Schaft, A.J.: L2-Gain and Passivity Techniques in Nonlinear Control. Springer-Verlag (2000)
13. van der Schaft, A.J.: On balancing of passive systems. In Proceedings of the European Control Conference, Kos, pp. 4173–4178, 2–5 July 2007
14. Scherpen, J.: Balancing for nonlinear systems. Ph.D. thesis, University of Twente (1994)
15. Sorensen, D.: Passivity preserving model reduction via interpolation of spectral zeros. Systems Control Lett. **54**, 347–360 (2004)
16. Willems, J.C.: Dissipative dynamical systems. Arch. Ration. Mech. Analysis **45**, 321–393 (1972)
17. Zhou, K., Doyle, J., Glover, K.: Robust and Optimal Control. Prentice-Hal (1996)

Chapter 15
Coupling of Numerical and Symbolic Techniques for Model Order Reduction in Circuit Design

Oliver Schmidt, Thomas Halfmann and Patrick Lang

Abstract In this paper, we will give a short motivation for the use of symbolic methods for model order reduction and present some of these techniques. Furthermore, we will discuss the hierarchy usually at hand for circuit design and present a workflow for the exploitation of this hierarchy. By applying it to a simple circuit example, we will show how model order reduction (MOR) methods can be profitably employed.

Keywords Model reduction · Symbolic model reduction · Hierarchical model reduction · Coupled symbolic-numerical model reduction · Structure preservation · Structure-exploiting model reduction · Structure-preserving model reduction · Analog circuit · Electrical circuit

15.1 Motivation

First, we will give a motivation for symbolic model order reduction methods (MOR methods) by using the example of the folded-cascode operational amplifier (see also [3]) depicted in Fig. 15.1. The original design showed an instability in the amplifier's small-signal behavior at a frequency of approximately 10 MHz (cf.

O. Schmidt (✉) · T. Halfmann · P. Lang
Fraunhofer ITWM, Fraunhofer-Platz 1, 67663 Kaiserslautern, Germany
e-mail: oliver.schmidt@itwm.fraunhofer.de

T. Halfmann
e-mail: thomas.halfmann@itwm.fraunhofer.de

P. Lang
e-mail: patrick.lang@itwm.fraunhofer.de

P. Benner et al. (eds.), *Model Reduction for Circuit Simulation*,
Lecture Notes in Electrical Engineering, 74, DOI: 10.1007/978-94-007-0089-5_15,

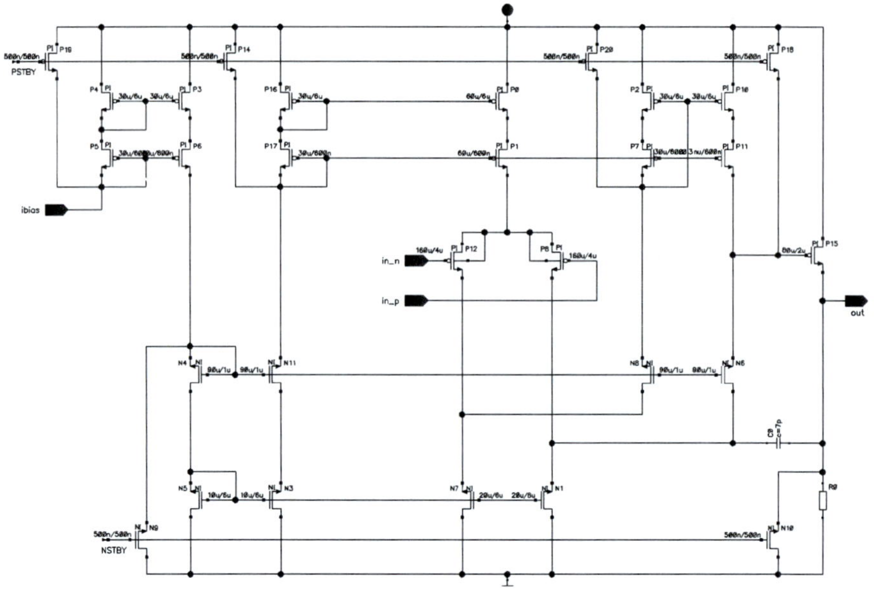

Fig. 15.1 Folded-cascode operational amplifier

Fig. 15.2, solid curves), visible by a resonance peak, which was caused by a pair of complex conjugate parasitic poles.

As a classic approach, one would have performed parameter variations and numerical simulations in a trial-and-error fashion to solve the underlying instability problem. But due to the fairly high number of parameters, this was not feasible here. However, a symbolic analysis of the amplifier allowed a computation of its transfer function. Since the exact symbolic transfer function consisted of more than $5 \cdot 10^{19}$ terms, a symbolic approximation was necessary, that further allowed a derivation of the formula for the dominant pole pair depicted in Fig. 15.3.

By interpretation of this formula, one detected that one of the circuit's components—a Gate-Source capacitance—had dominant influence with respect to the unstable behavior. By adding an additional capacitance, the original capacitance value could be modified such that the resonance could be damped (cf. frequency analysis in Fig. 15.2, dashed curves) and a stable behavior after a redesign of the amplifier was obtained (cf. transient analysis in Fig. 15.4, dotted curve).

It should be mentioned here that in this case symbolic MOR was used to improve the original design of the circuit and not the reduced model itself.

15.2 Symbolic Techniques

What in particular does nonlinear symbolic analysis aim for? Usually we start at a netlist description that is translated into an equation system (*differential-algebraic*

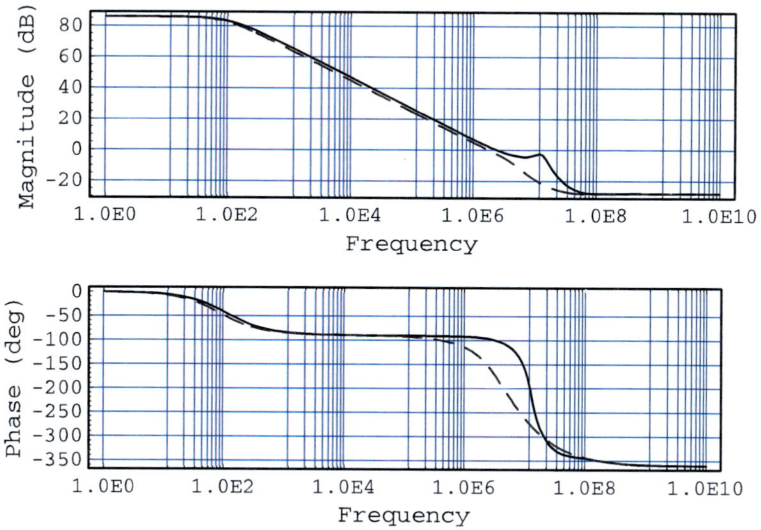

Fig. 15.2 The amplifier's transfer function: original (*solid*) and with an additional capacitance (*dashed*)

$$s_p{}^{1,2} == -\frac{(CC0 + CL)\ gm\$MN6}{2CC0CL} \pm \frac{\sqrt{Cgs\$MP15gm\$MN6\ (Cgs\$MP15\ (CC0 + CL)^2\ gm\$MN6 - 4CC0^2\ CLgm\$MP15\ )}}{2CC0Cgs\$MP15CL}$$

Fig. 15.3 Approximated formula for the dominant pole pair of the amplifier's transfer function

Fig. 15.4 Stable behavior (*dotted*) after redesign of the originally unstable behavior (*solid*)

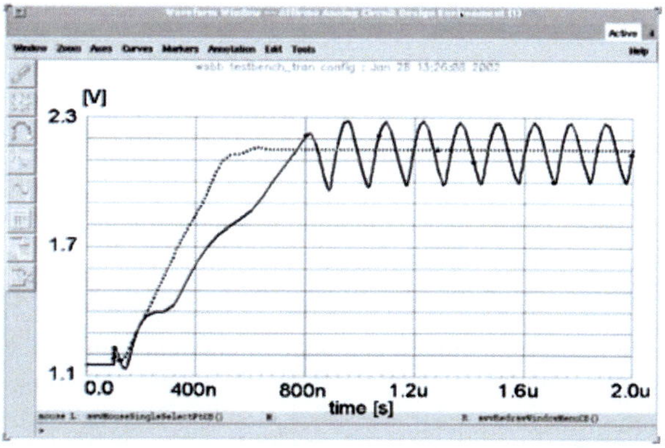

equations, DAE) via standard graph theoretical methods like *modified nodal analysis* (MNA) or *sparse tableau analysis* (STA) (see e.g. [13]). Based on a *reference solution* of the original system, we can symbolically reduce it to a *simpler* form with less equations, less terms, less derivatives, and so on [14]. This simpler form represents a behavioral model for the original system, which can also be translated into a hardware description language (HDL) and then simulated using a circuit simulator.

The general workflow is as follows (see [11, 14]); according to a numerical analysis $A \in \{\text{AC analysis, DC analysis, transient analysis}\}$ and an input u we have a reference solution v_F for the output v of the original DAE system F, as well as a user-given error bound ε and an error function E to compute the actual error. This reference solution v_F is given by pairs of sampling points and interpolation values, hence the accuracy in comparison to the exact mathematical solution v depends on the numerical solver's choice of the step size. The complexity of F is reduced by iteratively applying reduction techniques R and comparing the (numerical) solution v_G of the so far reduced system G to the reference solution v_F. As long as the error is within the given bound, the performed reduction step is allowed, otherwise it is rejected.

In the following, four symbolic techniques will be explained [14]. They are implemented in *Analog Insydes* [18], an add-on for *Mathematica* [17]. Analog Insydes is developed by Fraunhofer ITWM in Kaiserslautern, Germany.

15.2.1 Algebraic Manipulations

The first symbolic reduction technique we want to present is a purely algebraic manipulation. Here, certain variables are eliminated and substituted in the remaining set of equations. Additionally, independent blocks of equations are removed, e.g.:

$$F: \left.\begin{matrix} \cdots \\ y = f(x) \\ 0 = g(x, y) \\ \cdots \end{matrix}\right\} \Rightarrow \left\{\begin{matrix} \cdots \\ 0 = g(x, f(x)) \\ \cdots \end{matrix}\right. : G$$

Obviously, this method is mathematically exact. Despite this, we have to be careful, since applying this method may lead to equations including a huge number of terms.

15.2.2 Branch Reductions

Using branch reductions, unused branches of piecewise-defined functions are detected and removed. Piecewise-defined functions occur, e.g., in the device model equations of transistors or diodes. Usually these components only work in a certain operating region that is contained, e.g., in the interval $[a, b]$, then branch reductions are possible:

$$f(x) = \begin{cases} f_1(x) & x < a \\ f_2(x) & a \leq x \leq b \\ f_3(x) & x > b \end{cases} \quad \Rightarrow f(x) = f_2(x) \text{ for all } x.$$

Here, f is to be considered as a function that is contained in at least one term in at least one of the equations of the DAE system F.

15.2.3 Term Substitutions

Here, an equation, which typically consists of a sum of products,[1] is considered, where certain terms are replaced by appropriate constant values, e.g. adequate average values. This technique can also be applied to different levels, e.g. when some of the summands are functions and their arguments themselves contain sums, and so on.

In the following example, the *j*th equation F_j in the equation system F is a sum of terms t_i, where the *k*th term t_k will be replaced by the constant κ:

$$F_j : \sum_{i=1}^{N} t_i(x) = 0 \Rightarrow G_j : \sum_{i=1, i \neq k}^{N} t_i(\widetilde{x}) + \kappa = 0.$$

This method obviously is not an exact one, since it alters the solution x of the original equation system. The solution of the approximated system is denoted by $\widetilde{x}$.

15.2.4 Term Reductions

This technique is a special case of term substitutions, since certain terms are replaced by zero, which means that they are cancelled:

$$F_j : \sum_{i=1}^{N} t_i(x) = 0 \Rightarrow G_j : \sum_{i=1, i \neq k}^{N} t_i(\widetilde{x}) = 0.$$

Concerning all the above techniques, we consider that as long as the error that is caused by applying these methods is within the user-given error bound, the corresponding reductions are valid.

15.3 Hierarchical Systems

Within the last few years the growing miniaturization of integrated circuits and increasing integration density of circuit components have led to nanoelectronic structures. It turns out that physical effects like thermic or electro-magnetic interactions cannot be neglected anymore. One also observes a very fast growing number of parameters for the device models for a single transistor within the last

---

[1] In MNA contributions of components to the system typically are of the form $A^T \cdot f(Bx)$, where A and B are incidence matrices describing the circuit topology and f is a function describing the behavior of the corresponding component. The same expression holds for submodels.

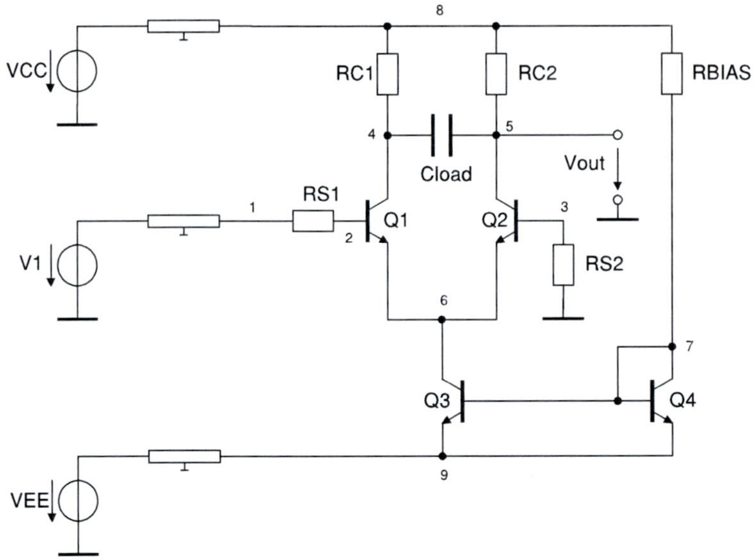

Fig. 15.5 Differential-amplifier circuit

years. This forces the usage of model descriptions based on *partial differential equations* (PDE) besides the netlist based models or those based on DAEs, especially for the semi-conductor components. Hence, one has to cope with *partial differential-algebraic equations* (PDAE), coupled systems of DAEs and PDEs.

Moreover, the circuit usually is hierarchical, as we have various subsystems coupled by a topology. Due to their respective accuracies, these subsystems have different mathematical descriptions. By using standard graph theoretical methods like MNA or STA (see [13]) to build up the describing equation system, this hierarchy is lost. So our goal is the exploitation of the circuit's hierarchical structure.

As an example, let us consider the differential-amplifier circuit depicted in Fig. 15.5. The voltage sources VCC and VEE denote the voltage supply for the amplifier circuit, whereas the voltage source V1 defines the input. The output is defined as the voltage-potential of node 5, denoted by Vout. The three sources are connected to the remaining circuit components via three transmission lines, for which we will use a discretized PDE model of the telegraph or transmission line equations with 20 line segments each (see e.g. [7] or [8], Chaps. 6 and 5).

15.3.1 Symbolic Reduction of the Entire Circuit

If we use MNA in node voltage formulation to set up the describing equation system for the transient circuit behavior, we obtain 167 equations with a total

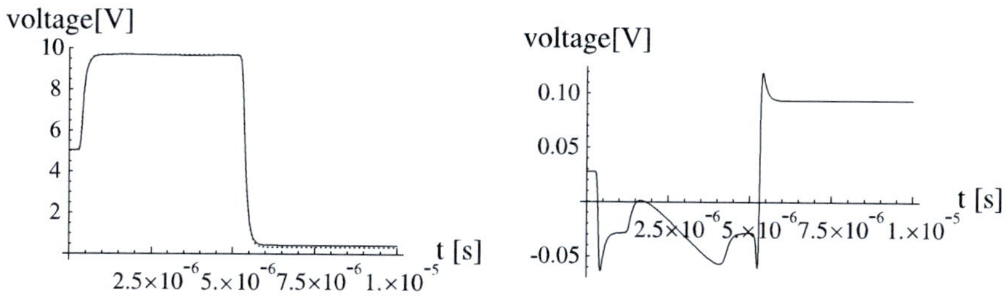

Fig. 15.6 *Left:* Solutions of the original system (*solid*) and the reduced system (*dotted*) allowing 2% error. *Right:* The corresponding error plot

number of 645 terms. For the input V1, we chose a sine-wave voltage excitation with a magnitude of 2 V and a frequency of 10^5 Hz. The supply voltages VCC and VEE are 12 V and −12 V, respectively. By using the symbolic reduction techniques presented in Sect. 15.2 and permitting an error of 2%[2] for the output Vout to reduce the entire system, we obtain a system consisting of 124 equations with 425 terms in total. For the reduction, a few hours are needed,[3] in which more than 95% of that time is spent for the computation of the transient ranking[4] (see e.g. [4, 14, 15]). Figure 15.6 shows the solution and the error of this reduced system in comparison to the solution of the original system.

According to these figures, our reduced system with only about 1% error is a very accurate approximation.

If we allow an error of 10% for the reduction of the original system, we only obtain 44 equations and a total of 284 terms. The time needed is almost the same because of the ranking computation mentioned above. However, the reduced equation complexity has been achieved at the expense of accuracy (cf. Fig. 15.7).

15.3.2 Exploitation of the Hierarchical Structure

What will happen, if we take an 'intuitive' hierarchy into account? Therefore consider certain structural patterns in the circuit's netlist: the differential pair of transistors (together with some resistors) in the upper box on the right side of Fig. 15.8, a current mirror in the lower right, three transmission lines in the middle, and the sources on the left.

---

[2] x% error here means $\frac{x}{100}$ of the maximum magnitude of the reference solution, considered on the entire time interval: $\frac{x}{100} \cdot \sup_{t\in[t_0,t_1]} ||v_F(t)||_\infty$.

[3] The computations were executed on a Dual Quad Xeon E5420 with 2.5 MHz and 16 GB RAM.

[4] A ranking is a trade-off between accuracy and efficiency in computation time that estimates the influence of a term in the equation system on the system's solution.

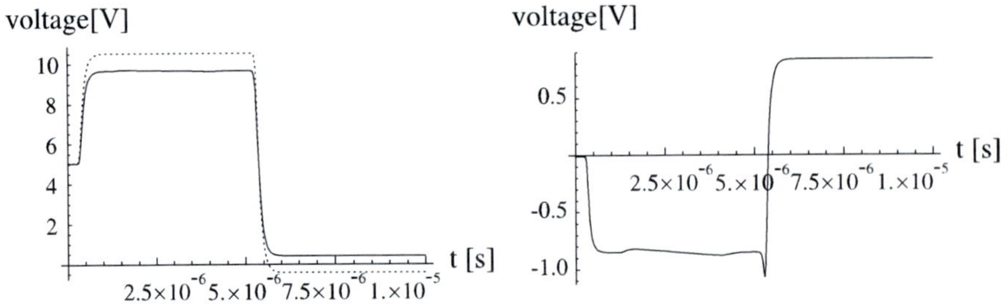

Fig. 15.7 *Left:* Solutions of the original system (*solid*) and the reduced system (*dotted*) allowing 10% error. *Right:* The corresponding error plot

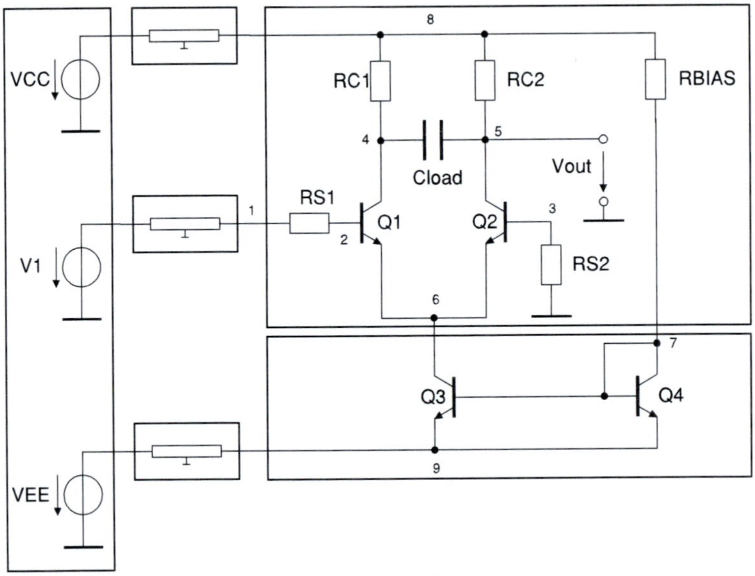

Fig. 15.8 Differential-amplifier circuit with its 'intuitive' hierarchy

As said before, by using e.g. MNA to set up the describing equation system for the entire circuit, the hierarchy information is lost, since the system contains equations with mixed parts from different subsystems. But if we consider the five subsystems one by one, we can transmit the hierarchy information into the describing set of equations and reduce them separately.

For the reduction of the three transmission lines, we translate the describing linear equation system into its state space form and apply Arnoldi's algorithm (see e.g. [5] or [1], Chaps. 10 and 11), a numerical approximation method based on projections on an appropriate Krylov space. For both the transmission lines that are connected to the voltage supplies VCC and VEE we iterate only one step,[5] for the

[5] One step provides sufficient accuracy here, since the sources are DC voltage sources.

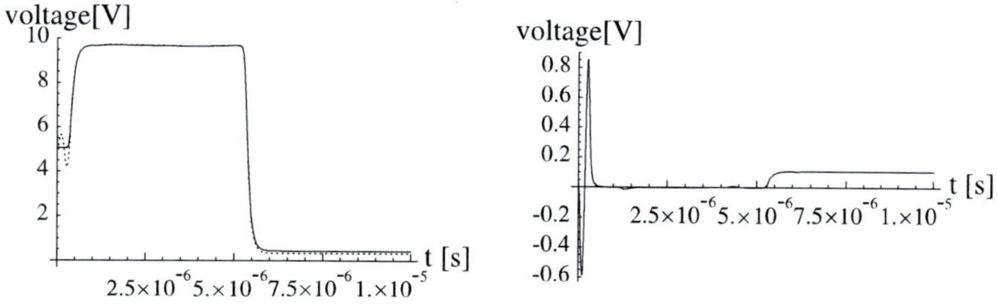

Fig. 15.9 *Left:* Solutions of the hierarchically reduced system (*dotted*) and the original system (*solid*) by applying the 3-step Arnoldi iteration to the input voltage transmission line. *Right:* The corresponding error plot

one connected to the input voltage V1 we perform three steps in the Arnoldi iteration.[6] Due to two additional equations and variables for internal port modeling, we thus can reduce the transmission line subsystems from 50 down to 8 resp. 4 equations.

The current mirror and the differential pair subsystems are reduced symbolically by using the presented techniques from Sect. 15.2. An exact description of how this is done is given in Sect. 15.4. In case of the current mirror subsystem, the symbolic simplification yields 9 instead of 16 equations with a total number of 20 instead of 59 terms. For the subsystem with the differential pair, we obtain a reduced system with 13 instead of 22 equations and 50 instead of 91 terms in total. For both of these reductions a 2% error bound[7] is permitted, but we observe almost no further reductions, if we allow 10% error instead. Note that all of the reductions mentioned above are computed within seconds. How these reduction steps—particularly in the symbolic case—can be carried out will be explained in the next section.

If we now plug together all the reduced subsystems to obtain a reduced form of the entire differential-amplifier circuit, the results are very accurate as one can see in Fig. 15.9. Instead of 167 equations and a total number of 645 terms, we only have 62 equations with 252 terms altogether. The error of the entire system compared to the original one is approximately 8%.

A higher dimensional projection space obtained by the Arnoldi iteration provides more accurate solutions, but of course also leads to a larger system. If, for example, we perform five iteration steps in the Arnoldi algorithm to reduce the transmission line that is connected to the input voltage V1 instead of only three, we can further improve our results and obtain a maximum error of the entire reduced

[6] We performed some experiments with a higher number of iteration steps, but the gain of additional accuracy in comparison to the size of the larger system was rather marginal for a number of iteration steps bigger than five.

[7] As we are reducing only the single subsystem, this error bound is not limiting the error of the entire circuit's output `Vout` to the given values (see Sect. 15.4 and Algorithm 1 for further details).

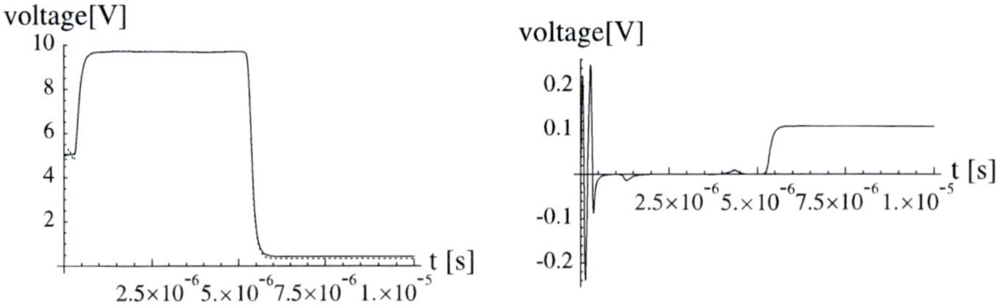

Fig. 15.10 *Left:* Solutions of the hierarchically reduced system (*dotted*) and the original system (*solid*) by applying the 5-step Arnoldi iteration to the input voltage transmission line. *Right:* The corresponding error plot

system of approximately 2% (Fig. 15.10). But then the reduced entire system consists of 66 equations and a total number of 396 terms.

15.3.3 Using the Reduced Model with Other Inputs

To check whether the reduced model obtained with our hierarchical approach works fine also for other inputs, we apply a pulse wave with a magnitude of 2 V, a sum of three pulses with magnitudes of 1, 2, and 3 V, respectively, and a sum of three sine functions

$$f(t) := 2 \cdot \sin(2\pi \cdot 10^5 \cdot t) + 2 \cdot \sin(2\pi \cdot 2 \cdot 10^5 \cdot t) + 1 \cdot \sin(2\pi \cdot 5 \cdot 10^5 \cdot t)$$

to V1. The graphs of these functions are shown in Fig. 15.11 on the time interval $I = [0\text{ s}, 10^{-5}\text{ s}]$.

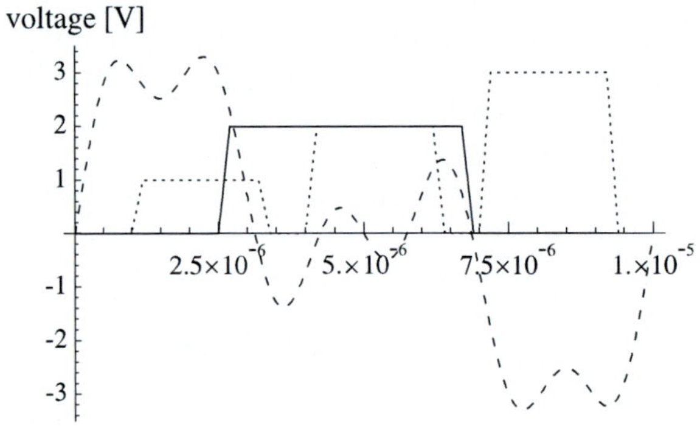

Fig. 15.11 Three other inputs `V1` to test the reduced model of the differential-amplifier circuit

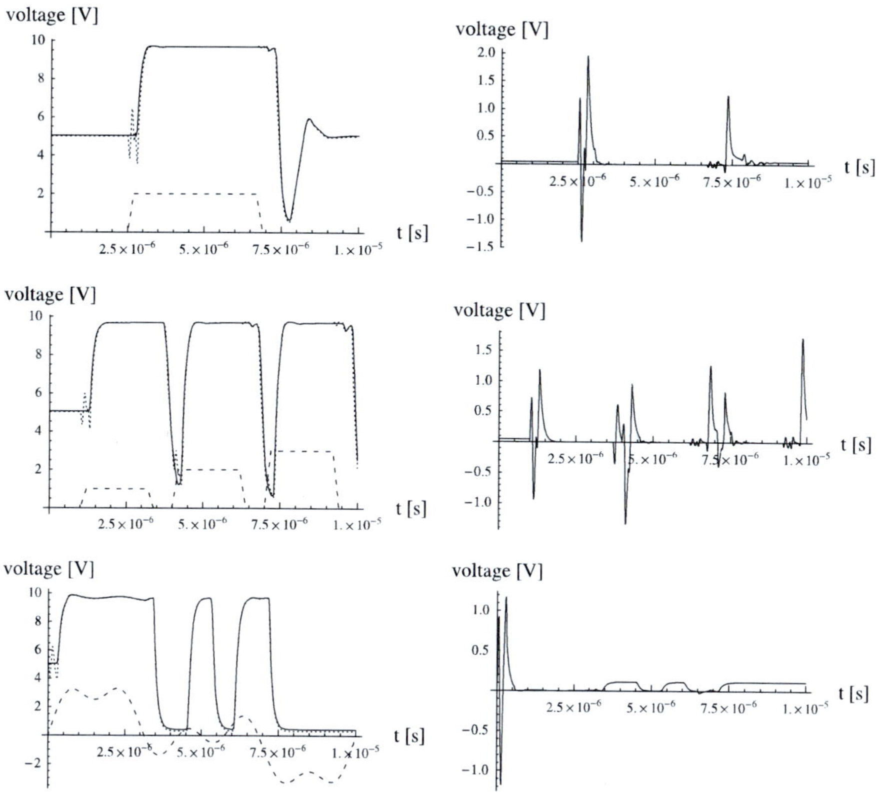

Fig. 15.12 *Left:* Simulation results of the original (*solid*) and the hierarchically reduced model (*dotted*), together with the corresponding input V1 (*dashed*). *Right:* The corresponding error plots

With these inputs, both the original differential-amplifier and our hierarchically reduced model[8] are simulated. The graphs of the simulations as well as the corresponding error plots are shown in Fig. 15.12.

According to these figures, the obtained model works quite well in all the test cases and therefore can be used as a behavioral model of the original differential amplifier in a certain domain.

15.3.4 Comparison of both Approaches

The results of the two different reduction approaches for the differential-amplifier circuit are listed in Table 15.1.

---

[8] We chose the more accurate model with 66 equations, see Fig. 15.10.

Table 15.1 Results of the two different reduction approaches for the differential-amplifier circuit

	Orig. system	Reduction of entire circuit		Reduction exploiting hierarchy	
Equations	167	124	44	66	62
Terms	645	425	284	396	252
(Permitted) error		2%	10%	~2%	~8%
Arnoldi iterations				(5/1)	(3/1)
Computation time		Few hours	Few hours	Within seconds	Within seconds

Comparing both approaches, we conclude that the one that exploits the hierarchical structure is the better choice, since it delivers much better results in less computation time; we obtained reduced systems with almost the same accuracy in the 2% error case, but only about half the number of equations. Performing only three steps in the Arnoldi iteration to reduce the transmission line connected to the input voltage V1, we still received a reduced overall system that fit the original (reference) solution quite well. On the other hand, the reduction of the whole circuit permitting an error of 10% led to a system that fully exploits the error bound.

15.4 Workflow for the Exploitation of the Hierarchy

Now we come to the workflow for the reduction of the particularly symbolic subsystems of the entire circuit. It is schematically shown in Fig. 15.13.

From the hierarchical segmentation, we have a set of interacting subsystems. Since we always need a reference solution to symbolically reduce a system, we need a 'closed circuit'. If we simply cut out a subsystem from its connecting structure, we will have no defined input/output behavior at its ports, i.e. we have no information about its current-voltage relation. However, the following algorithm provides a possibility for reducing such a subsystem (see also Fig. 15.13).

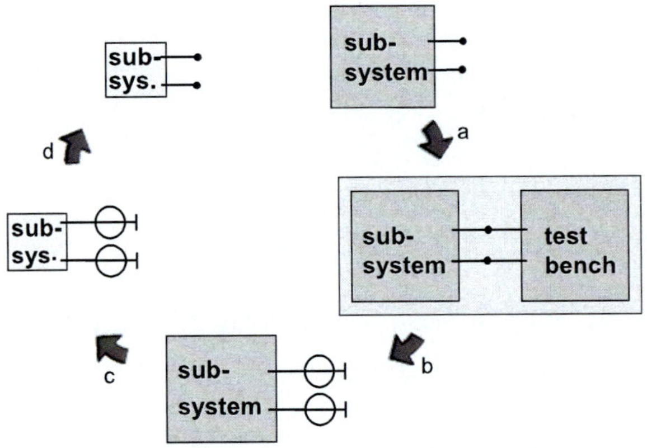

Fig. 15.13 Illustration of the workflow for the symbolic reduction of circuit subsystems

Algorithm 1 (Reduction of subsystems)

(a) connect the subsystem to a test bench[9] and record the voltages[10] at its ports
(b) remove the test bench and plug voltage sources to the subsystem's ports that deliver exactly the recorded voltages (as a data-based function) thus having the subsystems isolated
(c) now the closed circuit can be reduced[11] with appropriate (numerical or) symbolic methods presented in Sect. 15.2
(d) remove all the above sources after the reduction of the subsystem and finally obtain a reduced subsystem that serves as a behavioral model

This workflow has been applied to the subsystems of the differential-amplifier example and delivered the results described at the end of the previous section.

It should be mentioned here that this approach only controls the errors at the ports of the single subsystems. Thus, one cannot pose a statement about or guarantee a certain global error, i.e. the error on the output of the entire (reduced) circuit when replacing the original subsystems by the reduced ones. This could possibly be a point of future investigation (see also Sect. 15.6).

15.5 Comparison to Other Approaches

In [9], LTI coupled control systems are considered, which are coupled by input/output-relations. Moment matching approximation and balanced truncation, two numerical MOR methods, are used to reduce the order of the closed-loop systems. Moreover, structure-preserving model reduction approaches using balanced truncation or Krylov subspaces are presented and some error bounds are given. The internal inputs of the LTI subsystems are given by a linear combination of the other subsystem's internal outputs and the external input, the external output is a linear combination of all internal outputs of the subsystems.

In contrast, in this paper we are dealing with *nonlinear hierarchical* systems which are much less studied, but of great importance. We do not consider an approach that is based on inputs and outputs for the coupling of subsystems as adequate to model an electronic circuit, since the subsystems are *mutually* interacting with each other. Hence, we did not try to extend these approaches.

In [16], however, the author uses "variable sharing" and relations on system variables for modeling the behavior of the variables on the subsystems' external terminals to capture physical phenomena. This approach is applicable not only to

[9] A test bench is a simulation test environment. Here, one could take all the rest of the entire circuit as test bench.

[10] Of course, one could record the currents as well.

[11] By providing the *voltages* at all k ports, they work as the input to the subsystem. Then the *currents* flowing into the subsystem at the k ports are defined as the output and will be controlled by the specified error bound.

electronic circuits, but also to other systems. For example, in a hydraulic system, if two pipes are connected with each other, their pressure variables at the link are equalized while the sum of their flow variables is zero.[12] Thus the variables at the link are shared. Also in our opinion, this approach is appropriate to capture all relevant physical effects of the interacting subsystems, while viewing interconnections as output-to-input assignments introduces a signal transmission mechanism that is not part of the physics of an electronic circuit.

In our paper, we use a similar approach and, in addition, provide a sophisticated model reduction method for hierarchically structured nonlinear systems.

Of course there already exist some methods like POD for *nonlinear* model order reduction. However, we could not find any other approaches for nonlinear model order reduction techniques that explicitly exploit the hierarchy of the overall system. We found some other structure-preserving approaches [2, 6, 12], but these were applied only to *linear* systems. They are based on Krylov space approximations with a certain shape of the projecting matrices to preserve the structure of the original system in its reduced equivalent.

15.6 Summary and Future Work

In this paper, we considered two different approaches for the reduction of a differential-amplifier circuit. The first one reduced the entire circuit using methods as presented in Sect. 15.2 (cf. also [3, 14]). The second approach, introduced in this paper, exploits the circuit's hierarchy. Here we reduce the subsystems separately, then plug the reduced ones together, and thus obtain a reduced overall system.

The new hierarchy-exploiting approach turned out to be the better choice, since we obtained much better results with respect to computation time. Moreover, the accuracy was almost the same, but the number of equations could be halved.

The principal idea is to reduce the subsystems separately (either symbolically or numerically), each one allowing a certain approximation error ε_k. If we plug together the reduced subsystems, we obtain an approximation for the entire system with a certain error ε compared to the original overall system. If the total error ε or the degree of reduction appears not to be satisfactory, we restart the reduction of at least one appropriate subsystem with an altered error bound ε_k More details can be found in [10].

One goal of possible future investigation could be finding a functional relation between the errors ε_k and the error ε of the entire system. The challenge, probably realistic only for special systems, would be obtaining an estimate for the subsystem errors as function of the user-specified error ε.

Acknowledgements The work reported in this paper has been carried out within the project *SyreNe* (Systemreduktion für IC Design in der Nanoelektronik [19], Grant no. 03LAPAE6) which is supported by the German Federal Ministry of Education and Research (BMBF). The first

[12] The respective flows are directed towards the interior of each pipe.

named author wants to thank his thesis advisor G.-M. Greuel for additional support. Responsibility for the contents of this publication rests with the authors.

References

1. Antoulas, A.C.: Approximation of Large-Scale Dynamical Systems. SIAM, Philadelphia (2005)
2. Freund, R.W.: Structure-preserving model order reduction of RCL circuit equations. In: van der Vorst, H., Schilders, W., Rommes, J. (eds.) Model Order Reduction: Theory, Research Aspects and Applications, Mathematics in Industry. Springer, Berlin (2008)
3. Halfmann, T., Wichmann, T.: Overview of symbolic methods in industrial analog circuit design. Berichte des Fraunhofer ITWM, Nr. **44** (2003)
4. Halfmann, T., Wichmann, T.: Symbolic methods in industrial analog circuit design. In: Anile, A.M., Ali, G., Mascali, G. (eds.) Scientific Computing in Electrical Engineering. Springer, Berlin (2006)
5. Lassaux, G., Willcox, K.: Model reduction for active control design using multiple-point Arnoldi methods. AIAA Paper 2003-0616, Presented at 41st Aerospace Sciences Meeting & Exhibit, Reno, NV (2003) URL: http://www.hdl.handle.net/1721.1/3702
6. Li, R.-C., Bai, Z.: Structure-preserving model reduction using a Krylov subspace projection formulation. Comm. Math. Sci. **3**(2), 179–199 (2005)
7. Miri, A.M.: Ausgleichsvorgänge in Elektroenergiesystemen. Springer, Berlin (2000)
8. Reis, T.: Systems Theoretic Aspects of PDAEs and Applications to Electrical Circuits. Shaker, Aachen (2006)
9. Reis, T., Stykel, T.: A survey on model reduction of coupled systems. In: Schilders, W.H.A., van der Voorst, H.A., Rommes, J. (eds.) Model Order Reduction: Theory, Research Aspects and Applications, Mathematics in Industry, vol. 13, pp. 133–155. Springer, Berlin (2008)
10. Schmidt, O.: Structure-Exploiting Coupled Symbolic-Numerical Model Reduction For Electrical Networks. Ph.D. thesis, TU Kaiserslautern, Cuvillier, Göttingen (2010)
11. Sommer, R., Halfmann, T., Broz, J.: Automated behavioral modeling and analytical model-order reduction by application of symbolic circuit analysis for multi-physical systems. Simulation Modelling Practice Theory **16**, 1024–1039 (2008)
12. Villena, J.F., Schilders, W.H.A., Silveira, L.M.: Parametric structure-preserving model order reduction. In: Reis, R., Mooney, V., Hasler, P. (eds.) IFIP International Federation for Information Processing, vol. 291, VLSI-SoC: Advanced Topics on Systems on a Chip, pp. 69–88. Springer, Berlin (2009)
13. Vlach, J., Singhal, K.: Computer Methods for Circuit Analysis and Design, 2nd edn. Van Nostrand Reinhold, New York (1993)
14. Wichmann, T.: Symbolische Reduktionsverfahren für nichtlineare DAE-Systeme. Shaker, Aachen (2004)
15. Wichmann, T., Popp, R., Hartong, W., Hedrich, L.: On the simplification of nonlinear DAE systems in analog circuit design. In: Proceedings of CASC'99, Munich, Germany (1999)
16. Willems, J.C.: The behavioral approach to open and interconnected systems. IEEE Control Systems Magazine **27**, 46–99 (2007)
17. Wolfram, S.: The Mathematica Book, 5th edn. Wolfram Media Incorporated, Champaign (2003)
18. Analog Insydes website: URL: http://www.analog-insydes.de
19. SyreNe website: URL: http://www.syrene.org

Chapter 16
On Stability, Passivity and Reciprocity Preservation of ESVDMOR

Peter Benner and André Schneider

Abstract The reduction of parasitic linear subcircuits is one of many issues in model order reduction (MOR) for VLSI design. This issue is well explored, but recently the incorporation of subcircuits from different modelling sources into the circuit model has led to new structural aspects: so far, the number of elements in the subcircuits was significantly larger than the number of connections to the whole circuit, the so called pins or terminals. This assumption is no longer valid in all cases such that the simulation of these circuits or rather the reduction of the model requires new methods. In Feldmann and Liu (ICCAD '04: Proceedings of the 2004 IEEE/ACM International Conference on Computer-Aided Design, 88–92, 2004), Liu et al. (ICCAD '05: Proceedings of the 2005 IEEE/ACM International Conference on Computer-Aided Design, 821–826, 2004; Integr. VLSI J. **41**(2): 210–218, 2008) the extended singular value decomposition based model order reduction (ESVDMOR) algorithm is introduced as a way to handle this kind of circuits with a massive number of terminals. Unfortunately, the ESVDMOR approach has some drawbacks because it uses the SVD for matrix factorizations. In Benner (Mathematics in Industry 14, 2009), Schneider (Matrix decomposition based approaches for model order reduction of linear systems with a large number of terminals, 2008) the truncated SVD (TSVD) as an alternative to the SVD within the ESVDMOR is introduced. In this paper we show that ESVDMOR as well as

P. Benner (✉) · A. Schneider
Max Planck Institute for Dynamics of Complex Technical Systems, Computational Methods in Systems and Control Theory, Sandtorstr. 1, 39106 Magdeburg, Germany
e-mail: debenner@mpi-magdeburg.mpg.de

A. Schneider
e-mail: schneidera@mpi-magdeburg.mpg.de

P. Benner
Technische Universität Chemnitz, Fakultät für Mathematik, 09107 Chemnitz, Germany
e-mail: benner@mathematik.tu-chemnitz.de

P. Benner et al. (eds.), *Model Reduction for Circuit Simulation*,
Lecture Notes in Electrical Engineering, 74, DOI: 10.1007/978-94-007-0089-5_16,

the modified approach is stability, passivity, and reciprocity preserving under reasonable assumptions.

Keywords Model reduction · Model order reduction · Many terminals · Large number of inputs/outputs · Passivity · Stability · Reciprocity · Power grids

16.1 Introduction

MOR has become an important tool in the preprocessing of circuit simulation over the last decades. The original model which results from the mathematical modeling with the help of, e.g., modified nodal analysis [11] has to be simplified due to its complexity. One special part of this simplification for VLSI design is the MOR of parasitic linear interconnect circuits [7]. These circuits appear in form of substructures in the design of integrated circuits (ICs). They contain linear elements which do not necessarily have large influence on the result of the simulation.

There are applications in which the structure of these parasitic linear subcircuits has been changing recently in the following sense. So far, the number of elements in these interconnect circuits was significantly larger than the number of connections to the whole circuit, the so-called pins or terminals. This assumption is not true anymore in many cases. Power grid networks which supply elements of large circuits with energy are of this special form [22, 24]. Often these power grids are realized as an extra layer of elements in between the layers of transistors. The fact that they need connections to the elements for power supply explains the high number of pins. The same problem appears in connection with synchronization. In clock distribution networks, the clock signal is distributed from a common point to all the elements that need it for synchronization [14]. For simulating these special subcircuits, new methods are needed. In some applications, a lot of terminal signals behave similarly so that it is possible to compress the input-/output -matrices of the transfer function in such a way that the I/O behavior can be realized through a few so-called virtual inputs/outputs. As a consequence we deal with these virtual terminals, the number of which is much less than the original number of terminals. This allows the use of well known MOR methods like balanced truncation or Krylov subspace methods to reduce the number of inner nodes. The basic work was done a few years ago with the introduction of the (E)SVDMOR approach [6, 15–17, 23]. In this paper we review the existing ESVDMOR approach [6, 17] and discuss stability and passivity of the computed reduced-order model. As most linear subcircuits with a massive number of pins represent RC(L) circuits and contain no active devices, they are modelled as passive and thus stable systems. Therefore, it is generally important to compute reduced-order models that share these properties with the original model.

In the following section, we review the fundamentals of the ESVDMOR approach. We introduce the moments of a transfer function of the circuit describing system and show how to use the information in these moments in order to reduce the number of terminals. In this way we achieve a very compact model. The preservation of stability and passivity in the reduced-order models is the topic of Sect. 16.3. We introduce basic definitions and prove that the ESVDMOR approach is stability, passivity, and reciprocity preserving under certain conditions. A numerical example shows the passivity preservation. In Sect. 16.4, we sum up the results and outline future research activities.

16.2 The Extended SVDMOR Approach

We consider a linear time-invariant continuous-time descriptor system, which can be represented as

$$\begin{aligned} C\dot{x}(t) &= -Gx(t) + Bu(t), \quad x(0) = x_0, \\ y(t) &= Lx(t), \end{aligned} \tag{16.1}$$

with $C, G \in \mathbb{R}^{n \times n}$, $B \in \mathbb{R}^{n \times m_{\text{in}}}$, $L \in \mathbb{R}^{m_{\text{out}} \times n}$, $x(t) \in \mathbb{R}^n$ containing internal state variables, $u(t) \in \mathbb{R}^{m_{\text{in}}}$ the vector of input variables, e.g., the terminal currents, $y(t) \in \mathbb{R}^{m_{\text{out}}}$ being the output vector, e.g., the terminal voltages, $x_0 \in \mathbb{R}^n$ the initial value and n the number of state variables, called the *order* of the system. The original linear system (16.1) to be reduced has the following transfer function in frequency domain:

$$H(s) = L(sC + G)^{-1}B. \tag{16.2}$$

The number of inputs m_{in} and the number of outputs m_{out} are not necessarily equal. Further on we can define the ith block moment of (16.2) as

$$\mathbf{m_i} = L(-G^{-1}C)^i G^{-1}B, \quad i = 0, 1, \ldots \tag{16.3}$$

in terms of $\mathbf{m_i}$ as an $m_{\text{out}} \times m_{\text{in}}$ matrix

$$\mathbf{m_i} = \begin{bmatrix} m^i_{1,1} & m^i_{1,2} & \cdots & m^i_{1,m_{\text{in}}} \\ m^i_{2,1} & m^i_{2,2} & \cdots & m^i_{2,m_{\text{in}}} \\ \vdots & \vdots & \ddots & \vdots \\ m^i_{m_{\text{out}},1} & m^i_{m_{\text{out}},2} & \cdots & m^i_{m_{\text{out}},m_{\text{in}}} \end{bmatrix}. \tag{16.4}$$

Note that the moments in (16.3) are equal to the coefficients of the Taylor series expansion of (16.2) about $s = 0$. The expansion about $s = s_0 \neq 0$ leads to *frequency shifted moments* defined as

$$\mathbf{m_i}(s_0) = L(-(s_0C + G)^{-1}C)^i (s_0C + G)^{-1}B, \quad i = 0, 1, \ldots \tag{16.5}$$

In the ESVDMOR method [6, 17], the information of a combination of these moments to create a decomposition of (16.2) is used in the following way. For being able to allow terminal reduction for inputs and outputs separately, w.l.o.g. we use r different block moments forming two moment matrices, the input response matrix M_I and the output response matrix M_O, as follows:

$$M_I = \begin{bmatrix} \mathbf{m_0} \\ \mathbf{m_1} \\ \vdots \\ \mathbf{m_{r-1}} \end{bmatrix}, \quad M_O = \begin{bmatrix} \mathbf{m_0}^T \\ \mathbf{m_1}^T \\ \vdots \\ \mathbf{m_{r-1}}^T \end{bmatrix}. \tag{16.6}$$

Note that it is also possible to use different numbers of block moments to create M_I and M_O and that column k of M_I represents the coefficients (moments) of the series expansion of (16.2) at all outputs corresponding to input k. Similarly, each column k of M_O represents the coefficients of output k corresponding to all inputs. We assume the number of rows in each matrix to be larger than the number of columns. If not, r has to be increased.

Applying the SVD to these matrices, we can obtain a low rank approximation

$$M_I = U_I \Sigma_I V_I^T \approx U_{I_{r_i}} \Sigma_{I_{r_i}} V_{I_{r_i}}^T, \quad M_O = U_O \Sigma_O V_O^T \approx U_{O_{r_o}} \Sigma_{O_{r_o}} V_{O_{r_o}}^T, \tag{16.7}$$

where

- $U_I = \left[U_{I_{r_i}}, U_{I_{(rm_\text{out} - r_i)}} \right]$ is an $r \cdot m_\text{out} \times r \cdot m_\text{out}$ orthogonal matrix,
- $V_I = \left[V_{I_{r_i}}, V_{I_{(m_\text{in} - r_i)}} \right]$ is an $m_\text{in} \times m_\text{in}$ orthogonal matrix,
- $U_O = \left[U_{O_{r_o}}, U_{O_{(rm_\text{in} - r_o)}} \right]$ is an $r \cdot m_\text{in} \times r \cdot m_\text{in}$ orthogonal matrix,
- $V_O = \left[V_{O_{r_o}}, V_{O_{(m_\text{out} - r_o)}} \right]$ is an $m_\text{out} \times m_\text{out}$ orthogonal matrix,
- Σ_I and Σ_O are $r \cdot m_\text{out} \times m_\text{in}$ and $r \cdot m_\text{in} \times m_\text{out}$ diagonal matrices,

whereas

- $\Sigma_{I_{r_i}}$ and $\Sigma_{O_{r_o}}$ are $r_i \times r_i$ and $r_o \times r_o$ diagonal matrices,
- $V_{I_{r_i}}$ and $V_{O_{r_o}}$ are $m_\text{in} \times r_i$ and $m_\text{out} \times r_o$ isometric matrices that contain the dominant column subspaces of M_I and M_O,
- $U_{I_{r_i}}$ and $U_{O_{r_o}}$ are $rm_\text{out} \times r_i$ and $rm_\text{in} \times r_o$ isometric matrices that are not used any further. $r_i \leq m_\text{in}$ and $r_o \leq m_\text{out}$ are the numbers of significant singular values as well as the numbers of the reduced virtual input and output terminals. Due to the fact that the important information about the dependencies of the I/O-ports is hidden in the matrices $V_{I_{r_i}}^T$ and $V_{O_{r_o}}^T$, approximations of B and L using the factors in (16.7) lead to

$$B \approx B_r V_{I_{r_i}}^T \quad \text{and} \quad L \approx V_{O_{r_o}} L_r. \tag{16.8}$$

Here, $B_r \in \mathbb{R}^{n \times r_i}$ and $L_r \in \mathbb{R}^{r_o \times n}$ are computed by applying the Moore-Penrose pseudoinverse [18] (denoted by $(\cdot)^+$) of $V_{I_{r_i}}^T$ and of $V_{O_{r_o}}$ to B and L, respectively. In detail, that means

$$B_r = BV_{I_{r_i}} (V_{I_{r_i}}^T V_{I_{r_i}})^{-1} = BV_{I_{r_i}}^{T+} = BV_{I_{r_i}} \tag{16.9}$$

and

$$L_r = (V_{O_{r_o}}^T V_{O_{r_o}})^{-1} V_{O_{r_o}}^T L = V_{O_{r_o}}^{+} L = V_{O_{r_o}}^T L, \tag{16.10}$$

since $V_{I_{r_i}}$ and $V_{O_{r_o}}$ are isometric. As a consequence we get a new internal transfer function $H_r(s)$ by using the approximation

$$H(s) \approx \widehat{H}(s) = V_{O_{r_o}} \underbrace{L_r(G + sC)^{-1} B_r}_{:=H_r(s)} V_{I_{r_i}}^T. \tag{16.11}$$

This terminal reduced transfer function $H_r(s)$ can be further reduced to

$$\tilde{H}_r(s) = \tilde{L}_r(\tilde{G} + s\tilde{C})^{-1}\tilde{B}_r \approx H_r(s) = L_r(G + sC)^{-1}B_r \tag{16.12}$$

by some established MOR method, e.g., balanced truncation or a Krylov subspace method, see [2, 4, 20] for introductions to linear model reduction techniques. We end up with a very compact terminal reduced and reduced-order model $\tilde{H}_r(s)$, i.e.

$$H(s) \approx \widehat{H}(s) \approx \widehat{H}_r(s) = V_{O_{r_o}} \tilde{H}_r(s) V_{I_{r_i}}^T. \tag{16.13}$$

Note that the well-known SVDMOR approach [6] can be considered as a special case of ESVDMOR, using only one moment and one SVD, e.g. $r = 1$, and using $\mathbf{m}_0$ as moment

The ESVDMOR approach as described above is not appropriate for really large-scale circuit systems as it employs the full SVD of the moment matrices. Observing that the computation of the reduced-order model only needs the leading block-columns $U_{I_{r_i}}, V_{I_{r_i}}, U_{O_{r_o}}, V_{O_{r_o}}$ of the orthogonal matrices computed within the SVD, the expensive SVD can be replaced by a truncated SVD which can be computed cheaply employing sparsity of the involved matrices using Krylov subspace or Jacobi–Davidson methods [12, 13]. Efficient algorithms based on the first approach are suggested in [5, 21] while the Jacobi–Davidson version is under current investigation.

16.3 Stability, Passivity, and Reciprocity

In this section we establish some facts on the preservation of stability, passivity, and reciprocity in ESVDMOR reduced-order models.

16.3.1 Stability

In order to discuss stability of descriptor systems we need the following definition and lemma which can be found, e.g., in [4].

Definition 1 *The descriptor system* (16.1) *is called asymptotically stable if* $\lim_{t\to\infty} x(t) = 0$ *for all solutions* $x(t)$ *of* $C\dot{x}(t) = -Gx(t)$.

Lemma 1 *Consider a descriptor system* (16.1) *with a regular matrix pencil* $\lambda C + G$. *The following statements are equivalent*:

1. *System* (16.1) *is asymptotically stable.*
2. *All finite eigenvalues of the pencil* $\lambda C + G$ *lie in the open left half-plane.*

Using the results from Lemma 1 we are able to formulate the following theorem:

Theorem 1 *Consider an asymptotically stable system*(16.1) *with its transfer function* (16.2). *The ESVDMOR reduced-order system corresponding to* (16.13) *is asymptotically stable iff the inner reduction* (16.12) *is stability preserving.*

Proof It is obvious that none of the approximations (16.8), (16.11) and (16.13) change the eigenvalues of $\lambda C + G$. With Lemma 1 and the assumption that (16.12) is stability preserving it directly follows that the ESVDMOR approach is stability preserving. □

A possible stability preserving model reduction method that can be applied along the lines of Theorem 1 is balanced truncation for regular descriptor systems, see [3, 4].

16.3.2 Passivity

Showing that the ESVDMOR approach is passivity preserving turns out to be more difficult. First we note that a system is *passive* iff its transfer function is positive real [1]. The following definition of positive realness can be found, e.g., in [8].

Definition 2 The transfer function (16.2) is positive real iff the following three assumptions hold:

1. $H(s)$ has no poles in $\mathbb{C}_+ = \{s \in |\mathrm{Re}s > 0\}$, i.e. the system is stable if additionally there are no multiple poles on $i\mathbb{R}$,
2. $H(\bar{s}) = \overline{H(s)}$ for all $s \in \mathbb{C}$,
3. $\mathrm{Re}(x^H H(s)x) \geq 0$ for all $s \in \mathbb{C}_+$ and $x \in \mathbb{C}^m$.

For passive systems we have to assume that the number of inputs is equal to the number of outputs: $m_{\mathrm{in}} = m_{\mathrm{out}} = m$. Due to the fact that we often deal just with parasitic linear RLC circuits we furthermore assume $L = B^T$ such that

$$H(s) = B^T(sC + G)^{-1}B. \tag{16.14}$$

As a result of Modified Nodal Analysis (MNA) modeling the system has a well defined block structure [8] and we get

$$\begin{bmatrix} C_1 & 0 \\ 0 & C_2 \end{bmatrix}\dot{x} + \begin{bmatrix} G_1 & G_2 \\ -G_2^T & 0 \end{bmatrix} \quad x = \begin{bmatrix} B_1 \\ 0 \end{bmatrix} u, \qquad y = [B_1 \quad 0]x, \tag{16.15}$$

where G_1, C_1, C_2 are symmetric, $G_1, C_1 \geq 0$ (i.e., both matrices are positive semi-definite), and $C_2 > 0$, i.e., C_2 is positive definite. Moreover, as before, the matrix pencil $\lambda C + G$ is assumed to be regular. It is easy to see that under these assumptions, $H(s)$ is positive real and thus the system is passive [10].

Theorem 2 *Consider a passive system of the form* (16.15). *The ESVDMOR reduced system* (16.13) *is passive iff the inner reduction* (16.12) *is passivity preserving.*

Proof If we can show that $\widehat{H}_r(s)$ in (16.13) is positive real, we have shown that the reduced system is passive, see Definition 2. The moments of (16.15) are

$$\mathbf{m_i}(s_0) = B^T(-(s_0C + G)^{-1}C)^i(s_0C + G)^{-1}B, \tag{16.16}$$

with $0 \leq s_0 \in \mathbb{R}$ and $\det(s_0C + G) \neq 0$.

Following a technique which can be found, e.g., in [9] we define $J = \begin{bmatrix} I & 0 \\ 0 & -I. \end{bmatrix}$. The properties of G_1, C_1, and C_2 as well as the fact that $J = J^T$ and $J^2 = I$ lead to the following rules:

R1: $J = J^{-1}$,
R2: $JC = CJ$, hence $JCJ = C$,
R3: $(s_0C + G)^T = s_0C + JGJ = s_0JCJ + JGJ$, hence $(s_0C + G)^{-T} = (s_0JCJ + JGJ)^{-1} = J^{-1}(s_0C + G)^{-1}J^{-1} = J(s_0C + G)^{-1}J$,
R4: $B = JB$, and
R5: for every matrix X and Y, $(-X^{-1}Y)^i = X^{-1}(Y(-X)^{-1})^iX$ holds.

A straightforward calculation employing these rules shows that

$$\begin{aligned}
\mathbf{m_i}^T(s_0) &= (B^T(-(s_0C + G)^{-1}C)^i(s_0C + G)^{-1}B)^T \\
&= B^T(s_0C + G)^{-T}\{(-(s_0C + G)^{-1}C)^i\}^TB \\
&\overset{(R5)}{=} B^T(s_0C + G)^{-T}\{(s_0C + G)^{-1}(C(-(s_0C + G)^{-1}))^i(s_0C + G)\}^TB \\
&= B^T(s_0C + G)^{-T}(s_0C + G)^T\{(C(-(s_0C + G)^{-1}))^i\}^T(s_0C + G)^{-T}B \\
&= B^T\{(C(-(s_0C + G)^{-1}))^i\}^T(s_0C + G)^{-T}B \\
&\overset{(C=C^T)}{=} B^T((-(s_0C + G))^{-T}C)^i(s_0C + G)^{-T}B \\
&\overset{(R3)}{=} B^T(-J(s_0C + G)^{-1}JC)^i(J(s_0C + G)^{-1}J)B \\
&\overset{(R5)}{=} B^TJ(s_0C + G)^{-1}(JC(-J(s_0C + G)^{-1}))^i(s_0C + G)J(J(s_0C + G)^{-1}J)B
\end{aligned}$$

$$
\begin{aligned}
&\overset{(R1,R2)}{=} B^T J(s_0C+G)^{-1}(C(-(s_0C+G)^{-1}))^i(s_0C+G)(s_0C+G)^{-1}JB \\
&= B^T J(s_0C+G)^{-1}(C(-(s_0C+G)^{-1}))^i JB \\
&\overset{(R4)}{=} B^T(s_0C+G)^{-1}(C(-(s_0C+G)^{-1}))^i B \\
&\overset{(R5)}{=} B^T(-(s_0C+G)^{-1}C)^i(s_0C+G)^{-1}B \\
&= \mathbf{m_i}(s_0).
\end{aligned}
$$

It follows from (16.6) that $M_I = M_O$, such that

$$V_{I_{r_i}}^T = V_{O_{r_o}}^T = V_r^T \tag{16.17}$$

with the help of (16.7). It is easy to see that

$$\widehat{H}(s) = V_r B_r^T (G+sC)^{-1} B_r V_r^T, \tag{16.18}$$

with B_r analogously to (16.9). Hence, if the model reduction method used in (16.12) leads to a positive real transfer function of the reduced-order model, passivity is preserved. □

Remark 1 As a simple numerical example we show a RC circuit called *rc549*, which is also investigated in [21] and the Section by Antoulas/Lefteriu in this book. It is provided by the Infinoen Technologies AG, Munich, Germany. The order of the corresponding descriptor system is $n = 141$, the number of terminals is $m = 70$. Due to simplicity we use $H_{DC} = m_0(s=0)$ as basis for the matrices in (16.6) and we abstain from the reduction step in (16.12). Following Definition 2, we show that the Hermitian part of the transfer function on the imaginary axis is positive semi-definite, i.e., $H_H = H(j\omega) + H(j\omega)^* \geq 0$. Figure 16.1 shows the

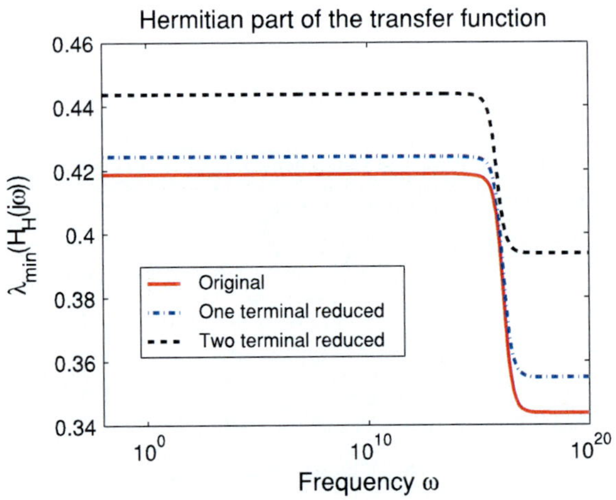

Fig. 16.1 Smallest eigenvalues of the hermitian part of the transfer function of *rc549* depending on the frequency ω

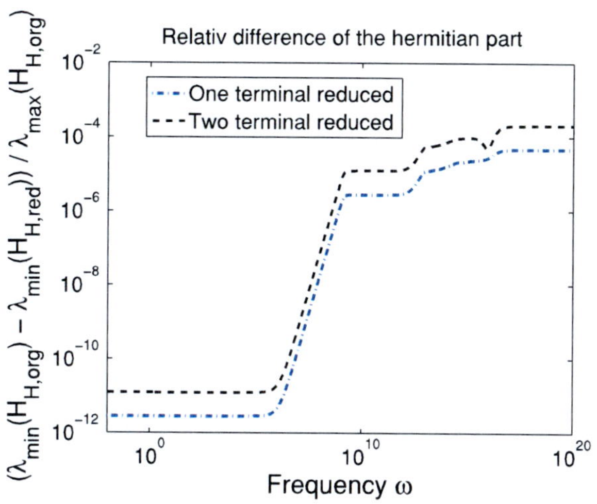

Fig. 16.2 Relative difference correlated to the largest eigenvalue of the original transfer function

smallest eigenvalue of H_H of the original transfer function and of the terminal reduced transfer functions $H_r(s)$. Although the frequency range is much larger than important in applications, this smallest eigenvalue grows depending on how much terminals are reduced. The relative difference of these smallest eigenvalues of the reduced systems to those of the original system is shown in Fig. 16.2. We recognize that this difference depends on the spectrum of the eigenvalues of $H(s)$ but does not play a too important role in the aspect of passivity preserving. Note, that we add numerically zero eigenvalues in (16.13) which do not destroy the positive semi-definiteness of the transfer function.

16.3.3 Reciprocity

Another important property of MOR methods is reciprocity preserving, which is a requirement for synthesis of the reduced order model as circuit. We assume the same setting as in Sect. 16.3.2. An appropriate definition can be found, e.g., in [19].

Definition 3 A transfer function (16.14) is *reciprocal* if there exists $m_1, m_2 \in \mathbb{N}$ with $m_1 + m_2 = m$, such that for $\Sigma_e = \operatorname{diag}(I_{m_1}, -I_{m_2})$ and all $s \in \mathbb{C}$ where $H(s)$ has no pole holds

$$H(s)\Sigma_e = \Sigma_e H^T(s).$$

The matrix Σ_e is called *external signature* of the system. A descriptor system is called reciprocal if its transfer function is reciprocal.

As a consequence, a transfer function of a reciprocal system has the form

$$H(s) = \begin{bmatrix} H_{11}(s) & H_{12}(s) \\ -H_{12}^T(s) & H_{22}(s) \end{bmatrix}, \tag{16.19}$$

where $H_{11}(s) = H_{11}^T(s) \in \mathbb{R}^{m_1,m_1}$ and $H_{22}(s) = H_{22}^T(s) \in \mathbb{R}^{m_2,m_2}$, which means that the transfer function is some kind of symmetric.

Theorem 3 *Consider a reciprocal system of the form* (16.15). *The ESVDMOR reduced system* (16.13) *is reciprocal iff the inner reduction* (16.12) *is reciprocity preserving.*

Proof Due to the reciprocity of the original system, the corresponding transfer function (16.14) has the structure given in (16.19). Equation 16.18 shows that none of the steps in ESVDMOR destroy this symmetric structure if (16.12) preserves reciprocity. □

16.4 Remarks and Outlook

We have reviewed the ideas of ESVDMOR and that this approach is stability, passivity, and reciprocity preserving under reasonable assumptions described in Sect. 16.3.2. As a direction of future research, it would be desirable to have a proof of passivity preservation for more general settings, e.g., a more general structure than in (16.15).

In [5] it is pointed out that an explicit computation of the moments in (16.3) would be numerically unstable and too expensive. In order to avoid this problem, a numerical solution by using the truncated SVD based on the implicitly restarted Arnoldi method is presented there. As an alternative, we also investigate the Jacobi–Davidson approach (JDSVD) [12] for computing the TSVD. In this context, the computation of the residual and the solution of the correction equation within JDSVD are based on iterative methods. This offers a large potential for increased efficiency in ESVDMOR due to the usual structures of circuit equations that can be exploited here.

An automatic error control which regulates the number of significant singular values as well as the best choice of the decomposition method and the number of used moments are other interesting topics for future research.

Acknowledgements The work reported in this paper was supported by the German Federal Ministry of Education and Research (BMBF), grant no. 03BEPAE1. We also want to thank our industrial project partners Infineon Technologies AG and NEC Laboratories Europe, IT Research Division, NEC Europe Ltd. Responsibility for the contents of this publication rests with the authors.

References

1. Anderson, B., Vongpanitlerd, S.: Network Analysis and Synthesis. Prentice Hall, Englewood Cliffs, New Jersey (1973)
2. Antoulas A.C.: Approximation of Large-Scale Dynamical Systems. SIAM Publications, Philadelphia, PA (2005)

3. Benner, P.: Advances in balancing-related model reduction for circuit simulation. In: Roos, J., Costa, L.R.J. (eds.) Scientific Computing in Electrical Engineering SCEE 2008, Mathematics in Industry, vol. 14. Springer-Verlag, Berlin pp. 469–482 (2010)
4. Benner, P., Mehrmann, V., Sorensen, D.: Dimension reduction of large-scale systems, Lecture Notes in Computational Science and Engineering, vol.45. Springer-Verlag, Berlin (2005)
5. Benner, P., Schneider, A.: Model order and terminal reduction approaches via matrix decomposition and low rank approximation. In: J. Roos, L.R.J. Costa (eds.) Scientific Computing in Electrical Engineering SCEE 2008, Mathematics in Industry, vol. 14. Springer-Verlag, Berlin pp. 523–530 (2010)
6. Feldmann, P., Liu, F.: Sparse and efficient reduced order modeling of linear subcircuits with large number of terminals. In: ICCAD '04: Proceedings of the 2004 IEEE/ACM International Conference on Computer-Aided Design, pp. 88–92. IEEE Computer Society, Washington, DC (2004)
7. Freund, R.W.: Passive reduced-order models for interconnect simulation and their computation via Krylov-subspace algorithms. In: DAC '99: Proceedings of the 36th Annual ACM/IEEE Design Automation Conference, pp. 195–200. ACM, New York, NY (1999)
8. Freund, R.W.: SPRIM: structure-preserving reduced-order interconnect macromodeling. In: ICCAD '04: Proceedings of the 2004 IEEE/ACM International Conference on Computer-aided Design, pp. 80–87. IEEE Computer Society, Washington, DC (2004)
9. Freund, R.W.: On Padé-type model order reduction of J-Hermitian linear dynamical systems. Linear Algebra and its Applications **429**(10), 2451 – 2464 (2008) Special Issue in honor of Richard S. Varga
10. Freund, R.W., Jarre, F.: An extension of the positive real lemma to descriptor systems. Optim. Methods Softw. **19**(1):69–87 (2004)
11. Günther, M., Feldmann, U., ter Maten, E.J.W.: Modelling and discretization of circuit problems. In: W.H.A. Schilders, E.J.W. ter Maten (eds.) Numerical Methods in Electromagnetics, Special Volume of Handbook of Numerical Analysis, vol. XIII, pp. 523–659. Elsevier Science BV, Amsterdam (2005)
12. Hochstenbach, M.E.: A Jacobi–Davidson type SVD method. SIAM J. Sci. Comput. **23**(2):606–628 (2001)
13. Lehoucq, R., Sorensen, D., Yang, C.: ARPACK user's guide. Solution of large-scale eigenvalue problems with implicitly restarted Arnoldi methods. Software—Environments—Tools, 6. p.142. SIAM, Society for Industrial and Applied Mathematics (1998)
14. Lin, Z., Carpenter, A., Ciftcioglu, B., Garg, A., Huang, M., Hui, W.: Injection-locked clocking: A low-power clock distribution scheme for high-performance microprocessors. IEEE Trans. VLSI Syst. **16**(9):1251–1256 (2008)
15. Liu, P., Tan, S.X.D., Li, H., Qi, Z., Kong, J., McGaughy, B., He, L.: An efficient method for terminal reduction of interconnect circuits considering delay variations. In: ICCAD '05: Proceedings of the 2005 IEEE/ACM International Conference on Computer-Aided Design, pp. 821–826. IEEE Computer Society, Washington, DC (2005)
16. Liu, P., Tan, S.X.D., Yan, B., Mcgaughy, B.: An extended SVD-based terminal and model order reduction algorithm. In: Proceedings of the 2006 IEEE International Behavioral Modeling and Simulation Workshop, pp. 44–49 (2006)
17. Liu, P., Tan, S.X.D., Yan, B., McGaughy, B.: An efficient terminal and model order reduction algorithm. Integr. VLSI J. **41**(2):210–218 (2008)
18. Penrose, R.: A generalized inverse for matrices. Mathematical Proceedings of the Cambridge Philosophical Society **51**(03):406–413 (1955). doi: 10.1017/S0305004100030401
19. Reis, T.: Circuit synthesis of passive descriptor systems—a modified nodal approach. Int. J. Circ. Theor. Appl. **38**(1):44–68 (2010)
20. Schilders, W., van der Vorst, H., Rommes, J.: Model Order Reduction: Theory, Research Aspects and Applications. Springer-Verlag, Berlin (2008)

21. Schneider, A.: Matrix decomposition based approaches for model order reduction of linear systems with a large number of terminals. Diplomarbeit, Chemnitz University of Technology, Faculty of Mathematics, Germany (2008)
22. Singh, J., Sapatnekar, S.: Congestion-aware topology optimization of structured power/ground networks. IEEE Trans. CAD **24**(5), 683–695 (2005)
23. Tan, S.X.D., He, L.: Advanced Model Order Reduction Techniques in VLSI Design. Cambridge University Press, New York (2007)
24. Tan, S.X.D., Shi, C.J.R., Lungeanu, D., Lee, J.C., Yuan, L.P.: Reliability-constrained area optimization of VLSI power/ground networks via sequence of linear programmings. In: Proceedings of the ACM/IEEE Design Automation Conference, pp. 78–83 (1999)

Chapter 17
Model Order Reduction of Nonlinear Systems in Circuit Simulation: Status and Applications

Michael Striebel and Joost Rommes

Abstract In this paper we review the status of existing techniques for nonlinear model order reduction by investigating how well these techniques perform for typical industrial needs. In particular the Trajectory Piecewise Linear-method and the Proper Orthogonal Decomposion approach are taken under consideration.

Keywords Model order reduction · Nonlinear systems · Proper orthogonal decomposition · Piece-wise linearisation

17.1 Introduction

The dynamics of electrical circuits can generally be described by a nonlinear, first order, differential-algebraic equation (DAE) of the form:

$$\frac{d}{dt}\mathbf{q}(\mathbf{x}(t)) + \mathbf{j}(\mathbf{x}(t)) + \mathbf{B}\mathbf{u}(t) = \mathbf{0};\ \mathbf{y}(t) = \mathbf{L}^T\mathbf{x}(t), \tag{17.1}$$

where $\mathbf{x}(t) \in \mathbb{R}^n$ represents the unknown vector of circuit variables at time $t \in \mathbb{R}$; $\mathbf{q}, \mathbf{j} : \mathbb{R}^n \to \mathbb{R}^n$ describe the contribution of reactive and nonreactive elements, respectively; $\mathbf{B} \in \mathbb{R}^{n\times m}$ distributes the input excitation $\mathbf{u} : \mathbb{R} \to \mathbb{R}^m$ and $\mathbf{L} \in \mathbb{R}^{n\times q}$ maps the state $\mathbf{x}$ to the system response $\mathbf{y}(t) \in \mathbb{R}^q$. We interpret (17.1) as

M. Striebel (✉)
Bergische Universität Wuppertal, Wuppertal, Germany
e-mail: striebel@math.uni-wuppertal.de

J. Rommes
NXP Semiconductors, Eindhoven, The Netherlands
e-mail: joost.rommes@nxp.com

P. Benner et al. (eds.), *Model Reduction for Circuit Simulation*,
Lecture Notes in Electrical Engineering, 74, DOI: 10.1007/978-94-007-0089-5_17,

description of a subcircuit. Thus the input $\mathbf{u}$ and the output $\mathbf{y}$ are terminal voltages and terminal currents that are injected and extracted linearly.

The dimension n of the unknown vector $\mathbf{x}(t)$ is of the order of the number of elements in the circuit, which can easily reach hundreds of millions. Therefore, one may solve the network equations (17.1) by means of computer algebra in an unreasonable amount of time only.

Model order reduction (MOR) aims to replace the original model (17.1) by a system

$$\frac{d}{dt}[\tilde{\mathbf{q}}(\mathbf{z}(t))] + \tilde{\mathbf{j}}(\mathbf{z}(t)) + \tilde{\mathbf{B}}\mathbf{u}(t) = \mathbf{0}; \quad \tilde{\mathbf{y}}(t) = \tilde{\mathbf{L}}^T\tilde{\mathbf{x}}(t), \tag{17.2}$$

with $\mathbf{z}(t) \in \mathbb{R}^r$; $\tilde{\mathbf{q}}, \tilde{\mathbf{j}} : \mathbb{R}^r \to \mathbb{R}^r$ and $\tilde{\mathbf{B}} \in \mathbb{R}^{r\times m}$ and $\tilde{\mathbf{L}} \in \mathbb{R}^{r\times q}$, which can compute a system response $\tilde{\mathbf{y}}(t) \in \mathbb{R}^q$ that is sufficiently close to $\mathbf{y}(t)$ given the same input signal $\mathbf{u}(t)$, but in much less time.

In the following we recall the fundamental problem that arises in nonlinear model order reduction. We overview different nonlinear model order reduction techniques and their field of application. For two of these methods, proper orthogonal decomposition (POD) [12, 14, 16] and trajectory piecewise linearization (TPWL) [15, 21] we will go more into details and report observations we made when applying them to different test examples.

17.2 Linear Versus Nonlinear Model Order Reduction

So far most research effort was spent on developing and analysing MOR techniques suitable for linear problems. For an overview on these methods we refer to [1].

When transferring approaches from linear MOR, especially projection based methods, fundamental differences emerge. Considering a linear system

$$\begin{aligned} \mathbf{C}\frac{d}{dt}\mathbf{x}(t) + \mathbf{G}\mathbf{x}(t) + \mathbf{B}\mathbf{u}(t) = 0, \quad &\text{with } \mathbf{C}, \mathbf{G} \in \mathbb{R}^{n\times n}, \\ \mathbf{y}(t) = \mathbf{L}^T\mathbf{x}(t), \quad & \end{aligned} \tag{17.3}$$

usually the state $\mathbf{x}(t)$ is approximated in a lower dimensional space of dimension $r \ll n$, spanned by basis vectors which we subsume in $\mathbf{V} = (\mathbf{v}_1, \ldots, \mathbf{v}_r) \in \mathbb{R}^{n\times r}$:

$$\mathbf{x}(t) \approx \mathbf{V}\mathbf{z}(t), \quad \text{with}\, \mathbf{z}(t) \in \mathbb{R}^r. \tag{17.4}$$

The reduced state $\mathbf{z}$ is defined by a reduced dynamical system that arises from projecting (17.3) on a test space spanned by the columns of $\mathbf{W}$. There, $\mathbf{W}$ and $\mathbf{V}$ are chosen biorthonormal, i.e., $\mathbf{W}^T\mathbf{V} = \mathbf{I}_{r\times r}$. The Galerkin projection[1] yields

[1] Most frequently $\mathbf{V}$ is constructed to be orthogonal, such that $\mathbf{W} = \mathbf{V}$ can be chosen.

$$\begin{aligned}\tilde{\mathbf{C}}\frac{d}{dt}\mathbf{z}(t) + \tilde{\mathbf{G}}\mathbf{z}(t) + \tilde{\mathbf{B}}\mathbf{u}(t) &= 0, \\ \mathbf{y}(t) &= \tilde{\mathbf{L}}^T\mathbf{z}(t),\end{aligned} \tag{17.5}$$

with $\tilde{\mathbf{C}} = \mathbf{W}^T\mathbf{C}\mathbf{V}$, $\tilde{\mathbf{G}} = \mathbf{W}^T\mathbf{G}\mathbf{V} \in \mathbb{R}^{r\times r}$ and $\tilde{\mathbf{B}} = \mathbf{W}^T\mathbf{B} \in \mathbb{R}^{r\times m}$, $\tilde{\mathbf{L}} = \mathbf{V}^T\mathbf{L} \in \mathbb{R}^{r\times p}$. The system matrices $\tilde{\mathbf{C}}, \tilde{\mathbf{G}}, \tilde{\mathbf{B}}, \tilde{\mathbf{L}}$ of this reduced substitute model are of smaller dimension and constant, i.e., need to be computed only once.

Applying the same technique directly to the nonlinear system means obtaining the reduced formulation (17.2) by defining $\tilde{\mathbf{q}}(\mathbf{z}) = \mathbf{W}^T\mathbf{q}(\mathbf{V}\mathbf{z})$ and $\tilde{\mathbf{j}}(\mathbf{z}) = \mathbf{W}^T\mathbf{j}(\mathbf{V}\mathbf{z})$. Clearly, $\tilde{\mathbf{q}}$ and $\tilde{\mathbf{j}}$ map from $\mathbb{R}^r$ to $\mathbb{R}^r$.

Multistep methods to solve (17.2) cause at each time point t_l a nonlinear equation

$$\alpha\tilde{\mathbf{q}}(\mathbf{z}_l) + \tilde{\beta} + \tilde{\mathbf{j}}(\mathbf{z}_l) + \tilde{\mathbf{B}}\mathbf{u}(t_l) = \mathbf{0}, \tag{17.6}$$

that defines $\mathbf{z}_l$ which is the approximation of $\mathbf{z}(t_l)$. In the above equation α is the integration coefficient of the method and $\tilde{\beta} \in \mathbb{R}^r$ contains history from previous timesteps. Newton techniques that are used to solve (17.6), usually require an update of the system's Jacobian matrix in each iteration ν:

$$\tilde{\mathbf{J}}_l^{(\nu)} = \left(\alpha\frac{\partial\tilde{\mathbf{q}}}{\partial\mathbf{z}} + \frac{\partial\tilde{\mathbf{j}}}{\partial\mathbf{z}}\right)\Bigg|_{\mathbf{z}=\mathbf{z}_l^{(\nu)}} = \mathbf{W}^T\left[\alpha\frac{\partial\mathbf{q}}{\partial\mathbf{x}} + \frac{\partial\mathbf{j}}{\partial\mathbf{x}}\right]\Bigg|_{\mathbf{x}^{(\nu)}=\mathbf{V}\mathbf{z}_l^{(\nu)}}\mathbf{V}. \tag{17.7}$$

The evaluation of the reduced system, i.e., $\tilde{\mathbf{q}}$ and $\tilde{\mathbf{j}}$, necessitates in each step the back projection of the argument $\mathbf{z}$ to its counterpart $\mathbf{V}\mathbf{z}$ followed by the evaluation of the functions $\mathbf{q}$ and $\mathbf{j}$ and the projection to the reduced space with $\mathbf{W}$ and $\mathbf{V}$.

Consequently, with respect to computation time no reduction will be obtained unless additional measures are taken or other strategies are pursued.

17.3 Some Nonlinear MOR Techniques

In MOR for linear systems especially methods based on Krylov subspaces [7] and balanced realization [13] are well understood and highly elaborated. In the following, we shortly describe how these approaches are adapted to nonlinear problems and give references for further reading.

17.3.1 Krylov Subspace Methods in Nonlinear MOR

In linear MOR, Krylov subspace methods construct order reduced models of the system (17.3) such that the moments of the transfer functions of full and reduced problem match up to a certain order. The transfer function $\mathbf{H} : \mathbb{C} \to \mathbb{C}^p$ is defined by the linear equation $\mathbf{H}(s) = \mathbf{L}^T(s\mathbf{C} + \mathbf{G})^{-1}\mathbf{B}$.

As there is no direct counterpart of the transfer function for the nonlinear problem (17.1), Krylov techniques are applied to systems derived from the general problem:

$$\text{bilinear system:} \quad \frac{d}{dt}\widehat{\mathbf{x}}(t) + \widehat{\mathbf{A}}\widehat{\mathbf{x}}(t) + \widehat{\mathbf{N}}\widehat{\mathbf{x}}(t)\mathbf{u}(t) + \widehat{\mathbf{B}}\mathbf{u}(t) = \mathbf{0}, \tag{17.8a}$$

$$\text{LPTV system:} \quad \frac{d}{dt}[\mathbf{C}(t)\mathbf{x}(t)] + \mathbf{G}(t)\mathbf{x}(t) + \mathbf{B}\mathbf{u}(t) = \mathbf{0}. \tag{17.8b}$$

The problem (17.8a) arises from expanding $\dot{\mathbf{x}}(t) + \mathbf{f}(\mathbf{x}(t)) + \mathbf{B}\mathbf{u}(t) = \mathbf{0}$ around an equilibrium point. Linear periodically time varying systems of type (17.8b), with matrices $\mathbf{C}(t), \mathbf{G}(t)$ that are periodic with period T, one gets when linearizing the system (17.1) around a periodic steady state solution with $\mathbf{x}^0(t + T) = \mathbf{x}^0(t)$.

For bilinear systems, Volterra-series expansion and multimoment expansions are the key to apply Krylov subspace based MOR. For further reading we refer to [6].

For LPTV systems, a time varying system function $\mathbf{H}(s, t)$ can be defined, where $s \in \mathbb{C}$ corresponds to the frequency of the input and $t \in [0, T)$ is used to capture the system's variation. The transfer function $\mathbf{H}$ is defined by a differential equation. Choosing an expansion point s, $\mathbf{H}(s, t)$ can be determined using time- or frequency samples on $[0, T)$. We refer to [9] and the references therein.

The techniques mentioned do not directly work on the nonlinear problem (17.1) of dimension n. The bilinear system that is reduced actually has a dimension of $n + n^2 + n^3 + \cdots$, depending on the order of expansion. Similar, the system in the LPTV case has dimension $p \cdot n$ with p being the number of time- or frequency-samples used to approximate $\mathbf{H}(s, t)$.

17.3.2 Balanced Truncation in Nonlinear MOR

The energy $L_c(\mathbf{x}_0)$ that is needed to drive a system to a given state $\mathbf{x}_0 \in \mathbb{R}^n$ and the energy $L_o(\mathbf{x}_0)$ the system provides to observe the state $\mathbf{x}_0$ it is in, are the main terms in balanced truncation. A system is called balanced if states that are hard to reach are also hard to observe and vice versa, i.e., $L_c(\mathbf{x})$ large implies $L_o(\mathbf{x})$ small. Truncation, i.e., reduced order modeling is then done by eliminating these states.

For linear problems L_c and L_o are connected directly to the reachability and observability Gramians $\mathbf{P}$ and $\mathbf{Q}$ that are computed from Lyapunov equations, involving the system matrices of the linear system (17.3). Balancing is reached by transforming the state space such that $\mathbf{P}$ and $\mathbf{Q}$ are simultaneously diagonalised. The entries $\sigma_1, \ldots, \sigma_n$ on the diagonal are called Hankel singular values. From the basis that arises from the transformation only those basis vectors that correspond to large Hankel singular values are kept. The main advantage of this approach is that there exists an a priori computable error bound for the truncated system.

In transferring balanced truncation to nonlinear problems, three main tracks can be recognized. Energy consideration is the common ground for the three directions.

In [8] the energy functions arise from solving Hamilton-Jacobi differential equations. Similar to the linear case, a, now state dependent, state-space transformation is derived such that L_c and L_o are formulated as quadratic form with diagonal matrix. The magnitude of the entries, the singular value functions, are then basis to truncation again. As the Hamilton-Jacobi system has to be solved and the varying state-space transformations have to be computed, it is an open issue, how the theory could be applied in a computer environment.

In sliding interval balancing [20], the nonlinear problem is first linearized around a nominal trajectory, giving a linear time varying system like (17.8b). At each state finite time reachability and observability Gramians are defined and approximated by truncated Taylor series expansion. Analytic calculations, basically the series expansions, connect the local balancing transformation smoothly. This necessary step is the limiting factor for this approach in circuit simulation.

Finally, balancing is also applied to bilinear systems (17.8b). Here the key tools are so called algebraic Gramians arising from generalised Lypunov equations. However, the algebraic Gramians can serve only to get bounds for L_c and L_o.

For further details we refer to [3, 5] and the references therein.

17.4 TPWL and POD

In view of high dimensional problems in circuit simulation and feasability in a computational environment, trajectory piecewise linearization (TPWL) [15] and proper orthogonal decomposition (POD) [14] are the most promising approaches for the time being.

17.4.1 Trajectory Piecewise Linearization

The idea of TPWL [15] is to reproduce the typical behaviour of the full nonlinear system (17.1) by a varying combination of a set of order reduced linear models.

For this purpose a training input $\mathbf{u}(t)$ for $t \in [t_0, t_e]$ is chosen that drives the system into typical states, i.e., situations. A transient simulation with the chosen input yields a collection of points $\bar{\mathbf{x}}_0, \ldots, \bar{\mathbf{x}}_N$, approximating the trajectory $\mathbf{x}(t_i)$ at timepoints $t_0 < t_1 < \cdots < t_N = t_e$. On the trajectory, points $\{\mathbf{x}_1^{\text{lin}}, \ldots, \mathbf{x}_s^{\text{lin}}\} \subset \{\bar{\mathbf{x}}_0, \ldots, \bar{\mathbf{x}}_N\}$ are chosen around which the nonlinear functions $\mathbf{q}$ and $\mathbf{j}$ are linearized. The linear models, that are all of dimension n, are reduced individually. This delivers local reduced subspaces. A common subspace of dimension $r \ll n$, which is represented by the columns of $\mathbf{V} \in \mathbb{R}^{n \times r}$, is constructed that describes the primary information of all local subspaces. All linear models are projected on this space. Finally a mechanism, a weighting $w_i(\mathbf{Vz}) \in [0, 1]$ for $i = 1, \ldots, s$ is introduced to decide which linear submodel is valid in a certain situation. And the full system is replaced by

$$\sum_{i=0}^{s} w_i(\mathbf{Vz}) \left[\mathbf{V}^T \mathbf{C}_i \mathbf{V} \frac{d}{dt} \mathbf{z} + \mathbf{V}^T \mathbf{G}_i \mathbf{Vz} + \mathbf{V}^T (\mathbf{j}(\mathbf{x}_i^{\text{lin}}) - \mathbf{G}_i \mathbf{x}_i^{\text{lin}}) \right] + \mathbf{V}^T \mathbf{Bu}(t) = \mathbf{0}, \tag{17.9}$$

that determines $\mathbf{z} \in \mathbb{R}^r$ such that $\mathbf{Vz} \approx \mathbf{x}$. Here $\mathbf{C}_i = \frac{\partial \mathbf{q}}{\partial \mathbf{x}}\Big|_{\mathbf{x}=\mathbf{x}_i^{\text{lin}}}$ and $\mathbf{G}_i = \frac{\partial \mathbf{j}}{\partial \mathbf{x}}\Big|_{\mathbf{x}=\mathbf{x}_i^{\text{lin}}}$.

The formulation (17.9) is taken from [21]. We find a slightly different setup in [15].

Selection of linearization points: The first crucial point in TPWL is to decide, when to add a linear substitute for the nonlinear problem automatically during the training simulation. Too many linear models could make the final model slow, too few could make it inaccurate.

In literature we find two different strategies to decide upon adding a new model:

- In [15] the author suggests to check at each accepted timepoint t during simulation for the relative distance of the current state $\bar{\mathbf{x}}$ to all yet existing i linearization states $\mathbf{x}_1^{\text{lin}}, \ldots, \mathbf{x}_i^{\text{lin}}$. If the minimum exceeds the bound $\delta > 0$, i.e.,

$$\min_{1 \le j \le i} \left(\frac{\|\bar{\mathbf{x}} - \mathbf{x}_j^{\text{lin}}\|}{\|\mathbf{x}_j^{\text{lin}}\|} \right) \ge \delta, \tag{17.10}$$

  $\bar{\mathbf{x}}$ becomes the $(i + 1)$st linearization point. Accordingly, a new linear model, arising from linearizing around $\bar{\mathbf{x}}$ is added to the collection.
- In [21] the mismatch of nonlinear and linear system motivates the creation of a new linearization point and an additional linear model: during training both the nonlinear and the lastly constructed linear system are computed in parallel. If the difference of the two approximations to the true solution at a timepoint t_n becomes too large, a new linear model is created from linearizing the nonlinear system around the state the system was in at the previous timepoint t_{n-1}.

Reducing the linear models: Basically, any linear MOR technique can be applied to the s linear systems. In [15] e.g., a Krylov-subspace method is used. However, the effect of reducing linear submodels on the overall accuracy in time domain simulation and computational work is an open question. Also, the additional source terms $\mathbf{j}(\mathbf{x}_i^{\text{lin}}) - \mathbf{G}_i \mathbf{x}_i^{\text{lin}}$ in (17.9) and observations from experiments (see Sect. 17.5) give rise to the question, if and how smoothness of the solution can be guaranteed.

Determination of the weights: After having done the training, a set of linear, reduced, models are available to replace the nonlinear problem. In (17.9) this is done by a convex sum of the models at hand. To have minimum complexity one aims at having to deal with a combination of just a small number of linear submodels. Hence, the weighting function is chosen such that only a few (in the ideal case just one) dominant linear model is chosen:

$$w_i(\mathbf{x}) = e^{-\frac{\beta}{m}\cdot\|\mathbf{x}-\mathbf{x}_i^{\text{lin}}\|}, \quad \text{for } i = 1, \ldots, s, \text{ with } m = \min_{1\leq j\leq s} \|\mathbf{x} - \mathbf{x}_j^{\text{lin}}\|. \tag{17.11}$$

The constant β adjusts how abrupt the change of models is. A typical value is $\beta = 25$. To guarantee a convex combination, the weights are normalized such that $\sum_i w_i(\mathbf{x}) = 1$.

17.4.2 Proper Orthogonal Decomposition and Adaptions

Similar to TPWL, a training input $\bar{\mathbf{u}}$ is chosen and a transient simulation is run. During this numerical benchmark integration of the nonlinear problem (17.1) on the training interval $[t_0, t_e]$ K snapshots of the state $\mathbf{x}(t)$ are collected in a snapshot matrix

$$\mathbf{X} = (\bar{\mathbf{x}}_1, \ldots, \bar{\mathbf{x}}_K) \in \mathbb{R}^{n\times K}. \tag{17.12}$$

The snapshots, i.e., the columns of $\mathbf{X}$, span a space of dimension $k \leq K$. We search for an orthonormal basis $\{\mathbf{v}_1, \ldots, \mathbf{v}_k\}$ of this space that is optimal in the sense that the time-averaged error that is made when the snapshots are expanded in the space spanned by just $r<k$ basis vectors to $\tilde{\mathbf{x}}_{r,i}$,

$$\langle\|\mathbf{x} - \tilde{\mathbf{x}}_r\|_2^2\rangle \quad \text{with the averaging operator} \quad \langle\mathbf{f}\rangle = \frac{1}{K}\sum_{i=1}^{K}\mathbf{f}_i \tag{17.13}$$

is minimized. This least squares problem is solved by computing the eigenvalue decomposition of the state covariance matrix $\frac{1}{K}\mathbf{X}\mathbf{X}^T$ or, equivalently by the singular value decomposition (SVD) of the snapshot matrix

$$\mathbf{X} = \mathbf{UST} \quad \text{with} \quad \mathbf{U} \in \mathbb{R}^{n\times n}, \mathbf{T} \in \mathbb{R}^{K\times K} \quad \text{and} \quad \mathbf{S} = \left(\begin{array}{ccc|c} \sigma_1 & & & \\ & \ddots & & \mathbf{0}_{n\times(K-n)} \\ & & \sigma_n & \end{array}\right), \tag{17.14}$$

where $\mathbf{U}$ and $\mathbf{T}$ are orthogonal and the singular values satisfy $\sigma_1 \geq \sigma_2 \geq \cdots \sigma_n \geq 0$. Note, that we assumed $K > n$ in (17.14).

The matrix $\mathbf{V} \in \mathbb{R}^{n\times r}$ whose columns span the reduced subspace is now build from the first r columns of $\mathbf{U}$, where the truncation r is chosen based on energy considerations [14].

For $\mathbf{V}$, it holds $\mathbf{V}^T\mathbf{V} = \mathbf{I}_{r\times r}$. Therefore, Galerkin projection as described above can be applied to create a reduced system (17.2). However, as mentioned in Sect. 17.2 the cost for evaluating the nonlinear functions $\mathbf{q}$, $\mathbf{j}$ is not reduced.

Missing point estimation (MPE) [2], adapted POD [19], the reduced-basis method [10] and Discrete Empirical Interpolation Method (DEIM) [4] offer techniques to overcome this problem. Basically, besides constructing the r dimensional subspace of $\mathbb{R}^n$ spanned by the columns of $\mathbf{V}$, the number n of rows is reduced, i.e., $\mathbf{V} \in \mathbb{R}^{n\times r}$ is replaced by $\mathbf{P}_g^T\mathbf{V} = \bar{\mathbf{V}} =\in \mathbb{R}^{g\times r}$ with a selector matrix $\mathbf{P}_g \in \{1,0\}^{n\times g}$. In this way, the n-dimensional function $\tilde{\mathbf{q}}$ (similar $\tilde{\mathbf{j}}$) is replaced by some $\mathbf{P}_g^T\tilde{\mathbf{q}} = \bar{\mathbf{q}} : \mathbf{R}^n \to \mathbf{R}^g$.

17.5 Numerical Examples

For testing purposes, a time-simulator, has been implemented in `octave`. The underlying DAE integration scheme used here is CHORAL [11], a Rosenbrock-Wanner type of method, adapted to circuitry problems.[2]

Our main questions for investigating MOR techniques for nonlinear problems, intending to use these approaches on an industrial scale, are:

- How sensitive is the reduction w.r.t. heuristics?
- How sensitive are reduced order models w.r.t. to changing input signals?
- Are there classes of circuits where the reduction does not work?

We stress that we do not intend to let the methods appear in bad light but rather put our results to discussion in order to find explanations and improvements. The authors are open for discussion on the presented work.

17.5.1 Testcases

The nonlinear transmission line in Fig. 17.1 is taken from [15]. It has dimension 100 and contains nonlinear diodes whose currents are modelled by $\imath_d(v) = \exp(40\cdot v) - 1$, where v is the voltage drop between the diodes' terminals.

The inverter chain depicted in Fig. 17.2 is taken from [21]. It has dimension 104 with a transistor whose drain-source current is modelled by $\imath_{\mathrm{ds}} = 2\cdot 10^{-4}\cdot \max(u_g - u_s - 1.0, 0)^2 - \max(u_g - u_d - 1.0, 0)^2$, where u_g, u_d, u_s denote the gate-, drain- and source-voltage, respectively.

2 Similar results are obtained from a matlab-implementation using ode15s as integration scheme.

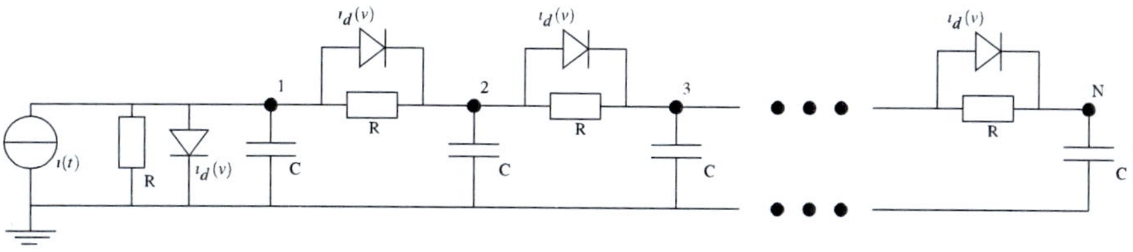

Fig. 17.1 Nonlinear transmission line

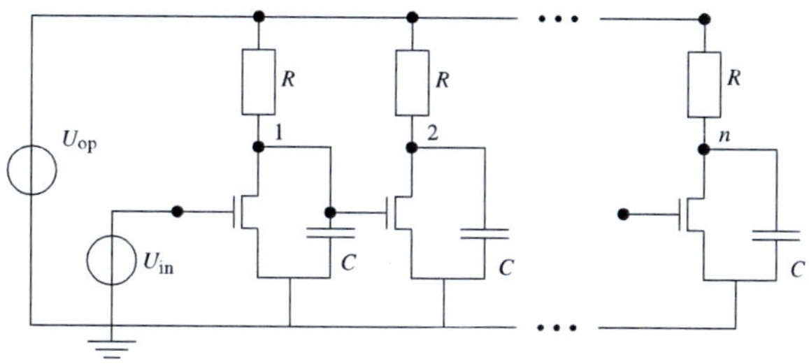

Fig. 17.2 Inverter chain [21]

17.5.2 Testruns

Nonlinear transmission line: As in [15] for extracting a model a shifted Heaviside function was used as training input. Resimulation is done both with the training input and with a cos-function on the interval $[t_{start}, t_{end}] = [0, 10]$:

$$u_{train}(t) = H(t-3) = \begin{cases} 0 & t<3 \\ 1 & t\geq 3 \end{cases}, \quad u_{resim}(t) = \frac{1}{2}\left(1 + \cos\left(\frac{2\pi}{10}t\right)\right).$$

The TPWL-model was extracted with the Arnoldi-method as suggested in [15], leading to a order reduced model of dimension 10. For choosing linearization points, the strategy (17.10) with $\delta = 0.00167$ has been tested. With this setting, 20 linear models are constructed. Also the extended strategy described in [21] is implemented, but does not show much different results for the transmission line. A more detailed discussion on the model extraction and statistics on which models are chosen can be found in [18].

Concerning the first two questions posed in the beginning of this section, i.e., the influence of heuristics and the impact of changing input signals, we turn to the graphs shown in Fig. 17.3. On the left part we see what happens when we feed the TPWL-model (17.9) with the 20 models just created and weights being determined according to (17.11) applying both the maximum norm and the Euclidean norm. In the top graph we see that in neither case the trajectory of the voltage at node 1 can follow the true solution correctly. The kinks, evident especially when taking the 2-norm for measuring the distance, suggest that there are some difficulties with choosing the weight, i.e., switching the responsible substitute model. Indeed, as the

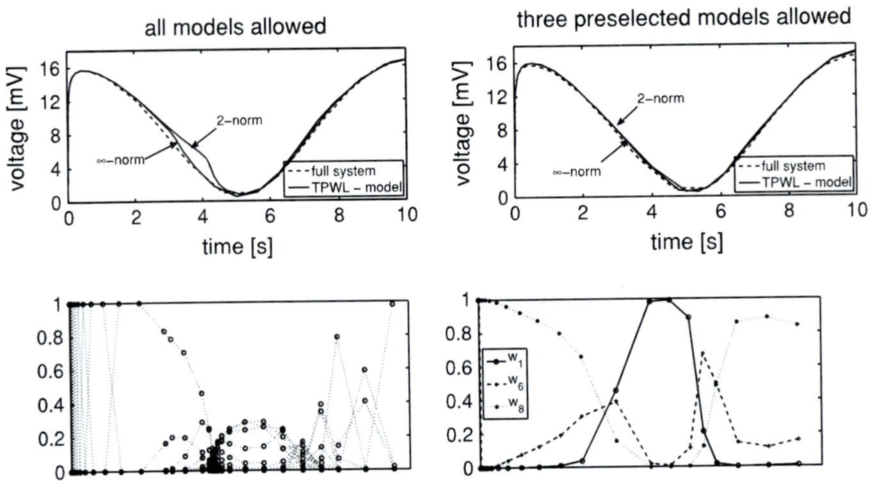

Fig. 17.3 Transmission line: different settings—trajectories (*top*) and weights (*bottom*)

lower picture shows, the importance of models is fluttering and unlike predicted we have to deal with several models of equivalent importance in (17.9).

The right column of Fig. 17.3 shows trajectories that are closer to the reference solution. This improvement was achieved by letting the TPWL model (17.9) choose from not the full set of 20 submodels but only three of them. However, this tuning[3] was done manually, i.e., testing different settings. Using different models or increasing the number of models to choose from, steers the system again to the situation depicted on the left of Fig. 17.3. We note that by changing δ in (17.10), a smaller number of linear submodels is constructed but these were not able to represent the full problem sufficiently accurate.

By the observations made, TPWL appears to be very sensitive to heuristics and implementation details.

Also a POD model as well as a POD model that has been modified with the DEIM algorithm [4] has been extracted. In both cases the state-space was reduced to dimension 4. Applying the DEIM algorithm $\mathbf{q}$ and $\mathbf{j}$ of dimension 100 where replaced by $\bar{\mathbf{q}}, \bar{\mathbf{j}}$ of dimension 5. We abandon to show a corresponding graph for the POD/POD-DEIM case as for node 1 the trajectories cannot be distinguished by inspection.

Table 17.1 gathers the performance of the models, measured in time consumption and L_2-norm of the error of the reconstruction of the node voltage at node 1 and, exemplary, at node 50. As predicted in Sect. 17.2, the plain POD-projection creates a reduced model that is even more expensive than the full system. The DEIM adapted POD model is superior to TPWL as no decision about

[3] The models used arise from linearization around the states the system was in during training at $t \in \{0.0, 3.01943, 3.04244\}$.

Table 17.1 Transmission line: performance of nonlinear MOR techniques

	Time for simulation with cos input [s]	L_2-norm of error reconstructing node 1 [10^{-4}]	node 50 [10^{-4}]
Full problem	4.16		
TPWL all models, 2- (inf-) norm	3.79 (3.23)	26.50 (10.72)	0.76 (0.29)
TPWL preselected, 2- (inf-) norm	2.88 (3.28)	10.86 (7.86)	0.44 (0.47)
POD model	4.34	3.68	9.18
POD–DEIM model	1.47	3.76	9.82

which model to use has to be made. Concerning the errors, the third column reflects the observations from Fig. 17.3. However, being interested in a different output, e.g., the voltage at node 50, the rating of the approaches can change and the TPWL- are superior to the POD-models. Note, that here the output, i.e., **L** in (17.1), is not taken into account in reduction, neither for TPWL nor POD.

Inverter chain: As suggested in [21], a piecewise linear voltage source

$$U_{\text{in}} = u(t) \text{ with } u(0) = 0, u(5\text{ns}) = 0, u(10\text{ns}) = 5, u(15\text{ns}) = 5, u(17\text{ns}) = 0.$$

was used for training purposes. Resimulation is done with both the training signal and a signal where the input impulse repeats after each 30 ns. From Fig. 17.4 it is obvious that both TPWL- and POD-model can not accurately reproduce the behaviour of the nonlinear system. The trajectories produced by both models—TPWL reduced to dimension 50, POD to dimension 28—wiggle notably. And the TPWL-model even misses every other impulse running through the system.

The inadequacy of the presented MOR techniques for the inverter chain becomes also obvious from the performance of the approaches in comparison to the work necessary for treating the full problem as displayed in Table 17.2.

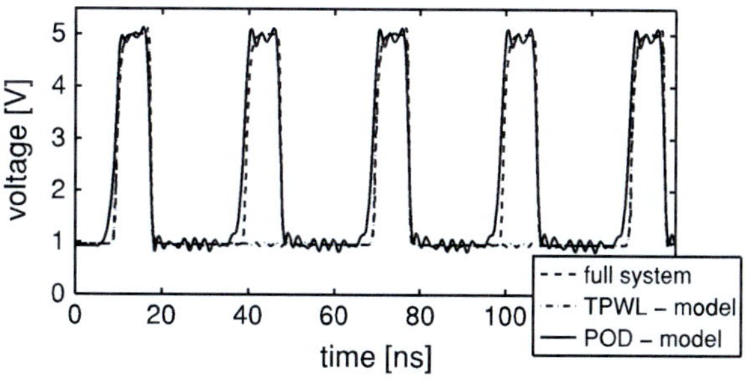

Fig. 17.4 Inverter chain: repeated pulse, inverter 6

Table 17.2 Inverter chain: performance of nonlinear MOR techniques

	Repeated impulse	Accepted (rejected) timesteps
Full problem	677.5 s	1074(182)
TPWL model	510.8 s	1437(824)
POD model	829.4 s	1276(104)

The reason for this behaviour might be found in the class of system the inverter chain belongs to. Here, a signal passes through the system activating at each time point a different set of elements, i.e., states and leaving the others untouched until they are activated. Hence, we are concerned rather with a redundancy in time than one in space. Here multirate methods (e.g., [17]) might be used to accelerate computation.

17.6 Discussion and Outlook

Although a variety of theoretically appealing MOR techniques for nonlinear problems exists, TPWL and POD seem to be the most practicable approaches. However, as our investigations show, care has to be taken when these techniques are being applied, as the approaches seem to be very sensitive to various heuristics. Furthermore, as seen with the inverter chain there are circuit problems where model order reduction is not viable. For TPWL there seems to be space for improvement in selecting linearization points and combining the submodels. Here it could be especially interesting to investigate the impact of the sharp, non-smooth, switching of the submodels on the smoothness of the final solution. The POD adaptions, originating from MOR for PDEs, deserve closer attention to adapt the ideas suitable for specially structured right-hand-sides to the rather complicated mixed structures in circuit problems.

Acknowledgements The work presented is funded by the Marie-Curie Transfer-of-Knowledge project O-MOORE-NICE! under call Identifier FP6-2005-Mobility-3.

References

1. Antoulas, A.C.: Approximation of Large-Scale Dynamical Systems. SIAM, Philadelphia (2005)
2. Astrid, P., Weiland, S., Willcox, K., Backx, T.: Missing point estimation in models described by proper orthogonal decomposition. IEEE Trans. Automat. Contr. **53**(10), 2237–2251 (2008)
3. Benner, P., Damm, T.: Lyapunov equations, energy functionals and model order reduction. Submitted for publication
4. Chaturantabut, C., Sorensen, D.C., Rice, U.: Discrete empirical interpolation for nonlinear model reduction. Tech. Rep. TR09–05, CAAM (2009)
5. Condon, M., Ivanov, R.: Nonlinear systems—algebraic gramians and model reduction. COMPEL: Int J Comput Mathe Elect Electron Eng **24**(1), 202–219 (2005)
6. Condon, M., Ivanov, R.: Krylov subspaces from bilinear representations of nonlinear systems. COMPEL: Int. J. Comput. Math. Electr. Electron. Eng. **26**(2), 399–406 (2007)
7. Freund, R.W.: Krylov-subspace methods for reduced-order modeling in circuit simulation. J. Comput. Appl. Math. **123**(1–2), 395–421 (2000)
8. Fujimoto, K., Scherpen, J.M.A.: Singular value analysis and balanced realizations for nonlinear systems. In: Schilders, W.H.A., van der Vorst, H.A., Rommes, J. (eds.) Model Order Reduction: Theory, Research Aspects and Applications. pp. 251–272 Springer, Berlin (2008)

9. Gad, E., Nakhla, M.: Efficient model reduction of linear periodically time-varying systems via compressed transient system function. IEEE Trans. Circ. Syst. **52**(6), 1188–1204 (2005)
10. Grepl, M., Maday, Y., Nguyen, N., Patera, A.: Efficient reduced-basis treatment of nonaffine and nonlinear partial differential equations. Math. Model. Numer. Anal. **41**(3), 575–605 (2007)
11. Günther, M.: Simulating digital circuits numerically—a charge-oriented ROW approach. Numer. Math. **79**, 203–212 (1998)
12. Holmes, P., Lumley, J.L., Berkooz, G.: Turbulence, Coherent Structures, Dynamical Systems and Symmetry. Cambridge Monographs on Mechanics. Cambridge University Press, Cambridge (1996)
13. Moore, B.: Principal component analysis in linear systems: controllability, observability, and model reduction. IEEE Trans. Automat. Contr. **26**(1), 17–32 (1981)
14. Pinnau, R.: Model reduction via proper orthogonal decomposition. In: Schilders, W., van der Vorst, H., Rommes, J. (eds.) Model Order Reduction: Theory, Applications, and Research Aspects, pp. 95–109. Springer, Berlin (2008)
15. Rewieński, M.J.: A trajectory piecewise-linear approach to model order reduction of nonlinear dynamical systems. Ph.D. thesis, Massachusetts Institute of Technology (2003)
16. Sirovich, L.: Turbulence and the dynamics of coherent structures part I-III. Q. App. Math. **45**(3), 561–571, 573–590 (1987)
17. Striebel, M., Bartel, A., Günter, M.: A multirate ROW-scheme for index-1 network equations. Appl. Numer. Math. **59**(3–4), 800–814 (2009)
18. Striebel, M., Rommes, J.: Model order reduction of nonlinear systems in circuit simulation: status, open issues, and applications. CSC Preprint 08-07, Chemnitz University of Technology, http://www.tu-chemnitz.de/mathematik/csc/index.php (2008)
19. Verhoeven, A., ter Maten, J., Striebel, M., Mattheij, R.: Model order reduction for nonlinear IC models. CASA-Report 07–41, TU Eindhoven (2007). To appear in IFIP2007 conference proceedings
20. Verriest, E.I.: Time variant balancing and nonlinear balanced realizations. In: Schilders, W.H.A., van der Vorst, H.A., Rommes, J. (eds.) Model Order Reduction: Theory, Research Aspects and Applications. pp. 243–250 Springer Berlin (2008)
21. Voß, T., Pulch, R., ter Maten, J., El Guennouni, A.: Trajector piecewise linear approach for nonlinear differential-algebraic equations in circuit simulation. In: Ciuprina, G., Ioan, D. (eds.) Scientific Computing in Electrical Engineering—SCEE 2006, pp. 167–173. Springer (2007)

Chapter 18
An Approach to Nonlinear Balancing and MOR

Erik I. Verriest

Abstract For linear systems we recall the notion of sliding interval balancing (SIB). It is shown that subsystems conserve stability. The truncated SIB realization bounds the optimal approximation in an input–output sense. Nonlinear balancing is approached by applying SIB on the linear variational system.

Keywords Balanced realization · Gramians · Lyapunov equation · Model reduction

18.1 Static Versus Dynamic Approximation

The problem considered in this chapter is the reduction of a system represented in state space form by $\dot{x} = f(x, u)$ and output $y = h(x)$ by a system of lower dimension, say $\dot{\xi} = \widehat{f}(\xi, u)$, and output given by $\widehat{h}(\xi)$. We get some queues from the approximation of *parameterized space curves*, where each fixed $\omega \in \Omega$, $\phi(\cdot, \omega) : [0, 1] \rightarrow \mathbb{R}^n$ corresponds to a curve in $\mathbb{R}^n$ with line coordinate t. If one considers all curves in Ω, according to some measure μ, the problem amounts to finding a submanifold M of suitable dimension $p < n$ such that $\hat{\phi}(\cdot, \cdot) \in M$ while minimizing

$$\int_0^1 \int_{\omega\in\Omega} \|\phi(t, \omega) - \hat{\phi}(t, \omega)\|^2 \, \mathrm{d}\mu(\omega) \, \mathrm{d}t. \tag{18.1}$$

If M is a subspace, this is known as the Galerkin projection method. For linear dynamical systems, it has been proposed to approximate the response of the state

E. I. Verriest (✉)
Georgia Institute of Technology, School of ECE, Atlanta, GA 30332-0250, USA
e-mail: erik.verriest@ece.gatech.edu

P. Benner et al. (eds.), *Model Reduction for Circuit Simulation*,
Lecture Notes in Electrical Engineering, 74, DOI: 10.1007/978-94-007-0089-5_18,

to an impulse input in the above sense. Clearly, this neglects any state to output properties the system may have. See for instance [1, 6]. In order to obtain a better balance between the input-to-state and state-to-output properties of a system, the notion of a balanced realization was introduced by Moore [3].

18.2 Gramians for Linear Systems and Applications

In this section we first derive some structural properties for analytic time varying systems, and prove a stability conservation property. These properties will be utilized in what follows to derive structural properties of smooth nonlinear systems via the linear perturbational system along a nominal trajectory, as this perturbational system is time varying in general.

18.2.1 Analytic Time Varying Systems

Let $\delta > 0$ be given, and consider a linear time varying system $(A(t), B(t), C(t))$. The finite time sliding interval reachability and observability Gramians are defined by

$$\mathcal{R}(t,\delta) = \int_{t-\delta}^{t} \Phi(t,\tau)B(\tau)B'(\tau)\Phi'(t,\tau)\mathrm{d}\tau. \tag{18.2}$$

$$\mathcal{O}(t,\delta) = \int_{t}^{t+\delta} \Phi'(\tau,t)C'(\tau)C(\tau)\Phi(\tau,t)\mathrm{d}\tau. \tag{18.3}$$

where $\Phi(t,\tau)$ is the transition matrix, satisfying for all t and $\tau, \frac{\partial}{\partial t}\Phi(t,\tau) = A(t)\Phi(t,\tau), \Phi(\tau,\tau) = I$. The Gramians respectively satisfy the time variant Lyapunov equations (where $\tau_+ = t+\delta$ and $\tau_- = t-\delta$)

$$\begin{aligned}\dot{\mathcal{R}}(t,\delta) =& A(t)\mathcal{R}(t,\delta) + \mathcal{R}(t,\delta)A'(t) + B(t)B'(t) - \Phi(t,\tau_-)B(\tau_-)B'(\tau_-)\Phi'(t,\tau_-)\\ -\dot{\mathcal{O}}(t,\delta) =& A'(t)\mathcal{O}(t,\delta) + \mathcal{O}(t,\delta)A(t) + C'(t)C(t) - \Phi'(\tau_+,t)C'(\tau_+)C(\tau_+)\Phi(\tau_+,t).\end{aligned}$$

For *analytic* systems, we have the convergent Taylor series

$$\Phi(t,\tau)B(\tau) = \sum_{i=0}^{\infty} \frac{(t-\tau)^i}{i!}\left[(A-\mathbf{D})^i B\right]_t \tag{18.4}$$

$$\Phi'(\tau,t)C'(\tau) = \sum_{i=0}^{\infty} \frac{(\tau-t)^i}{i!}\left[(A'+\mathbf{D})^i C'\right]_t \tag{18.5}$$

where $\mathbf{D}$ is the differentiation operator (acting to the right).

Defining local reachability and observability matrices:

$$\mathbf{R}_\infty(t) = [B, (A - \mathbf{D})B, (A - \mathbf{D})^2 B, \ldots]_t \tag{18.6}$$

$$\mathbf{O}'_\infty(t) = [C', (A' + \mathbf{D})C', (A' + \mathbf{D})^2 C', \ldots]_t \tag{18.7}$$

the Gramians can be computed via termwise integration, and by convergence of the Taylor series (for analytic systems) approximated by their truncation:

$$\mathcal{R}(t-\delta, t) = \mathbf{R}_\infty(t)\Delta_m(\delta)\mathbf{R}'_\infty(t) = \mathbf{R}_n(t)\Delta_m(\delta)\mathbf{R}'_n(t) + \mathrm{O}(\delta^{n+1}) \tag{18.8}$$

$$\mathcal{O}(t, t+\delta) = \mathbf{O}'_\infty(t)\Delta_p(\delta)\mathbf{O}_\infty(t) = \mathbf{O}'_n(t)\Delta_p(\delta)\mathbf{O}_n(t) + \mathrm{O}(\delta^{n+1}) \tag{18.9}$$

where $\Delta_m(\cdot)$ is a block matrix, with ijth block

$$[\Delta_m(\delta)]_{ij} = \frac{\delta^{i+j-1}}{(i-1)!(j-1)!(i+j-1)} I_m. \tag{18.10}$$

These Gramians are weighting matrices for energies and uncertainties [7].

18.2.2 Sliding Interval Balancing

In order to obtain a uniform measure for the relative importance of subspaces (w.r.t. input and output), a state space transformation $T(\cdot)$ is found such that the new input to state and state to output maps are symmetric, i.e., $\mathcal{R}_t = \mathcal{O}_t$ and both are diagonal. The rationale behind this is that in balanced coordinates, for each coordinate direction (more general, for each subspace) the degrees of reachability and observability are equal. Using the fact that the Gramians transform under the similarity $T(\cdot)$ as

$$\mathcal{R}_t \to T(t)\mathcal{R}_t T(t)' \tag{18.11}$$

$$\mathcal{O}_t \to T(t)^{-T}\mathcal{O}_t T(t)^{-1} \tag{18.12}$$

it was shown that, for analytic systems, a diffeomorphic transformation $T(\cdot)$, pointwise defined, exists that brings the original time varying system to the balanced form. The resulting realization is called *sliding interval balanced* (SIB) and the common Gramian, $\Lambda(t) = \operatorname{diag}[\lambda_1(t), \ldots, \lambda_n(t)]$, with $\lambda_i(t) > \lambda_{i+1}(t)$, the *canonical Gramian*. Note that the $\lambda_i^2(t)$ are the eigenvalues of the product $\mathcal{O}_t\mathcal{R}_t$. Because of the requisite ordering, this can only be described piecewise if *umbilic* crossover points exist. For details, see [8, 13].

When balancing was first introduced by Moore (1980), only LTI realizations were considered, and balancing in the sense of Moore corresponds to SIB for $\delta \to \infty$, *asymptotic stability* being implied. A balanced realization is obtained (it is not unique) by determining a T_{bal} which simultaneously diagonalizes and equalizes the Gramian matrices which satisfy the *algebraic Lyapunov equations*

$$A\mathcal{R} + \mathcal{R}A' + BB' = 0 \tag{18.13}$$

$$A'\mathcal{R} + \mathcal{O}A + C'C = 0. \tag{18.14}$$

18.2.3 Properties of SIB(δ) for LTI Systems

It was shown in [11] that for siso systems, the SIB(δ) realizations are sign symmetric, i.e. for some signature matrix, S, we have $A' = SAS$, $Sb = c'$. Conversely, any sign symmetric realization is SIB. Likewise, any SIB is M-balanced, which implies also a time weighted measure for energies and/or uncertainties. A system with reachability and observability matrix, respectively $\mathbf{R}$ and $\mathbf{O}$, is M-balanced, where $M = M' > 0$ if the M-Gramian matrices $\mathcal{R}_M = \mathbf{R}M\mathbf{R}'$ and $\mathcal{O} = \mathbf{O}'M\mathbf{O}$ are equal and diagonal [11].

Once a balanced realization is determined, a reduced order model is determined by *projection of dynamics* (POD). This method is known as balanced truncation.

$$\left(\begin{matrix}\begin{bmatrix} A_{11} & A_{12} \\ A_{21} & A_{22}\end{bmatrix} & \begin{bmatrix} B_1 \\ B_2 \end{bmatrix} \\ \begin{bmatrix} C_1 & C_2 \end{bmatrix} & \end{matrix}\right) \overset{\text{POD}}{\longrightarrow} \left(\begin{matrix}\begin{bmatrix} A_{11} & 0 \\ 0 & 0\end{bmatrix} & \begin{bmatrix} B_1 \\ 0 \end{bmatrix} \\ \begin{bmatrix} C_1 & 0 \end{bmatrix} & \end{matrix}\right) \longrightarrow \begin{pmatrix} A_{11} & B_1 \\ C_1 & \end{pmatrix}$$

We prove here a new property, which generalizes the stability result of Moore.

Theorem 1 *All truncated SIB-realization of an asymptotically stable, minimal system* (A, B, C) *are asymptotically stable for sufficiently large* δ.

Proof In balanced coordinates, the canonical Gramian satisfies

$$A\Lambda + \Lambda A' + BB' - \mathrm{e}^{A\delta}BB'\mathrm{e}^{A'\delta} = 0 \tag{18.15}$$

$$A'\Lambda + \Lambda A + C'C - \mathrm{e}^{A'\delta}C'C\mathrm{e}^{A\delta} = 0. \tag{18.16}$$

Adding, gives for A_s, the symmetric part of A,

$$A_s\Lambda + \Lambda A_s + Q(\delta) = 0 \tag{18.17}$$

where $Q(\delta) = \frac{1}{2}\left[(BB' + C'C) - \mathrm{e}^{A\delta}BB'\mathrm{e}^{A'\delta} - \mathrm{e}^{A'\delta}C'C\mathrm{e}^{A\delta}\right]$. By asymptotic stability of A, there exists $\delta_0 > 0$, such that for all $\delta > \delta_0$, $\|\mathrm{e}^{A\delta}\| < 1$. Consequently, $Q(\delta) > 0$, a property inherited by any principal submatrix of $Q(\delta)$. In particular, if (A_r, B_r, C_r) is a balanced truncation, and Λ_r and Q_r are the consistent truncations of Λ and $Q(\delta)$, we get

$$(A_r)_s\Lambda_r + \Lambda_r(A_r)_s + Q_r(\delta) = 0. \tag{18.18}$$

Minimality implies $\Lambda > 0$. By the strict positive definiteness of Λ_r, the solution of this Lyapunov equation for A_s, can be represented in the form

$$(A_r)_s = -\int_0^\infty e^{-\Lambda_r t} Q_r(\delta) e^{-\Lambda_r t}\, dt. \tag{18.19}$$

Since the real parts of the eigenvalues of A_r are bounded by the minimal and maximal eigenvalues of its symmetric part $(A_r)_s$, stability follows. □

Remarks

1. The balanced truncation of an SIB(δ) realization is in general not balanced (SIB or otherwise) unless $\delta = \infty$. This is analogous to the discrete LTI case.
2. Equation 18.17 and thus also (18.19) remain valid for time variant SIB realizations, if Q is replaced by the appropriate $Q(t, \delta)$. Then the balanced truncation gives *pointwise*

$$[A_r(t)]_s = -\int_0^\infty e^{-\Lambda_r(t)\tau} Q_r(t,\delta) e^{-\Lambda_r(t)\tau}\, d\tau. \tag{18.20}$$

If for some δ it holds that $Q(t, \delta) > 0$ for all t, then the spectrum of $A_r(t)$ lies in the open left half plane. It is known [15] that if the largest eigenvalue of $[A_r(t)]_s$ satisfies

$$\lim_{t\to\infty} \int_{t_0}^{t} \lambda_{\max}(t)\, dt = -\infty \tag{18.21}$$

then the null solution of the SIB-truncated system is asymptotically stable, and uniformly asymptotically stable if property (18.21) holds uniformly with respect to t_0. If the above integral is merely bounded by $M(t_0)$, the null solution is stable and uniformly stable if M does not depend on t_0.

18.3 Metric Properties of Balanced Truncation

In what sense is the balanced truncation close to the original system? In the LTI case, it is known that the H_∞ norm of the difference between the transfer functions is upper bounded by twice the sum of the neglected singular values [4]. In the time varying and general nonlinear case, this metric cannot be used. Willems [14] argued that a natural distance between two controllable LTI systems should be based on the distance between their L_2 behaviors, which are distances between subspaces. A convenient choice is therefore the *gap-metric*. This inspired us to pose the approximation problem as an LQ problem in the state space.

18.3.1 A Natural Problem and a Simplification

Assume a system is given in state space form by the triple (*A*, *B*, *C*). Let its initial condition be $x_0 \in \mathbb{R}^n$. The objective of model reduction is to model the input–output behavior of this system by one with a reduced state space, say of dimension *r*. This means that we postulate a form

$$\dot{\hat{x}} = A_r\hat{x} + B_r\hat{u} \tag{18.22}$$

$$\hat{y} = C_r\hat{x}. \tag{18.23}$$

with dim $\hat{x} = r < n$, and with $\hat{u}$ and $\hat{y}$ close to the given data (*u*, *y*) consistent with (*A*, *B*, *C*). In general $(\hat{u}, \hat{y}) \neq (u, y)$, as it is unlikely that the equations (18.23) are consistent with the given data. A reasonable objective function seems to be

$$J = \int_{\mathcal{T}} \|u - \hat{u}\|^2 + \|y - \hat{y}\|^2 \, \mathrm{d}t. \tag{18.24}$$

where $\mathcal{T}$ is the entire interval where the data is available. The optimization is over all systems (A_r, B_r, C_r), adjusted inputs $\hat{u}$, and initial state $\hat{x}_0$. The latter completely determine then $\hat{y}$. Note that *J* depends on the given data *u* and the initial condition x_0, the output *y* being determined by these. The above sketched problem has only a unique solution up to similarity: If $(A_r, B_r, C_r, \hat{u}, \hat{x}_0)$ is a solution then so is $(TA_rT^{-1}, TB_r, C_rT^{-1}, \hat{u}, T\hat{x}_0)$ for any $T \in G\ell_r(\mathbb{R})$.

In principle, this problem can be solved in two steps: First fix $\Sigma_r = (A_r, B_r, C_r)$ and optimize *J* over $\hat{u}$ and $\hat{x}_0$, resulting in an optimal performance $J(\Sigma_r)$. Next, minimize over all Σ_r. This may further be extended: $J(\Sigma_r|u, x_0)$ may be averaged over *several* input and initial state combinations, or even a continuum of them. Then optimize the average of these optima, $J_{\text{avg}}(\Sigma_r)$, over all Σ_r. The first step is a typical LQ problem, and its solution requires the solution of a Riccati equation, which we would like to avoid. Therefore we simplify the performance index and setup of the problem. Instead of simultaneously trading off the adjustment of *u* and *y*, let's stagger the penalties.

What to use as training data (*u*, *y*)? Of course this may be a mote point if there is no control over the system, as for instance when only causal relations are analyzed without being able to influence. If the input variables *u* can be freely chosen, then let it be the input of minimal energy ($\mathcal{E}_u = \int \|u\|^2 \, \mathrm{d}t$) which sets up the state x_0 from the zero state, say in subinterval $\mathcal{T}_-$, and which is zero after that in subinterval $\mathcal{T}_+$, with $\mathcal{T}_- \cup \mathcal{T}_+ = \mathcal{T}$. Hence $\mathcal{E}_u = \int_{\mathcal{T}} \|u - \hat{u}\|^2 \, \mathrm{d}t = \int_{\mathcal{T}_-} \|u - \hat{u}\|^2 \, \mathrm{d}t$.

The output energy, $\mathcal{E}_y = \int \|y\|^2 \, \mathrm{d}t$ is only measured after that, again for reasons of mathematical simplicity. Furthermore if $u = 0$, we shall insist that then also $\hat{u}$ must be zero. Letting $\mathcal{T}_- = (-\delta, 0)$ and $\mathcal{T}_+ = (0, \delta)$ we arrive at

$$J = \int_{-\delta}^{0} \|u - \hat{u}\|^2 \, \mathrm{d}t + \int_{0}^{\delta} \|y - \hat{y}\|^2 \, \mathrm{d}t. \tag{18.25}$$

Note that the input u is parameterized by the state x_0 set up at time zero. In turn, this state determines the output y completely, so that the optimal J in (18.25) is also parameterized by x_0. Because of the LQ nature of the problem, J scales quadratically in the norm $\|x_0\|$. Unless special weights in the state space are intended, it suffices to let x_0 live on the unit ball in $\mathbb{R}^n$. For $\delta \to \infty$, the problem touches on the optimal Hankel norm reduction [2].

18.3.2 Solution

Given Σ, and x_0, the input u which reaches x_0 from the zero state in time δ is given by

$$u(t) = B' \mathrm{e}^{-A't} \mathscr{R}^{-1} x_0 \tag{18.26}$$

where $\mathscr{R}$ is the reachability Gramian over $(-\ \delta, 0)$. Let the reduced system Σ_r be steered by input $\hat{u}$, and set up a state $\hat{x}_0$ (also from its zero state). It follows then that the second integral in (18.25) is expressible as a quadratic form in $\hat{x}_0$

$$\Psi(\hat{x}_0) = \int_{0}^{\delta} \|y - \hat{y}\|^2 \, \mathrm{d}t = \int_{0}^{\delta} \|C\mathrm{e}^{At} x_0 - C_r \mathrm{e}^{A_r t} \hat{x}_0\|^2 \, \mathrm{d}t.$$

Hence,

$$J = \int_{-\delta}^{0} \|u - \hat{u}\|^2 \, \mathrm{d}t + \Psi(\hat{x}_0). \tag{18.27}$$

It is further known that the specific training input u has energy and $\mathscr{E}_u = x_0' \mathscr{R}^{-1} x_0$, and that likewise $\mathscr{E}_y = x_0' \mathscr{O} x_0$. The reduced LQ optimization problem is then restated as the minimization of

$$\int_{-\delta}^{0} \|u - \hat{u}\|^2 \, \mathrm{d}t + x_0' \mathscr{O} x_0 - 2 x_0' \mathscr{O}_s \hat{x}_0 + \hat{x}_0' \widehat{\mathscr{O}} \hat{x}_0, \tag{18.28}$$

where $\mathscr{O}_s$ is the mixed observability Gramian defined by

$$\mathscr{O}_s = \int_{0}^{\delta} \mathrm{e}^{A't} C' C_r \mathrm{e}^{A_r t} \, \mathrm{d}t. \tag{18.29}$$

It satisfies the Sylvester equation

$$A'\mathcal{O}_s + \mathcal{O}_s A_r + C'C_r - \mathrm{e}^{A'\delta} C'C \mathrm{e}^{A_r\delta} = 0. \tag{18.30}$$

The following theorem provides a bound to the above problem solution in terms of the SIB(δ) realization.

Theorem 2 *If the SIB(δ) canonical Gramian is partitioned as* $\begin{bmatrix} \Lambda_1 & \\ & \Lambda_2 \end{bmatrix}$ *with* $\|\Lambda_2\|/\|\Lambda_1\| = \varepsilon$ *and* $\varepsilon < 1$, *then the balanced truncation will yield a mismatch* $J = O(\varepsilon)$ *for sufficiently large* δ.

Proof The problem of minimizing (18.28) is a standard LQ problem, with Hamiltonian

$$H = \|u - \hat{u}\|^2 + \lambda'(A_r\hat{x} + B_r\hat{u}) \tag{18.31}$$

and terminal manifold

$$\Psi = \hat{x}_0'\widehat{\mathcal{O}}\hat{x}_0 - 2x_0'\mathcal{O}_s\hat{x}_0 + x_0'\mathcal{O}x_0. \tag{18.32}$$

It follows from the optimality conditions and Euler–Lagrange equations that

$$(I + \widehat{\mathcal{R}}\widehat{\mathcal{O}})\hat{x}_0 = (\mathcal{R}_s\mathcal{R}^{-1} + \widehat{\mathcal{R}}\mathcal{O}_s')x_0. \tag{18.33}$$

Substituting in

$$u - \hat{u} = B_r'\mathrm{e}^{-A_r't}[\widehat{\mathcal{O}}\hat{x}_0 - \mathcal{O}_s'x_0]. \tag{18.34}$$

"squaring up" and integrating, we get after some manipulations $J(x_0)$ as a quadratic form with weight matrix

$$\begin{aligned} &\mathcal{R}^{-1}[\mathcal{R}_s'\widehat{\mathcal{O}} - \mathcal{R}\mathcal{O}_s][I + \widehat{\mathcal{R}}\widehat{\mathcal{O}}]^{-1}\widehat{\mathcal{R}}[I + \widehat{\mathcal{O}}\widehat{\mathcal{R}}]^{-1}[\widehat{\mathcal{O}}\mathcal{R}_s - \mathcal{O}_s'\mathcal{R}]\mathcal{R}^{-1} + \mathcal{O} \\ &- \mathcal{O}_s[I + \widehat{\mathcal{R}}\widehat{\mathcal{O}}]^{-1}[\mathcal{R}_s'\mathcal{R}^{-1} + \widehat{\mathcal{R}}\mathcal{O}_s'] - [\mathcal{R}^{-1}\mathcal{R}_s' + \mathcal{O}_s\mathcal{R}][I + \widehat{\mathcal{O}}\widehat{\mathcal{R}}]^{-1}\mathcal{O}_s' \\ &+ [\mathcal{R}^{-1}\mathcal{R}_s' + \mathcal{O}_s\mathcal{R}][I + \widehat{\mathcal{O}}\widehat{\mathcal{R}}]^{-1}\widehat{\mathcal{R}}^{-1}[I + \widehat{\mathcal{R}}\widehat{\mathcal{O}}]^{-1}[\mathcal{R}_s'\mathcal{R}^{-1} + \widehat{\mathcal{R}}\mathcal{O}_s']. \end{aligned} \tag{18.35}$$

where we introduced the mixed reachability Gramian

$$\mathcal{R}_s = \int_0^\delta \mathrm{e}^{A_r t}B_r B'\mathrm{e}^{A't}\,\mathrm{d}t. \tag{18.36}$$

If the system Σ is in SIB(δ)-balanced form with canonical Gramian Λ partitioned as given, then take for Σ_r its SIB(δ)-balanced truncation. As there is no guarantee for its optimality, we shall use this reduced model to establish an upper bound on J. The truncation of an SIB(δ) realization is not balanced. However, if Σ is asymptotically stable, it follows from Theorem 1 that the chosen Σ_r is asymptotically stable for sufficiently large δ. Let then δ be such that $\|\mathrm{e}^{A_r\delta}\| < \sqrt{\varepsilon}$, and thus, using the relations (18.15) and (18.16), the δ-reachability Gramian $\mathscr{R}_r$ of Σ_r satisfies the order bound

$$A_r(\Lambda_1 - \mathscr{R}_r) + (\Lambda_1 - \mathscr{R}_r)A_r' = O(\varepsilon). \tag{18.37}$$

It follows that $\Lambda_1 - \mathscr{R}_r = O(\varepsilon)$, and similarly for the observability Gramians. A similar argument then establishes the ε closeness of $\mathscr{O}_s$ and $\mathscr{R}_s$ respectively to $[\Lambda_1, 0]'$ and $[\Lambda_1, 0]$. Substituting these approximations in (18.35) shows that this non-optimal J is a term of order ε away from $x_0'\begin{bmatrix} 0 & \\ & \Lambda_2 \end{bmatrix}x_0$, which is itself of order ε. □

Remark The same argument as used in the proof shows that in fact, the δ-Gramians are order ε close to the infinite time Gramians. Hence, for stable A, one could work with realizations balanced in the sense of Moore. Such a realization will then approximately be SIB(δ) balanced as well if δ is sufficiently large. The benefit of the SIB-realizations is that they may be computed via direct integration, and they also shed light on the time varying and nonlinear extensions. Theorem 2 motivates the use of the balanced truncation as the reduced order model.

18.4 Nonlinear Model Reduction

The discussion in the previous sections set the stage for an extension of the balanced truncation for nonlinear system. In order to avoid any pathological cases, the class of systems is restricted to analytic system, and for simplicity, we limit this further to siso systems. Let thus the system

$$\dot{x} = \bar{f}(x, u) \tag{18.38}$$

$$y = h(x) \tag{18.39}$$

be given, with $\bar{f}$ and h analytic. This implies that convergent Taylor series exist.

Let further a class of "typical" initial states be given, say some domain $V \subset X$. The rationale behind this is that if V is not too large, one could focus attention on some nominal input, u_{nom}, say defined with respect to the system with initial condition x_0, an average over V. Deviations form the nominal input are assumed "small". This implies that the system can be "recentered", by incorporating this u_{nom} directly in the model to yield

$$\dot{x} = \bar{f}(x, u_{\text{nom}} + \tilde{u}) \stackrel{\triangle}{=} f(x) + g(x)\tilde{u} \tag{18.40}$$

$$y = h(x) \tag{18.41}$$

Consider the linear variational system in the tangent space moving with the nominal state along the nominal trajectory. In [7], and references therein, it was shown that the instantaneous reachability and observability matrices can be defined associated with system (18.41), with the use of the gradient, d, Lie derivative, L_f, and Lie product $[f,g]=\frac{\partial g}{\partial x}f-\frac{\partial f}{\partial x}g$ operations, with $\mathrm{ad}_f^{k+1}g=[f,\mathrm{ad}_f^k g]$, and $\mathrm{ad}_f^0 g=g$.

Remarks

1. If the system has strict relative degree n at x_0, i.e., (see for instance [5]),

$$
\begin{aligned}
L_g L_f^i h(x) &\equiv 0, \quad \forall x \in U(x_0),\ i=0,\ldots,n-2 \\
L_g L_f^{n-1} h(x_0) &\neq 0,
\end{aligned}
$$

   then $g, \mathrm{ad}_f g, \ldots, \mathrm{ad}_f^{n-1} g$ are linearly independent at x_0. For linear systems the relative degree is the pole zero excess.
2. A system may fail to have relative degree at some points.

Definition 1 The nonlinear local reachability and observability matrices are respectively

$$
\mathbf{R}_\infty(x) = [g, \mathrm{ad}_{-f}g, \ldots, \mathrm{ad}_{-f}^{n-1}g, \ldots]_x \tag{18.42}
$$

$$
\mathbf{O}'_\infty(x) = [(\mathrm{d}h)',\ (\mathrm{d}L_f h)', \ldots,\ (\mathrm{d}L_f^{n-1}h)', \ldots] \tag{18.43}
$$

Denote their truncated (n-component) versions by

$$
\begin{aligned}
R_n(f,g) &= [g, \mathrm{ad}_{-f}g, \ldots, \mathrm{ad}_{-f}^{n-1}g] \\
O_n(f,h) &= \begin{bmatrix} \mathrm{d}h \\ \mathrm{d}L_f h \\ \vdots \\ \mathrm{d}L_f^{n-1}h \end{bmatrix}.
\end{aligned}
$$

Definition 2 The local reachability and observability Gramians are respectively

$$
\mathscr{R}_\delta(x) = \int_0^\delta (\mathrm{e}^{\tau\,\mathrm{ad}_{-f}}g)_x (\mathrm{e}^{\tau\,\mathrm{ad}_{-f}}g)_x^T \,\mathrm{d}\tau \tag{18.44}
$$

$$
\mathscr{O}_\delta(x) = \int_0^\delta (\mathrm{d}\mathrm{e}^{\tau L_f}h)_x^T (\mathrm{d}\mathrm{e}^{\tau L_f}h)_x \,\mathrm{d}\tau. \tag{18.45}
$$

Motivation for these definitions stems from the linear variational system, which has time varying reachability matrices and Gramians. When mapped back to the original nonlinear model, they have the form given in the definitions. The exponential of the operators appearing in the Gramians is then defined via the series expansion. The Gramians (18.44, 18.45) satisfy *Lyapunov-like equations* (locally at x) [7]:

$$L_f\mathscr{R}_\delta(x) = \left(\frac{\partial f}{\partial x}\right)_x \mathscr{R}_\delta(x) + \mathscr{R}_\delta(x)\left(\frac{\partial f}{\partial x}\right)_x' + g_x g_x' - \left(e^{\delta\,\mathrm{ad}_{-f}}g\right)_x\left(e^{\delta\,\mathrm{ad}_{-f}}g\right)_x'$$

$$-L_f\mathscr{O}_\delta(x) = \left(\frac{\partial f}{\partial x}\right)_x' \mathscr{O}_\delta(x) + \mathscr{O}_\delta(x)\left(\frac{\partial f}{\partial x}\right)_x + (\mathrm{d}h)_x'(\mathrm{d}h)_x - (\mathrm{d}e^{\delta L_f}h)_x'(\mathrm{d}e^{\delta L_f}h)_x.$$

18.4.1 Local Balancing

Locally, at $X \in Y$, the balancing transformation $T(x)$ can be found from the Gramians $(\mathscr{R}_\delta(x), \mathscr{O}_\delta(x))$. The elements of the canonical Gramian $\Lambda(x)$, now defined at each state x, are the positive square roots of the eigenvalues of the product $\mathscr{O}_\delta(x)\mathscr{R}_\delta(x)$. If the system has strict relative degree n, the computation can be simplified by taking the first n terms in the series expansions and expressing Gramians as weighted "squares" of the reachability and observability matrices as suggested in (18.9).

Assume the system starts from rest ($x_{\mathrm{eq}} = 0$), and we want to analyze its input–output behavior. For small u, let U be the union of the reachable sets for all $t < t_f$, with t_f sufficiently small so that the entire space is not covered. U is some tubular neighborhood of the nominal trajectory. Let, for each $x \in U$, the corresponding canonical Gramian and local balancing transformation be known. In a patch where the dominant part of Λ has dimension r, the space can be foliated into the span of the dominant directions, say X_r. Hence $U_{\mathrm{loc}} = (\mathrm{span}X_r + \mathrm{span}N(x)) \cap U$, where $N(x)$ is the $(n-r)$-dimensional space normal to the dominant subspace. By construction, different points (states) on a single leaf can be reached from each other, and can be discerned easily. However, it is hard to jump from leaf to leaf (move in the normal directions, or resolve states along the normal directions. In this sense, a single leaf suffices as approximation for the state space model, and thus the approximation of the input output behavior. Motivated by the linear case where the redundant subspace is collapsed, we want to postulate the existence of a Riemannian metric $\xi'\Lambda(x)\xi$ on the state manifold [12].

18.4.2 Global Model Reduction

A problem remains in that the reduction, given by a low dimensional manifold, is still embedded in the full space. Whereas the dominant submanifold is clearly identifiable, there still is no intrinsic lower dimensional set of state equations to make the reduced order model global.

Let's approach the problem from another starting point: Let $\widehat{\xi}$ be a diffeomorphism in a patch of the (full) state manifold. Denote the inverse of $\widehat{\xi}(x) = \xi$ by $\widehat{x}(\xi) = x$. Then the system (18.41) is represented by

$$\dot{\xi} = \frac{\partial \widehat{\xi}}{\partial x} f \bigg|_{\hat{x}(\xi)} + \frac{\partial \widehat{\xi}}{\partial x} g \bigg|_{\hat{x}(\xi)} u \triangleq \hat{f}(\xi) + \hat{g}(\xi) u \tag{18.46}$$

$$y = h(\hat{x}(\xi)) \triangleq \hat{h}(\xi). \tag{18.47}$$

Locally, the diffeomorphism is determined by the Jacobian $J(x) = \frac{\partial \xi}{\partial x}$. Of course, if we want the ξ coordinates to be balanced in some global sense, we would naturally require that $J(x)$ corresponds to the local balancing transformation, $T(x)$. This implies that $\xi(x)$ should be an integral of $T(x)$. However, the system

$$\frac{\partial \xi}{\partial x} = T(x) \tag{18.48}$$

is a Mayer–Lie type system of partial differential equations, which has no solution unless the Frobenius conditions hold

$$\frac{\partial T_{ij}(x)}{\partial x_k} - \frac{\partial T_{ik}(x)}{\partial x_j} = 0 \tag{18.49}$$

In [9], a method was suggested based on the approximation of the local balancing transformation by an integrable Jacobian (Mayer–Lie interpolation).

So, let us now assume that the system (18.47) is a good *approximation* of (18.41) in balanced coordinates. Following the philosophy in Sect. 18.3 project the state ξ to $[\xi_r', 0]'$ where $\dim \xi_r = r < n$, and discard the equations of the rate of change in the last $n - r$ components of ξ. The resulting set gives a global (actually only in a patch) reduced approximation of dimension r:

$$\xi_r = \hat{f}([\xi_r', 0]') + \hat{g}([\xi_r', 0]')u \triangleq \hat{F}(\xi_r) + \hat{G}(\xi_r)u \tag{18.50}$$

$$y = \hat{h}([\xi_r', 0]') \triangleq \hat{h}(\xi_r). \tag{18.51}$$

18.4.3 Comments

As all computations of observability and reachability matrices and the computation of the local balancing transformation must be done symbolically, it cannot be expected that truly very large scale systems will easily be reduced unless the system have a specific simplifying structure. Such an example was explored in [10]. However, from the point of view of sensitivity analysis with respect to parameters in a nominal system, the method may turn out to be viable.

Acknowledgments The author thanks SCD, ESAT at KULeuven, Leuven, B-3001, Belgium for the generous support during his sabbatical semester, while this document was prepared.

References

1. Astolfi, A.: Model reduction by moment matching for nonlinear systems. In: Proceedings of the 47th Conference on Decision and Control, Cancun, Mexico (2008)
2. Glover, K.: Optimal Hankel-norm approximations of linear multivariable systems and their ℓ_∞-error bounds. Int. J. Control, **39**(6), 115–193 (1984)
3. Moore, B.: Principal Component Analysis in Linear Systems: Controllability, Observability and Model Reduction. IEEE Trans. Automat. Contr. **26**(1), 17–32 (1981)
4. Pernebo, L., Silverman, L.: Model reduction via balanced state space representation. IEEE Trans. Automat. Contr. **27**, 382–387 (1982)
5. Sastry, S.: Nonlinear Systems. Springer-Verlag, New York (1999)
6. Schilders, W.H.A., van der Vorst, H., Rommes, J.: Model Order Reduction. Springer-Verlag, Heidelberg (2008)
7. Verriest, E.I.: Time variant and nonlinear balanced realizations. In: Schilders, W.H.A., van der Vorst, H., Rommes, J. (eds.) Model Order Reduction, pp. 213–250. Springer-Verlag (2008)
8. Verriest, E.I.: On balanced realizations for time variant linear systems. Ph.D. Dissertation, Stanford University (1980)
9. Verriest, E.I.: Nonlinear balancing and Mayer—lie interpolation. In: Proceedings of the 43rd IEEE Conference SSST-04, Atlanta, GA 180–184 (2004)
10. Verriest, E.I.: Local nonlinear balancing for structural information with an application in super collider particle dynamics. In: Proceedings of the 46th IEEE Conference on Decision and Control, New Orleans, LA, pp. 4614–4619 (2007)
11. Verriest, E.I., Gray, W.S.: Nonlinear balanced realizations. In: Proceedings of the 43rd IEEE Conference on Decision and Control, Paradise Island, Bahamas, pp. 1164–1169 (2004)
12. Verriest, E.I., Gray, W.S.: Geometry and topology of the state space via balancing. In: Proceedings of the 17th International Symposium on Mathematical Theory of Networks and Systems, Kyoto, Japan, pp. 840–848 (2006)
13. Verriest, E.I., Kailath, T.: On generalized balanced realizations. IEEE Trans. Auto. Control. **28**(8), 833–844 (1983)
14. Willems, J.C.: Model reduction in the GAP metric. SyReNe Workshop, Universität Hamburg, Hamburg, Germany, October 30–31 2008
15. Willems, J.L.: Stability Theory of Dynamical Systems. Nelson, London (1970)